Advances in Water, Air and Soil Pollution Monitoring, Modeling and Restoration

Advances in Water, Air and Soil Pollution Monitoring, Modeling and Restoration

Editors

Alina Barbulescu
Lucica Barbes

MDPI

Basel • Beijing • Wuhan • Barcelona • Belgrade • Novi Sad • Cluj • Manchester

Editors
Alina Barbulescu
Civil Engineering
Transilvania University
of Brașov
Brașov
Romania

Lucica Barbes
Applied Sciences
Ovidius University
of Constanța
Constanța
Romania

Editorial Office
MDPI
St. Alban-Anlage 66
4052 Basel, Switzerland

This is a reprint of articles from the Special Issue published online in the open access journal *Toxics* (ISSN 2305-6304) (available at: www.mdpi.com/journal/toxics/special_issues/0I5O0Y56F0).

For citation purposes, cite each article independently as indicated on the article page online and as indicated below:

Lastname, A.A.; Lastname, B.B. Article Title. *Journal Name* **Year**, *Volume Number*, Page Range.

ISBN 978-3-7258-0906-6 (Hbk)
ISBN 978-3-7258-0905-9 (PDF)
doi.org/10.3390/books978-3-7258-0905-9

Contents

About the Editors

Alina Barbulescu

Alina Bărbulescu received a Ph.D. in Mathematics from Al. I. Cuza University of Iași, Romania, in Cybernetics and Econometrics from the Academy of Economics Studies of Bucharest, Romania, and in Civil Engineering from the Technical University of Civil Engineering of Bucharest. She received a Habilitation in Civil Engineering, as well as Cybernetics and Econometrics. Currently, she is a Professor at the Transilvania University of Brașov, Romania. Bărbulescu's main research fields are environmental pollution, hydrological modeling, applied statistics and time series analysis. Recently, most research has been developed within the frameworks of UEFISCDI Romania, Horizon Europe, or projects funded by the Zayed University Research Center, the United Arab Emirates. Her publishing activity includes over 200 articles (over half indexed in Web of Science), 32 books and chapters. She was the editor of 22 Special Issues of international conferences and is a member of the editorial boards of different journals (three indexed in Web of Science).

Lucica Barbes

Lucica Barbeş received a Ph.D. in Chemical Engineering from the Politehnica University of Bucharest, Romania, in 2003. She is currently an Associate Professor at the Ovidius University of Constanţa, Romania. She is the author or co-author of 5 books, holds 1 patent (2003), and has published over 70 research articles, of which over 40 are included in the WOS, Scopus or ACS database. She has earned the Marie Curie grant from the French National Centre for Scientific Research (2003–2004), and the individual mobility training staff Lifelong Learning Programme – Erasmus at universities in Italy (2011) and Spain (2012). She was an invited speaker at the Interdisciplinary Nanoscience Center, University of Aarhus, Denmark (2006). She has participated in more than 40 national and international conferences. She serves on the Reviewer Board of *Water* and other highly reputed international journals. She is vice-president of the Chemistry Society of Romania - Constanta Branch, and an active member of the Romanian Society of Chemical Engineering, the Romanian Association of Food Industry Specialists (2010 – 2016), and the Internationale Gesellschaft fur Warenwissenschaft und Technologie, Austria (2011–2016). Her research interests include the following: indicators and integrated environmental monitoring; emerging pollutants in the air, water and soil; water and wastewater treatment; marine fouling; VOC monitoring and analysis; spectral analyses (FTIR, IRRAS); techniques of enzyme immobilization on different metallic surfaces; and conventional and alternative fuel combustion.

Preface

The pollution of air, water and soil is constantly increasing, becoming a global issue. Biodegradation modifies pollutants' structure to the molecular level, generating more waste that must be monitored, reduced and controlled.

The purpose of this Special Issue was to assess air, water and soil pollution employing advanced methods, and apply the findings to possible mitigation measures. The published articles provide an overview of the actual research stage in the field, aiming to emphasize the pollution risks and impact on people's health and environment.

Alongside the solutions to the practical problems of cleaning the water, air and soil, the study topics directly answer questions relating to selecting different tools that best emphasize the environmental quality changes and their impact on society's future.

Alina Barbulescu and Lucica Barbes
Editors

toxics

Editorial

Advances in Water, Air and Soil Pollution Monitoring, Modeling and Restoration

Alina Bărbulescu [1], Lucica Barbeş [2,3,*] and Cristian Ştefan Dumitriu [4,*]

1 Department of Civil Engineering, Transilvania University of Braşov, 5 Turnului Str., 500152 Braşov, Romania; alina.barbulescu@unitbv.ro
2 Department of Chemistry and Chemical Engineering, "Ovidius" University of Constanţa, 124 Mamaia Bd., 900527 Constanţa, Romania
3 Doctoral School of Biotechnical Systems Engineering, Politehnica University of Bucharest, 313, Splaiul Independenţei, 060042 Bucharest, Romania
4 Faculty of Mechanical Engineering and Robotics in Constructions, Technical University of Civil Engineering, Calea Plevnei 59, 021242 Bucharest, Romania
* Correspondence: lucille.barbes2020@gmail.com (L.B.); cristian.dumitriu@utcb.ro (C.Ş.D.)

Global pollution demands continuous attention and concerted efforts to reduce its effects. Every day, our planet faces increasing pressure from various sources, including industrial activities, urbanization, agriculture, and waste generation. In this context, the articles featured in this Special Issue shed light on the multifaceted nature of environmental pollution and provide innovative approaches for its monitoring and modeling, proposing solutions for restoration. Several articles delve into the complexity of pollution assessment, providing insights into the impact of pollutants on environmental health and human wellbeing. Studies focusing on degradation processes emphasize the importance of understanding pollution's ecological consequences. Therefore, a key theme of these investigations is the urgent need for effective mitigation measures to address environmental restoration. Moreover, the articles provide valuable guidance for policy makers, practitioners, and researchers.

As academic editors, we are particularly excited to see this collection's diverse topics. We hope that the presented discoveries will inspire further interdisciplinary collaboration and innovative solutions to the challenges posed by environmental pollution. Eleven papers were selected for inclusion in this issue after the peer review process of the twenty-three submitted manuscripts. The complexity of this Special Issue lies in interpreting the multifaceted interactions between various environmental parameters and developing effective pollution monitoring and management strategies.

In the article "Polycyclic Aromatic Hydrocarbons (PAHs) in the Dissolved Phase, Particulate Matter, and Sediment of the Sele River, Southern Italy: A Focus on Distribution, Risk Assessment, and Sources", Montuori et al. present the findings on the concentrations and composition of PAHs in the Sele River, Italy. Low-molecular-weight (LMW) PAH levels were notably elevated in water samples, while high-molecular-weight (HMW) PAHs were predominant in sediment samples. Analysis of the PAHs' diagnostic ratio indicates that the primary sources were pyrolytic, suggesting a significant contribution from vehicle emissions and combustion processes. The concentrations of numerous individual PAHs at various sites surpassed environmental risk limits (ERLs) and threshold effect levels (TELs) [1], occasionally resulting in adverse environmental impacts. However, the toxic equivalent concentration (TEQ) of carcinogenic PAHs shows a definite carcinogenic risk in the Sele River basin. Hence, continuous monitoring of Sele River waters is imperative as PAH contamination could affect aquatic ecosystems.

The article "Occurrence and Distribution of Persistent Organic Pollutants (POPs) from Sele River, Southern Italy: Analysis of Polychlorinated Biphenyls and Organochlorine Pesticides in a Water–Sediment System" investigates the pollution characteristics, spatiotemporal

check for updates

Citation: Bărbulescu, A.; Barbeş, L.; Dumitriu, C.Ş. Advances in Water, Air and Soil Pollution Monitoring, Modeling and Restoration. *Toxics* **2024**, *12*, 244. https://doi.org/10.3390/toxics12040244

Received: 15 March 2024
Accepted: 24 March 2024
Published: 26 March 2024

variation, source, and potential ecological risk of PCBs and OCPs in the Sele River, including their contribution to the Tyrrhenian Sea. Sediment samples exhibited higher levels of these contaminants compared to those in the water bodies, DP, and SPM, indicating that suspension processes and sedimentation are the primary mechanisms at work in the Sele River. The data showed that industrial processes were the primary source of PCBs. Risk assessment revealed elevated PCB risk factors at the mouth of the Sele River and 500 m south, while levels were lower at other sites. In contrast, OCP ratios were generally lower, with most analytes showing a risk quotient (RQ) below 1. Consequently, regular monitoring of pollution in the Sele River and its estuary is necessary to evaluate ecological risks over time. These findings enhance our understanding of Sele River water quality and inform environmental monitoring, applications of sediment quality guidelines, and ecological risk assessments [2]. It is expected that establishing a comprehensive database for various pollution factors and including more emerging contaminants in river ecosystem risk assessments will be crucial. Moreover, this study's results will aid in preventing future contamination of the Sele River's water system by PCBs and OCPs, thereby strengthening prevention and pollution control measures against future risks. The results will help policy makers identify high-risk pollutant areas, improve environmental protection regulations, and raise public awareness of their importance.

The research study "Health Risk Assessment of PAHs from Estuarine Sediments in the South of Italy" introduces, for the first time, an evaluation of the carcinogenic risk posed to human health by dermal and ingestion exposure to polycyclic aromatic hydrocarbons (PAHs) present in sediments within the primary surface water streams of the Campania Region, located in southern Italy. It offers insights into the concentrations, spatial distribution, and composition profiles of PAHs found in sediments collected near the estuaries of the Sele, Sarno, and Volturno Rivers. The findings suggest that the risk of cancer resulting from oral exposure to PAHs in estuarine sediments [3]—quantified as incremental lifetime cancer risk (ILCR ingestion)—is low, unlike the risk associated with accidental skin exposure, which is moderate. The results underscore the need to continuously evaluate the carcinogenic risk to human health arising from dermal and oral exposure to PAHs and ongoing monitoring of PAH concentrations in surface water sediments within the Campania Region. Therefore, this study is a foundation for future investigations to comprehensively assess the carcinogenic risk to human health due to PAH exposure to inform pollution prevention measures, ecological restoration strategies for rivers, and the preservation of our overall wellbeing.

In their work titled "Modeling the Chlorine Series from the Treatment Plant of Drinking Water in Constanta, Romania", Bărbulescu and Barbeş introduced four alternative approaches for modeling monthly free chlorine residual concentration series from PCTP using decomposition, Holt–Winters, and SARIMA models. A key novelty lies in employing econometric models in engineering, thereby expanding upon previous studies on the water quality, which had primarily used statistical modeling [4,5]. Research in Romania has been limited in this field, with it being primarily experimental or presenting basic statistics without correlations. This article fills this gap in research, which is particularly crucial given the importance of monitoring chlorine concentration to avoid exceeding regulatory limits and potential public discontent due to changes in drinking water taste and smell. However, these models are recommended for short-term predictions without continuous updating. Automating chlorine concentration monitoring can improve dosage and forecasting accuracy. Additionally, future studies should consider incorporating risk factors and addressing water quality deterioration to ensure constant monitoring and intervention in the water treatment process.

The article "Assessing the Efficiency of a Drinking Water Treatment Plant Using Statistical Methods and Quality Indices" by Bărbulescu and Barbeș introduces various indicators utilized in a case study to assess the effectiveness of a water treatment plant. While individual indicators highlight efficiency concerning specific water parameters and underscore issues that may arise during particular periods or regarding specific parameters, cumulative indicators evaluate overall efficiency over time, considering all parameters. This study revealed that individual efficiencies are sensitive to fluctuations in effluent values compared to influent values, even if they fall within maximum allowable variation (MAV) limits. Consequently, cumulative indices can be significantly influenced when very low values contribute to their calculation. Weighted cumulative indices consistently differ from the average ones. However, given the significance of each water parameter and the imperative of maintaining high water quality standards, they must be considered.

The study paves the way for aligning evaluations of environmental pollution with sustainability objectives based on objective criteria [6–8]. Future research will explore opportunities to enhance the presented indices and establish a system that promptly implements necessary corrective measures upon issue detection. Additionally, a procedural framework must be devised to address the outliers' existence because these values introduce considerable biases in the indices computation.

The paper "Determinants Analysis Regarding Household Chemical Indoor Pollution" highlights the need for more comprehensive research on indoor household pollution among the general population. Despite being aware of the harmful effects of certain habits, it remains difficult for people to adopt behaviors that help reduce indoor pollution. Therefore, there is an urgent need for training programs that can target individuals with poor indoor habits, such as singles, smokers, and those with lower education, to help them improve their practices and minimize exposure to indoor pollutants [9]. Additionally, educational initiatives are needed to reinforce the importance of good practices among individuals who already exhibit positive attitudes and behaviors, such as those in committed relationships and non-smokers. Although there are behavior and attitude correction programs for highly educated youth, there is still a gap in translating this knowledge into practical measures to effectively address indoor chemical pollution.

The article "A New Method for Ecological Risk Assessment of Combined Contaminated Soil" indicates that the ecological risk assessment of combined polluted soil has traditionally relied on the risk screening value (RSV) of individual pollutants, but this approach has notable limitations. It overlooks the influence of soil properties and fails to consider interactions among different pollutants. This study evaluated the ecological risks associated with 22 soils from 4 smelting sites using toxicity tests involving soil invertebrates. In addition to the RSV-based assessment, a novel method was developed.

This new approach introduced a toxicity effect index (EI) to standardize the toxicity effects of various endpoints, enabling comparisons across different toxicity measures. Furthermore, an ecological risk probability assessment method (RP) was devised based on the cumulative probability distribution of EI. A significant correlation was observed between the RP based on EI and the Nemerow ecological risk index (NRI) [10] based on RSVs ($p < 0.05$). Additionally, the new method facilitates the visual representation of probability distributions for various toxicity endpoints, aiding risk managers in devising more effective risk management strategies to safeguard critical species. This innovative method is poised to integrate with a sophisticated dose–effect relationship prediction model constructed using machine learning algorithms, offering a fresh approach to the ecological risk assessment of combined contaminated soil.

The article "Evaluating the Contamination by Indoor Dust in Dubai" presents an analysis of metal enrichment levels in indoor dust collected from various locations in Dubai, utilizing multivariate statistics and pollution indices. The research addresses a significant gap in understanding indoor pollution caused by dust in a region prone to frequent dust storms. Results indicated that the highest enrichment factors (for Ca, Cu, Mg, and Fe) were attributed to soil lithology and industrial activities, particularly mining, with dust transportation over long distances during dust storms. Two novel pollution indices, CPI and AWI, were introduced and applied to assess contamination levels at observation sites. Classification of sites based on PLI, CPI, AWI, and the Nemerow index [11] differed from classifications based on raw data series, with two sites falling into distinct clusters in each classification. This study also suggests a promising research direction using different classification data sets. Notably, eliminating elements with concentrations significantly below warning limits from the data set resulted in more realistic classifications. The future of this research aims to develop a methodology for cross-validating clustering findings using supplementary selection criteria and decision trees. Employing various clustering algorithms on raw data series, pollution index series, and stability criteria will be crucial for identifying consistently similar series within the data set. Overall, this study provides valuable insights into indoor dust pollution in Dubai and lays the groundwork for further research, with potential implications for pollution management and public health.

The paper "Toxicity Risk Assessment Due to Particulate Matter Pollution from Regional Health Data: Case Study from Central Romania" presents the health implications of elevated levels of PM10 and PM2.5 above the average limits recommended by Romanian legislation and the World Health Organization (WHO) in the Central Region of Romania. The findings underscore the significant risk of prolonged exposure to airborne fine particulate matter, commonly found in urban areas, on cardiovascular health. According to the health impact assessment conducted in this study, adhering to the new WHO limits could yield substantial benefits in reducing mortality rates in the Central Region of Romania. Specifically, adopting these limits could reduce approximately 196 deaths on average and an increase in life expectancy by approximately 5.3 months due to lower PM2.5 levels. Furthermore, there could be a decrease of roughly 190 deaths on average, corresponding to a 3.5-month increase in life expectancy related to cardiovascular mortality.

These results highlight the urgent need to mitigate the health risks associated with pollutants' exposure [12,13] by implementing the new WHO-recommended limits in Romanian regulations. However, it is essential to acknowledge the limitations of this study, such as data gaps, especially regarding PM2.5, which may affect the accuracy of the PM2.5/PM10 ratio estimation. As a future direction, expanding the scope of this study to include other pollutants is crucial.

The paper "Prediction for Cyanobacterial Blooms Using Environmental Variable Selection and Data Resampling" introduce a series of processes to enhance the prediction accuracy of algal alert levels in the BJR by using observed data, feature selection techniques, and resampling methods to construct two machine learning models [14,15]. The primary objective of this study was to develop a prediction model for algal alert levels in reservoirs using readily available data from national monitoring stations. The proposed model, which incorporates feature selection and resampling methods, is anticipated to benefit engineers and decision makers in managing algal blooms in watershed areas, including inland weirs. This model will facilitate the development of effective strategies and regulations for constructing and operating these reservoirs.

The article titled "Application of Machine Learning in Modeling the Relationship between Catchment Attributes and Instream Water Quality in Data-Scarce Regions", highlights the efficacy of machine learning methods [16] in predicting and evaluating water quality parameters within a catchment area. Among these methods, the random forest (RF) model is the most effective, providing a robust tool for accurate and efficient water quality assessment. While certain models may exhibit shortcomings in specific criteria, a nuanced assessment using relative criteria such as accuracy (R^2) and mean absolute percentage error (MAPE) underscores the overall robustness of predictive models. Evaluation of R^2 values indicates satisfactory performance across all models except pH prediction. Despite slightly elevated MAPE values in five models (SAR, Na^+, SO_4, Cl^-, TDS), the primary research objective—understanding the significance of individual input variables within data constraints—was achieved. This accomplishment lays the groundwork for selecting and implementing optimal models from a broader spectrum of machine-learning techniques.

Integrating these research findings into decision-making processes offers transformative opportunities for strategic resource allocation and environmental impact mitigation. Furthermore, this integration empowers decision makers to adopt targeted strategies for promoting environmental sustainability, contributing to the broader objective of nurturing resilient water ecosystems. This approach signifies a practical pathway towards achieving a delicate balance between human activities and ecological preservation, actively promoting sustainable water ecosystems.

Finally, we sincerely thank the authors, reviewers, and editorial team for their invaluable contributions to this Special Issue. The research presented here will catalyze continued progress in environmental science and contribute to our ongoing efforts to safeguard our precious natural resources. Working together, we can strive towards a cleaner, healthier, and more sustainable planet for future generations.

Author Contributions: Conceptualization, A.B., L.B. and C.Ş.D.; methodology, A.B., L.B. and C.Ş.D.; validation, A.B., L.B. and C.Ş.D.; formal analysis, A.B., L.B. and C.Ş.D.; investigation, A.B., L.B. and C.Ş.D.; resources, A.B.; data curation, L.B. and C.Ş.D.; writing—original draft preparation, A.B. and L.B.; writing—review and editing, A.B. and C.Ş.D.; supervision, A.B.; project administration, A.B.; funding acquisition, A.B. All authors have read and agreed to the published version of the manuscript.

Funding: This research received no external funding.

Conflicts of Interest: The Guest Editor declares that there are no conflicts of interest or agreements with private companies that will prevent us working impartially in the editorial process.

List of Contributions:

1. Montuori, P.; De Rosa, E.; Di Duca, F.; De Simone, B.; Scippa, S.; Russo, I.; Sarnacchiaro, P.; Triassi, M. Polycyclic Aromatic Hydrocarbons (PAHs) in the Dissolved Phase, Particulate Matter, and Sediment of the Sele River, Southern Italy: A Focus on Distribution, Risk Assessment, and Sources. *Toxics* **2022**, *10*, 401. https://doi.org/10.3390/toxics10070401.
2. De Rosa, E.; Montuori, P.; Triassi, M.; Masucci, A.; Nardone, A. Occurrence and Distribution of Persistent Organic Pollutants (POPs) from Sele River, Southern Italy: Analysis of Polychlorinated Biphenyls and Organochlorine Pesticides in a Water–Sediment System. *Toxics* **2022**, *10*, 662. https://doi.org/10.3390/toxics10110662.
3. Di Duca, F.; Montuori, P.; Trama, U.; Masucci, A.; Borrelli, G.M.; Triassi, M. Health Risk Assessment of PAHs from Estuarine Sediments in the South of Italy. *Toxics* **2023**, *11*, 172. https://doi.org/10.3390/toxics11020172.
4. Bărbulescu, A.; Barbeş, L. Modeling the Chlorine Series from the Treatment Plant of Drinking Water in Constanta, Romania. *Toxics* **2023**, *11*, 699. https://doi.org/10.3390/toxics11080699.
5. Bărbulescu, A.; Barbeş, L. Assessing the Efficiency of a Drinking Water Treatment Plant Using Statistical Methods and Quality Indices. *Toxics* **2023**, *11*, 988. https://doi.org/10.3390/toxics11120988.
6. Montuori, P.; Gioia, M.; Sorrentino, M.; Di Duca, F.; Pennino, F.; Messineo, G.; Maccauro, M.L.; Riello, S.; Trama, U.; Triassi, M.; et al. Determinants Analysis Regarding Household Chemical Indoor Pollution. *Toxics* **2023**, *11*, 264. https://doi.org/10.3390/toxics11030264.
7. Wang, Q.; Wang, J.; Cheng, J.; Zhu, Y.; Geng, J.; Wang, X.; Feng, X.; Hou, H. A New Method for Ecological Risk Assessment of Combined Contaminated Soil. *Toxics* **2023**, *11*, 411. https://doi.org/10.3390/toxics11050411.
8. Nazzal, Y.; Bărbulescu, A.; Sharma, M.; Howari, F.; Naseem, M. Evaluating the Contamination by Indoor Dust in Dubai. *Toxics* **2023**, *11*, 933. https://doi.org/10.3390/toxics11110933.
9. Maftei, C.; Vaseashta, A.; Poinareanu, I. Toxicity Risk Assessment Due to Particulate Matter Pollution from Regional Health Data: Case Study from Central Romania. *Toxics* **2024**, *12*, 137. https://doi.org/10.3390/toxics12020137.
10. Kim, J.H.; Lee, H.; Byeon, S.; Shin, J.-K.; Lee, D.H.; Jang, J.; Chon, K.; Park, Y. Machine Learning-Based Early Warning Level Prediction for Cyanobacterial Blooms Using Environmental Variable Selection and Data Resampling. *Toxics* **2023**, *11*, 955. https://doi.org/10.3390/toxics11120955.
11. Kovačević, M.; Jabbarian Amiri, B.; Lozančić, S.; Hadzima-Nyarko, M.; Radu, D.; Nyarko, E.K. Application of Machine Learning in Modeling the Relationship between Catchment Attributes and Instream Water Quality in Data-Scarce Regions. *Toxics* **2023**, *11*, 996. https://doi.org/10.3390/toxics11120996.

References

1. Liu, Y.; Zarfl, C.; Basu, N.B.; Cirpka, O.A. Turnover and legacy of sediment-associated PAH in a baseflow-dominated river. *Sci. Total Environ.* **2019**, *671*, 754–764. [CrossRef] [PubMed]
2. Sun, C.; Zhang, J.; Ma, Q.; Chen, Y.; Ju, H. Polycyclic aromatic hydrocarbons (PAHs) in water and sediment from a river basin: Sediment-water partitioning, source identification and environmental health risk assessment. *Environ. Geochem. Health* **2016**, *39*, 63–74. [CrossRef]
3. Du, J.; Jing, C. Anthropogenic PAHs in lake sediments: A literature review (2002–2018). *Environ. Sci. Process. Impacts* **2018**, *20*, 1649–1666. [CrossRef] [PubMed]
4. Bucurica, I.A.; Dulama, I.D.; Radulescu, C.; Banica, A.L. Surface water quality assessment using electro-analytical methods and inductively coupled plasma mass spectrometry (ICP-MS). *Rom. J. Phys.* **2022**, *67*, 802.
5. Voinea, S.; Nichita, C.; Burchiu, E.; Diac, C.; Armeanu, I. Study case of potable water from wells in the metropolitan Bucharest area. Influences on human health–interdisciplinary lab. *Rom. Rep. Phys.* **2022**, *74*, 902.
6. Calotă, R.; Girip, A.; Savaniu, M.; Anica, I.; Glavă, G. Study on the heat transfer with regard to an off-grid vending machine having a low impact on the environment. *IOP Conf. Ser. Earth Environ. Sci.* **2023**, *1185*, 012004. [CrossRef]
7. Antonescu, N.N.; Stănescu, D.-P.; Calotă, R. CO_2 Emissions Reduction through Increasing H_2 Participation in Gaseous Combustible—Condensing Boilers Functional Response. *Appl. Sci.* **2022**, *12*, 3831. [CrossRef]
8. Bărbulescu, A.; Barbeş, L.; Dumitriu, C.Ş. Statistical Assessment of the Water Quality Using Water Quality Indicators—Case Study from India. In *Security and Sustainability. Advanced Sciences and Technologies for Security Applications*; Vaseashta, A., Mafte, C., Eds.; Springer: Cham, Switzerland, 2021; pp. 591–613.
9. Calotă, R.; Antonescu, N.N.; Stănescu, D.-P.; Năstase, I. The Direct Effect of Enriching the Gaseous Combustible with 23% Hydrogen in Condensing Boilers' Operation. *Energies* **2022**, *15*, 93733. [CrossRef]
10. Nazzal, Y.; Bou Orm, N.; Barbulescu, A.; Howari, F.; Sharma, M.; Badawi, A.; Al-Taani, A.A.; Iqbal, J.; El Ktaibi, F.; Xavier, C.M.; et al. Study of atmospheric pollution and health risk assessment. A case study for the Sharjah and Ajman Emirates (UAE). *Atmosphere*, 2021; 12, 1442.
11. Kowalska, J.B.; Mazurek, R.; Gąsiorek, M.; Zaleski, T. Pollution indices as useful tools for the comprehensive evaluation of the degree of soil contamination—A review. *Environ. Geochem. Health* **2018**, *40*, 2395–2420. [CrossRef] [PubMed]
12. Chiritescu, R.-V.; Luca, E.; Iorga, G. Observational study of major air pollutants over urban Romania in 2020 in comparison with 2019. *Rom. Rep. Phys.* **2024**, *76*, 702.
13. Dumitru, A.; Olaru, E.-A.; Dumitru, M.; Iorga, G. Assessment of air pollution by aerosols over a coal open-mine influenced region in Southwestern Romania. *Rom. J. Phys.* **2024**, *69*, 801.
14. Bourel, M.; Segura, A.M.; Crisci, C.; López, G.; Sampognaro, L.; Vidal, V.; Kruk, C.; Piccini, C.; Perera, G. Machine learning methods for imbalanced data set for prediction of faecal contamination in beach waters. *Water Res.* **2021**, *202*, 117450. [CrossRef] [PubMed]
15. Tahir, M.A.U.H.; Asghar, S.; Manzoor, A.; Noor, M.A. A classification model for class imbalance dataset using genetic programming. *IEEE Access* **2019**, *7*, 71013–71037. [CrossRef]
16. Haq, M.A.; Jilani, A.K.; Prabu, P. Deep Learning Based Modeling of Groundwater Storage Change. *Comput. Mater. Contin.* **2022**, *70*, 4599–4617.

toxics

Article

Toxicity Risk Assessment Due to Particulate Matter Pollution from Regional Health Data: Case Study from Central Romania

Carmen Maftei [1,*], Ashok Vaseashta [2,3,*] and Ionut Poinareanu [4,5,6]

1 Faculty of Civil Engineering, Transilvania University of Brasov, 900152 Brasov, Romania
2 Office of Research, International Clean Water Institute, Manassas, VA 20108, USA
3 Institute of Biomedical Engineering and Nanotechnologies, Faculty of Mechanical Engineering, Transport and Aeronautics, Ķīpsalas, LV1048 Rīga, Latvia
4 Faculty of Medicine, "Ovidius" University of Constanta, 900470 Constanta, Romania
5 Clinical Service of Pathology, "St. Apostol Andrei" Emergency County Hospital, 145 Tomis Blvd., 900591 Constanta, Romania
6 Faculty of Materials Science and Engineering, Transilvania University of Brasov, 500036 Brasov, Romania
* Correspondence: carmen.maftei@unitbv.ro (C.M.); prof.vaseashta@ieee.org (A.V.)

Abstract: Air pollution poses one of the greatest dangers to public well-being. This article outlines a study conducted in the Central Romania Region regarding the health risks associated with particulate matter (PM) of two sizes, viz., PM_{10} and $PM_{2.5}$. The methodology used consists of the following: (i) an analysis of the effects of PM pollutants, (ii) an analysis of total mortality and cardiovascular-related mortality, and (iii) a general health risk assessment. The Central Region of Romania is situated in the Carpathian Mountains' inner arch (consisting of six counties). The total population of the region under investigation is about 2.6 million inhabitants. Health risk assessment is calculated based on the relative risk (RR) formula. During the study period, our simulations show that reducing these pollutants' concentrations below the new WHO guidelines (2021) will prevent over 172 total fatalities in Brasov alone, as an example. Furthermore, the potential benefit of reducing annual $PM_{2.5}$ levels on total cardiovascular mortality is around 188 persons in Brasov. Although health benefits may also depend upon other physiological parameters, all general health indicators point towards a significant improvement in overall health by a general reduction in particulate matter, as is shown by the toxicity assessment of the particulate matter in the region of interest. The modality can be applied to other locations for similar studies.

Keywords: PM; risk assessment; central Romania; cardiovascular; health; indicators

Citation: Maftei, C.; Vaseashta, A.; Poinareanu, I. Toxicity Risk Assessment Due to Particulate Matter Pollution from Regional Health Data: Case Study from Central Romania. *Toxics* **2024**, *12*, 137. https://doi.org/10.3390/toxics12020137

Academic Editor: Chongjian Wang

Received: 15 November 2023
Revised: 2 February 2024
Accepted: 5 February 2024
Published: 7 February 2024

1. Introduction

Particulate matter (PM) is classified according to its diameters, viz., that with a diameter of 10 microns or less (PM_{10}), while fine particulate matter is defined as particles that are 2.5 microns or less in diameter ($PM_{2.5}$); thus, $PM_{2.5}$ comprises a portion of PM_{10}. The common sources of PM_{10} include manufacturing industries, construction, and fossil fuel combustion, such as diesel exhaust particles (DEP) and emissions from coal-burning stoves [1]. PM_{10} is inhalable into the lungs and can induce adverse health effects, while exposure to fine $PM_{2.5}$ aggravates cardiovascular disease (CVD), among other physiological effects. A recent study conducted by the authors [2] demonstrates that the key sectors responsible for polluting the air in Brasov with PM_{10} and $PM_{2.5}$ are commercial, institutional, and households (61.2% and 48.2%, respectively), manufacturing industries and construction (14.2% and 11.1%, respectively), transportation (11.8% and 10.6%, respectively), mineral products (12.3% and 29.6%, respectively), and energy production and distribution (~0.3%). In general, airborne PM varies widely in size, shape, and chemical composition. Particles are defined by their diameter for air quality regulatory purposes since a wide range of adverse health effects have been linked to air pollution exposure. A publication by the World Health Organization (WHO) [3] asserts that many health effects are associated with air pollution, such as mortality caused by chronic cardiovascular and respiratory diseases, chronic obstructive pulmonary disease (COPD), etc. A study reported by the Global Burden of Diseases (GBD) reveals that the global level of total death for the 1990–2019 period represented by cardiovascular disease (CVD) (GBD Compare, 2022) was at 32.84%, while, for the same period, at the Romanian level, the percent was 57.26%. Further studies conducted around the globe [4–6] show that air pollution, and especially exposure to $PM_{2.5}$, exacerbates CVD, resulting in a high mortality rate.

Several factors that generally contribute to CVD are certain lifestyles, such as obesity, alcohol consumption, and the use of tobacco [7], but several recent epidemiological studies show that air pollution could also contribute to an increased risk of CVD. In 2006, Pope et al. [8] showed that short-term exposure to $PM_{2.5}$ could be associated with acute ischemic heart disease (IHD) events. The research literature also suggests that increased PM concentrations are linked to higher rates of morbidity and mortality among EU countries. According to Orru [9], in Estonia, PM represents a public health concern, leading to an annual increase in estimated premature deaths, a decrease in life expectancy, and an increase in the number of hospitalizations. Exposure to $PM_{2.5}$ is a major cause of premature deaths worldwide, as compared to previous estimations of mortality due to this pollutant and is significantly higher (~50%) according to studies by Anenberg et al. [10]. The Aphekom project concludes that EU citizens are continuously exposed to air pollution of $PM_{2.5}$ exceeding the WHO limits, and it was concluded that life expectancy at age 30 could be reduced by 22 months [11].

Airborne particulate matter consists of a mixture of many chemical species, viz., solids, aerosols of small droplets of liquid, dry solid fragments, and solid cores with liquid coatings, particles of varying sizes, shapes, and chemical compositions, and may contain inorganic ions, metallic compounds, elemental carbon, organic compounds, and compounds from the earth's crust. The relative risk (*RR*) is a widely used function to estimate the health impact of different pollutants. Different studies suggest the logarithmic model recommended by WHO [8,10,12–16]. Concerning the relative risk (*RR*) for the long-term impact of $PM_{2.5}$, Pope [15] recommends a value of 1.06 (95% CI—confidence interval, which represents a range of estimates for an unknown parameter in frequentist statistics—1.02–1.11) per 10 $\mu g/m^3$ (total non-external causes mortality) and a value of 1.12 (95% CI 1.08–1.15) per 10 $\mu g/m^3$ (cardiovascular mortality) [16]. Related to the *RR* for the short impact of PM_{10}, WHO [17] recommends a value of 1.006 (95% CI 1.004–1.008) for all causes of mortality and all ages and a value of 1.009 (1.005–1.013) for cardiovascular disease. The Aphekom project [18] used for cardiovascular disease indicates a value of 1.006 (95% CI 1.004–1.008). The objective of this paper is to investigate the health risk assessment of PM_{10} and $PM_{2.5}$ for the Central Region of Romania.

The suspended aerosol mass, in addition to PM_{10} and $PM_{2.5}$, also contains new and emergent contaminants, such as nanoparticles and micro/nanoplastics. The nature and toxicity of suspended PMs, coupled with the presence of new and emergent contaminants, pose health impacts that can be very severe, especially for the very young, elderly, and people with immuno-compromised conditions. These health impacts include $PM_{2.5}$- and PM_{10}-induced airway inflammation, oxidative stress induced by polyaromatic hydrocarbons, covalent modifications of intracellular proteins/enzymes, the innate immune response, and inflammation by biologic compounds, adjuvant effects, suppression of normal defense mechanisms, suppression of oxygen transfer, and adverse impacts on the cardiovascular system and neurobehavior. Although a major source of PM is attributed to fossil fuel combustion and coal-burning stoves, the actual life cycle of suspended particulate matter, including new and emergent contaminants such as micro/nanoparticles [19] and micro/nanoplastics, is not well understood. The regional study conducted here can lead to mapping the health effects on a larger scale, and an integrated database may serve as the basis for expanded investigations into global health impacts due to PM and other suspended pollutants. Hence, this investigation lays the groundwork for developing some policies and guidelines regarding emission, exposure, and the use of associated personal protection equipment.

2. Study Area and Data Sets

The Central Region of Romania covers an area of 34,100 km^2 and is located in the Carpathian Mountains' inner arch (Figure 1). Parts of the three branches of the Carpathian Mountains, along with the hilly and depressed areas of the Transilvania Plateau form the relief, as shown in Figure 1. The hydrographic network is based on the Mures and Olt Rivers' tributaries. Natural and anthropic lakes, such as Balea (glacier), St. Ana (volcanic), and hypersaline lakes, complete the hydrography of the Central Region. The climate is temperate continental and varies according to altitude. From an administrative point of view, the Central Region consists of Brasov, Sibiu, Alba, Mures, Harghita, and Covasna counties (Figure 1). On average, the population is estimated at approx. 2.6 million inhabitants. The region's natural resources include natural gas, materials for construction (basalt, travertine marble), nonferrous metal, and numerous mineral springs. The vegetation consists of steppe; forests occupy about one-third of the territory (1185.1 thousand ha.). The economy has developed according to the resources of this area, such as industry (~32%), with the rest being allocated to agriculture, services, and tourism.

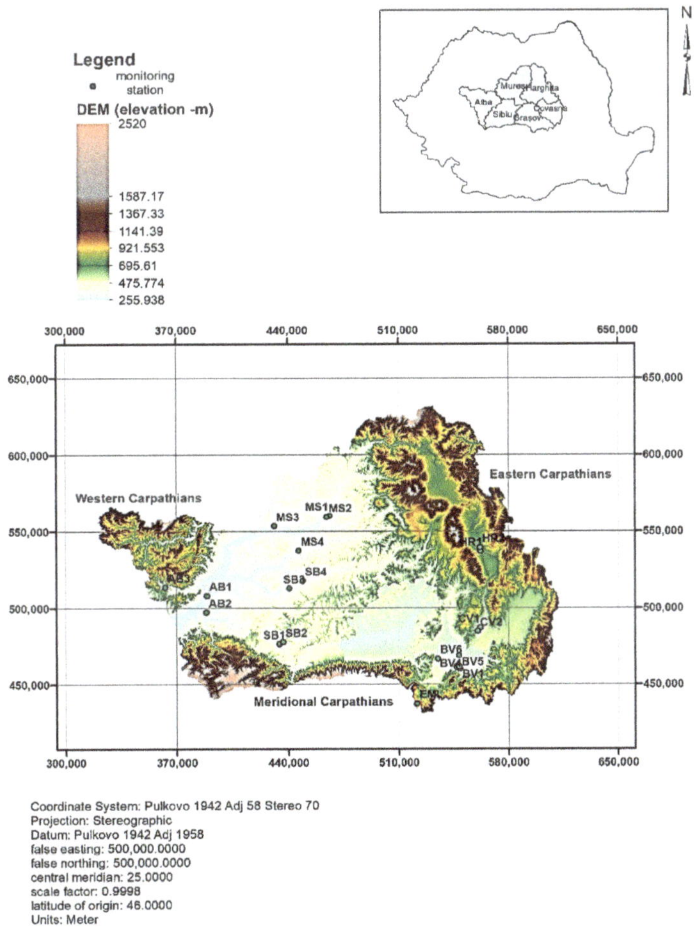

Figure 1. Elevation map of the Central Region of Romania and the location of monitoring stations.

In this study, we use two types of datasets, grouped by independent and dependent variables. The independent variables include PM_{10} and $PM_{2.5}$ as pollutants; the concentrations of these pollutants are obtained through the national air quality monitoring network [20] and correspond to daily data recorded between January 2012 and December 2021. The dependent variables include populations (ranging from age 30 to 85 years or more) and health data. Total mortality/morbidity and cardiovascular mortality data were collected from the National Institute of Statistics (NIS) over the same period (2012–2021) and within the same population age ranges. All data sets are available in the public domain without any attribution, and they do not contain any patient contact information and/or IDs.

The analysis of the status of principal air quality data and its correlation with the most affected area is based on the Romanian Network for Air Quality Monitoring, called Reţeaua Națională de Monitorizare automata a Calităţii Aerului (RNMCA), comprising 148 stations that survey air quality by measuring the concentration of principal pollutants. In the Central Region, there exist 22 stations distributed as follows: Brasov (BV)-seven, Sibiu (SB)-four, Alba Iulia (AB)-three, Mures (MS)-four, Harghita (HR)-two, Covasna (CV)-two, as presented in Figure 1 and Table 1. Unfortunately, the $PM_{2.5}$ pollutant is not measured at every station. In Table 1, we introduced "yes" (Y) or "no" (N) to indicate whether the monitoring station measures this specific pollutant or not. It should be noted that at the HR2 station, PM_{10} is not measured. The reference method for sampling and measuring PM fractions is the one provided in the standard SR EN 12341 (available at https://magazin.asro.ro/ro/standard/229855, accessed on 12 September 2023), viz., ambient air. This standardized method involves gravimetric measurement for the determination of the mass fraction of PM_{10} or $PM_{2.5}$ in suspended particles (in Romanian). All values are provided in $\mu g/m^3$.

Table 1. Measurements of $PM_{2.5}$ and PM_{10}, along with monitoring station details such as type, location, and elevation.

No.	County/Station Indicative	Village	Type	Elevation (m)	$PM_{2.5}$ (Y/N)
			Alba		
1	AB1	Alba Iulia	urban	246	N
2	AB2	Sebes	urban	256	N
3	AB3	Zlatna	urban	450	N
			Sibiu		
4	SB1	Sibiu	urban	430	Y
5	SB2	Sibiu	industrial	402	N
6	SB3	Copsa Mica	industrial	285	N
7	SB4	Medias	industrial	320	N
			Brasov		
8	BV1	Brasov	urban	593	N
9	BV2	Brasov	urban	593	Y
10	BV3	Brasov	urban	593	N
11	BV4	Sanpetru	suburban	560	N
12	BV5	Brasov	industrial	593	N
13	BV6	Codlea	urban	567	Y (Since 2022)
14	EMI	Fundata	suburban	1350	Y (Since 2022)

Table 1. *Cont.*

No.	County/Station Indicative	Village	Type	Elevation (m)	$PM_{2.5}$ (Y/N)
			Mures		
15	MS1	Targu Mures	urban	329	Y
16	MS2	Targu Mures	suburban	304	N
17	MS3	Ludus	suburban	270	N
18	MS4	Tarnaveni	suburban	284	N
			Covasna		
19	CV1	Sf Gheorghe	rural	522	N
20	CV2	Sf. Gheorghe	urban	563	N
			Harghita		
21	HR1	Miercurea Ciuc	industrial	710	Y (Since 2017)
22	HR2	Miercurea Ciuc	urban	689	Y (Since 2019)

It is important to mention the accepted limits according to Romanian legislation (Law #104, which translates to national legislation Directive 2000/60/EC, Water Framework Directive [21] and 2004/107/CE [22] provisions): the standard daily limit for PM_{10} is 50 $\mu g/m^3$, and 40 $\mu g/m^3$ is the annual limit; the standard annual limit for $PM_{2.5}$ is 25 $\mu g/m^3$, and the target until (and after) 2020 is 20 $\mu g/m^3$. The standard daily limit for $PM_{2.5}$ is not regulated. The World Health Organization (WHO) has established guidelines on outdoor (ambient) air pollution levels. These guidelines, initially established in 2005 and updated in 2021, offer the recommended limits for airborne particulate matter, as shown in Table 2.

Table 2. Safety limits for PM according to different air quality standards.

		$PM_{2.5}$ $\mu g/m^3$	PM_{10} $\mu g/m^3$
Romanian legislation and EU standards	annual average	25 (20 until (and after) 2020)	40
	24 h average	not regulated	50
WHO limits [3] 2005	annual average	10	20
	24 h average	25	50
WHO limits [23] 2021	annual average	5	15
	24 h average	15	45

3. Methods and Methodologies

Based on the data sets, the methodologies proposed for this study include the following: (i) analysis of the status of $PM_{2.5}$ and PM_{10} pollutants, (ii) analysis of total mortality and cardiovascular mortality, and (iii) health risk assessment. The analysis of PM_{10} and $PM_{2.5}$ data consists of the following: (i) evaluating yearly and monthly PM_{10} and $PM_{2.5}$ concentrations based on daily measurements; (ii) assessing descriptive statistics; (iii) assessing the number of days/year that exceed the limit value, as established by Romanian legislation and WHO requirements; and (iv) assessing the correlation between PM_{10} and $PM_{2.5}$. Descriptive statistics are conducted using a data analysis package available with MS Excel 365.

To analyze total mortality and cardiovascular mortality, two indicators are used: (i) the mortality rate and (ii) cardiovascular mortality. The two rates are calculated as the number of deaths by 100,000 inhabitants. Health risk assessment consists of two stages: (i) health risk assessment based on the Ostro methodology [14] for the short-term effect of PM_{10} or $PM_{2.5}$, (ii) health impact assessment based on the Pascal methodology [11] implemented in the "Aphekom" project and available on their website [24] for the long-term effect of $PM_{2.5}$. Detailed equations are available in the guidelines presented on the aforementioned website. According to Ostro et al. [14], the calculation of the number of deaths associated with exposure to PM_{10} (total non-external causes mortality) or/and to $PM_{2.5}$ (total mortality) is carried out based on the following equation:

$$N_{assigned} = AF \cdot N_{total} = \frac{(RR-1)}{RR} N_{total} \quad (1)$$

where N_{assigned} represents the number of deaths assigned to PM_{10} or $PM_{2.5}$ pollutants, *AF* the attributable fraction, N_{total} is the number total of deaths, and *RR* is the relative risk. The relative risk (*RR*) is calculated with the following formula:

$$RR = \exp[\beta(X - X_0)] \quad (2)$$

where X and X_0 represent the annual average of pollutant concentration and background concentration as baseline values, respectively (e.g., 10 $\mu g/m^3$), and β represents the concentration–response coefficient or the CFR coefficient. For short-term exposure of PM_{10}, Ostro et al. [14] proposed a value of 0.0008 for the CFR coefficient values. Anderson [17] estimated a β value of 0.00059 (+/−0.00019) for all ages and all-cause mortality, as recommended by WHO [3]. According to Pascal et al. (2013) [11], an impacted lifetime table is calculated using the following equation:

$$_nD_m^{impacted} = {_nD_m} \cdot e^{-\beta \cdot \Delta x} \quad (3)$$

where $_nD_m^{impacted}$ and $_nD_m$ are the total number of deaths in the age group starting at age n and covering m years for the impacted and baseline life tables, respectively; for the present study, m = 10 is considered to cover a ten-year interval. Δx is the decrease in the pollutant concentration in such a given scenario. In this study, a β value of 0.00059 (+/−0.00019) for PM_{10} and 0.000598 (+/−0.000299–0.000895) for $PM_{2.5}$ is used. Two scenarios are used: (1) where $PM_{2.5}$ yearly average is decreased to 5 $\mu g/m^3$, and (2) where the $PM_{2.5}$ yearly average is decreased to 10 $\mu g/m^3$. Concerning long-term exposure to $PM_{2.5}$, the coefficient used for total mortality is 0.005826, and for CVD, it is 0.011.

To obtain the spatial distribution of a parameter, a geographic information system (GIS) method was used. The Voronoi method (or Thiessen polygon) was used to assign a surface for each station. Following this, an individual weight was calculated and used to assign the number of total deaths to a polygon. The Thiessen polygon method assumes that the parameter of each point is the same as that of the adjacent station measurement. The investigation was conducted for the period 2011–2022. Concerning data series representing mortality by age group, the National Institute of Statistics (NIS) did not calculate those parameters for 2011.

4. Results and Discussion

In Table 2, we have presented some statistical information about the pollutants measured at the 22 stations and the number of days/years that are over the limits recommended by Romanian legislation and the World Health Organization [21–23]. As can be seen, additional data are needed for the time series data sets in Tables 1 and 3. During the study period, the multi-annual average concentration of PM_{10} varied between 9.41 $\mu g/m^3$ (at Fundata station) and 30.95 $\mu g/m^3$ (at Alba Iulia AB2). Generally, annual values of PM_{10} are not over the limit of 40 $\mu g/m^3$, as required by Romanian legislation and the European Directive (with two exceptions for Brasov, viz., BV3 and BV5). These findings are in line with the results obtained by other authors for Romania [25,26]. Both the daily limit value and annual limit value recommended by WHO were also used as benchmarks in this study (Tables 2 and 3a). Annual values of PM_{10} exceeded the annual safety limits required by WHO. The number of days over the daily limit (50 μg/m^3) varied between 24% (AB2) and 0.5% (EMI–Fundata–Brasov). The number of days over the daily limit required by WHO in 2021 (45 μg/m^3) varied between 18% (AB2) and 1% at the Fundata station. Concerning $PM_{2.5}$, the average value for the Central Region is 17.48 μg/m^3, continuing to surpass the thresholds recommended by the World Health Organization (WHO), which are set at 10 μg/m^3 and 5 μg/m^3, respectively [3,23]. The annual value of this pollutant is not over the limit of 25 μg/m^3 for Sibiu (SB1) but exceeds the limit for all the other stations. The annual limit of $PM_{2.5}$ concentrations exceeded for all stations, and the excess percentages were in the range of 11–100% (Table 3b).

Investigating the multi-annual monthly values of PM_{10} and $PM_{2.5}$, it was established that during the last spring period (May) and summer periods (June–August), the values of the two pollutants were lower than those observed in the rest of the year (Figure 2). These results are in agreement with the results obtained for Brasov by Maftei et al. [2] and for Cluj by Levei et al. [25]. In the autumn, winter, and earlier spring periods, the multi-annual monthly values are higher than the annual limits recommended by WHO in 2006 and 2021 [3,23]. Several studies indicate that variations in PM concentration throughout the seasons directly and adversely affect human health [27]. Only one exception exists in these datasets: the EMI station located in Fundata (Brasov County), where the variation of the PM_{10} pollutants shows a reverse trend. In January, February, November, and December, the multi-annual mean values are lower than the values registered during the rest of the year, when the multi-annual mean values are comparable with those in the rest of the monitoring stations. The influence of precipitation on PM_{10} was highlighted by Popescu LL et al. (2022) [28] through the daily measurements, while daily average $PM_{2.5}$ concentrations were less influenced.

To estimate the relationship between $PM_{2.5}$ (as a dependent variable) and PM_{10}, regression analysis using the Excel data analysis package is employed. Considering the values of the R-squared determination coefficient, the results show that between 63% (MS1 and SB1) and 91% (BV2), the $PM_{2.5}$ values fit the regression analysis model. The correlation coefficient of the linear relationship between the two variables is situated in the 0.79 and 0.95 range, which demonstrates a strong positive relationship. The F-significance value for all stations investigated is less than 0.05 (5%), which means that the null hypothesis is accepted, and the linear regression model fits the data well. Investigating the $PM_{2.5}/PM_{10}$ ratio, it is observed that this parameter varies within a narrow range (0.61 at SB1 station to 0.88 at MS1 station). The average ratio for the Central Region is 0.71, which is slightly over the value obtained by Bodor [12] for the same region (0.67) and in accordance with the value proposed by Pascal et al. (2013) in the Aphekom project [24].

The estimated population in the Central Region of Romania is approximately 2635 thousand inhabitants. The average population by county (Table 4) varies between 204,688 inhabitants (Covasna) and 551,685 inhabitants (Brasov). Generally, the population is decreasing (between −2.13% and −4.19%), with two exceptions: Brasov and Sibiu, where increases of 1.7 and 1.5%, respectively, were observed.

Table 3. Descriptive statistic on the annual values of PM_{10} data sets; no. of days over safety limits (see Table 2).

(a)																	
PM_{10} (2011–2021)—Statistics																	
No. crt.	Station Indicative	No. obsv.	Min. μg/m³	Max. μg/m³	Mean μg/m³	StD. μg/m³	5th μg/m³	95th μg/m³	No. of Days over Limit of 50 μg/m³		No. of Days over Limit of 45 μg/m³		No. of Year over Limit of 40 μg/m³		No. of Year over Limit of 20 μg/m³		No. of Year over Limit of 15 μg/m³	
									no	%	no	%	no	%	no	%	no	%
		1	2	3	4	5	6	7	8	9	10	11	12		13	14	15	16
Alba																	
1	AB1	3769	0.01	104.84	23.36	13.73	8.18	49.42	110	3%	361	10%	0	0	9	82%	11	100%
2	AB2	982	2.88	120.41	30.95	17.80	11.30	66.86	107	11%	179	18%	0	0	3	100%	3	100%
3	AB3	3601	0.38	74.03	20.92	12.24	7.45	47.87	51	1%	241	7%	0	0	6	55%	11	100%
Sibiu																	
4	SB1	3206	0.73	109	21.57	12.42	7.27	45.42	98	3.06%	171	5%	0	0	6	60.00%	10	100.00%
5	SB2	2813	0.27	102.97	12.75	11.69	1.58	35.56	49	1.74%	75	3%	0	0	2	18.18%	3	27.27%
6	SB3	3051	0.24	112.64	20.87	13.73	4.72	47.24	104	3.41%	214	7%	0	0	6	54.55%	8	72.73%
7	SB4	2845	1.27	125.36	22.74	15.14	5.45	49.86	142	4.99%	272	10%	0	0	7	70.00%	9	90.00%
Brasov																	
8	BV1	3557	1.82	179.23	26.09	16.87	7.63	55.81	252	7.08%	369	10%	0		11	100.00%	11	100.00%
9	BV2	1649	2.36	255.93	27.41	20.83	8.18	57.40	107	6.49%	181	11%	0		5	100.00%	5	100.00%
10	BV3	3710	1.27	216.48	30.81	19.37	10.22	64.14	368	9.92%	589	16%	1	0.02	11	100.00%	11	100.00%
11	BV4	3151	0.36	200.95	21.16	17.48	4.54	49.33	153	4.86%	205	7%	0		7	70.00%	10	100.00%
12	BV5	1167	0.43	272.09	22.80	22.65	5.63	56.21	74	6.34%	91	8%	2	0.17	6	75.00%	8	100.00%
13	BV6	249	5.09	80.36	22.06	13.01	7.73	46.68	11	4.42%	16	6%	0		1	100.00%	1	100.00%
14	EMI	726	0.73	66.86	9.48	8.00	1.45	23.89	4	0.55%	5	1%	0		0	0.00%	0	0.00%
Mures																	
15	MS1	2126	0.89	154.32	24.10	17.80	6.22	58.44	170	8.00%	231	11%	0		8	72.73%	9	81.82%
16	MS2	2438	0.52	263.71	25.19	18.98	5.79	60.47	210	8.61%	283	12%	0		8	72.73%	9	81.82%
17	MS3	881	2.05	110.36	22.43	13.21	7.26	48.39	40	4.54%	54	6%	0		3	75.00%	4	100.00%
18	MS4	2133	0.04	95.39	10.59	11.26	1.60	32.29	41	1.92%	57	3%	0		0	0.00%	3	33.33%
Covasna																	
19	CV1	2484	0.09	141.16	18.93	13.19	4.62	43.67	88	3.54%	118	5%	0		5	45.45%	8	72.73%
20	CV2	204	1.63	75.25	21.85	19.7	5.71	41.04	6	2.94%	8	4%	0		0	0.00%	1	100.00%

Table 3. *Cont.*

(a)

PM$_{10}$ (2011–2021)—Statistics

No. crt.	Station Indicative	No. obsv.	Min. µg/m³	Max. µg/m³	Mean µg/m³	StD. µg/m³	5th µg/m³	95th µg/m³	No. of Days over Limit of 50 µg/m³		No. of Days over Limit of 45 µg/m³		No. of Year over Limit of 40 µg/m³		No. of Year over Limit of 20 µg/m³		No. of Year over Limit of 15 µg/m³	
									no	%	no	%	no	%	no	%	no	%
		1	2	3	4	5	6	7	8	9	10	11	12		13	14	15	16
									Harghita									
21	HR1	3225	0.18	221.6	16.94	19.05	2.36	51.02	164	5%	197	6%	0		2	18%	8	73%
22	HR2	-	-	-	-	-	-	-	-	-	-	-	-		-	-	-	-

(b)

PM$_{2.5}$ (2011–2021)—Statistics

Station Indicative	No. obsv.	Min µg/m³	Max µg/m³	Mean µg/m³	StD. µg/m³	5th µg/m³	95th µg/m³	No. of Year over Limit of 25 µg/m³		No. of Year over Limit of 20 µg/m³		No. of Year over Limit of 10 µg/m³		No. of Year over Limit of 5 µg/m³		No. of Days over Limit of 25 µg/m³		No. of Days over Limit of 15 µg/m³	
								no	%	no	%	no	%	no	%	no	%	no	%
	1	2	3	4	5	6	7	8		9		10		11		12		13	14
								Sibiu											
SB1	2347.00	0.53	78.40	13.85	9.86	3.45	34.52	0	0%	1	11%	7	78%	9	100%	282	12%	791	34%
								Brasov											
BV2	3283	1.09	198.31	17.36	14.67	4.9	41	1	9%	3	27%	11	100%	11	100%	564	17%	1521	46%
								Mures											
MS1	1606	0.71	161.9	21.22	21.72	4	66.3	2	29%	3	43%	7	100%	7	100%	407	0.25	726	0.5
								Harghita											
HR1	697	0.18	154.99	15.68	19.79	2.3	52	2	40%	2	40%	4	80%	5	100%	99	14%	200	29%
HR2	194	2.17	78.8	19.3	14.51	4.9	52.1	1	100%	1	100%	1	100%	1	100%	43	22%	90	46%

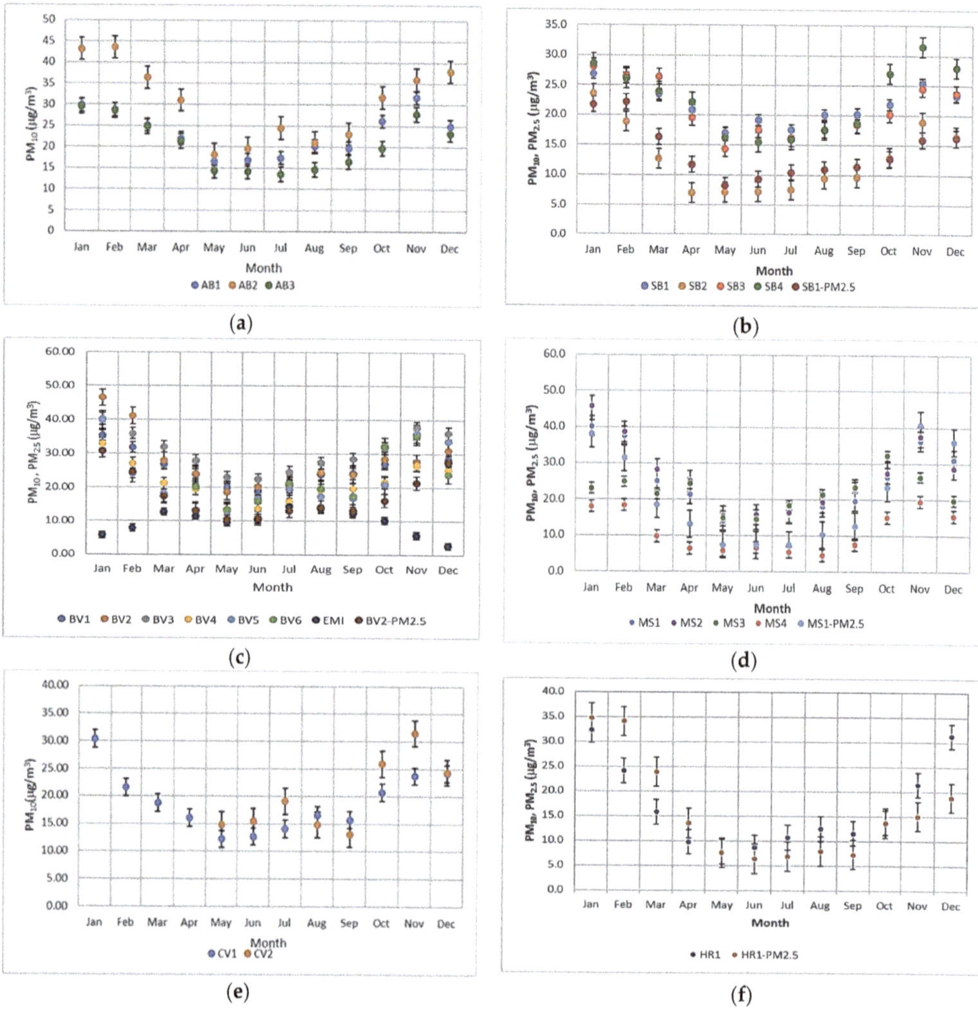

Figure 2. Evaluation of month-of-year mean concentrations of PM_{10} and $PM_{2.5}$ over multiple years for (**a**) Alba, (**b**) Sibiu, (**c**) Brasov, (**d**) Mures, (**e**) Covasna, and (**f**) Harghita.

Table 4. Average population for Central Region counties (2011–2021).

	Central Region	Alba	Brasov	Covasna	Harghita	Mures	Sibiu
average	2,635,986	380,571	632,764	228,459	333,280	595,829	465,083

The adult population by age group is presented in Figure 3. On average, more than 3% of the population around the Central Region was over the age of 80. The population aged 65 and above is between 13% (Sibiu) and 17% (Alba Iulia). Brasov has the highest population in the age group of 60–64 (6%). Moreover, the aging coefficient (its definition and explanation are beyond the scope of this article), physiologically, at the cellular level, is affected by the loss of specific regenerative and bioprotective mechanisms that occur over time in an organism due to exposure to PM. This coefficient varies between 19% (Sibiu) and 23% (Alba). Even while these percentages may seem negligible, the proportion of the elderly has increased significantly since 1960 (6.7%). The population aged under 30 is around 40%, with two exceptions: Covasna (20%) and Sibiu (24%). Moreover, Covasna has the lowest percentage in the age groups of 30–34, 35–39, and 40–45, which are ~7% each, respectively (Figure 3).

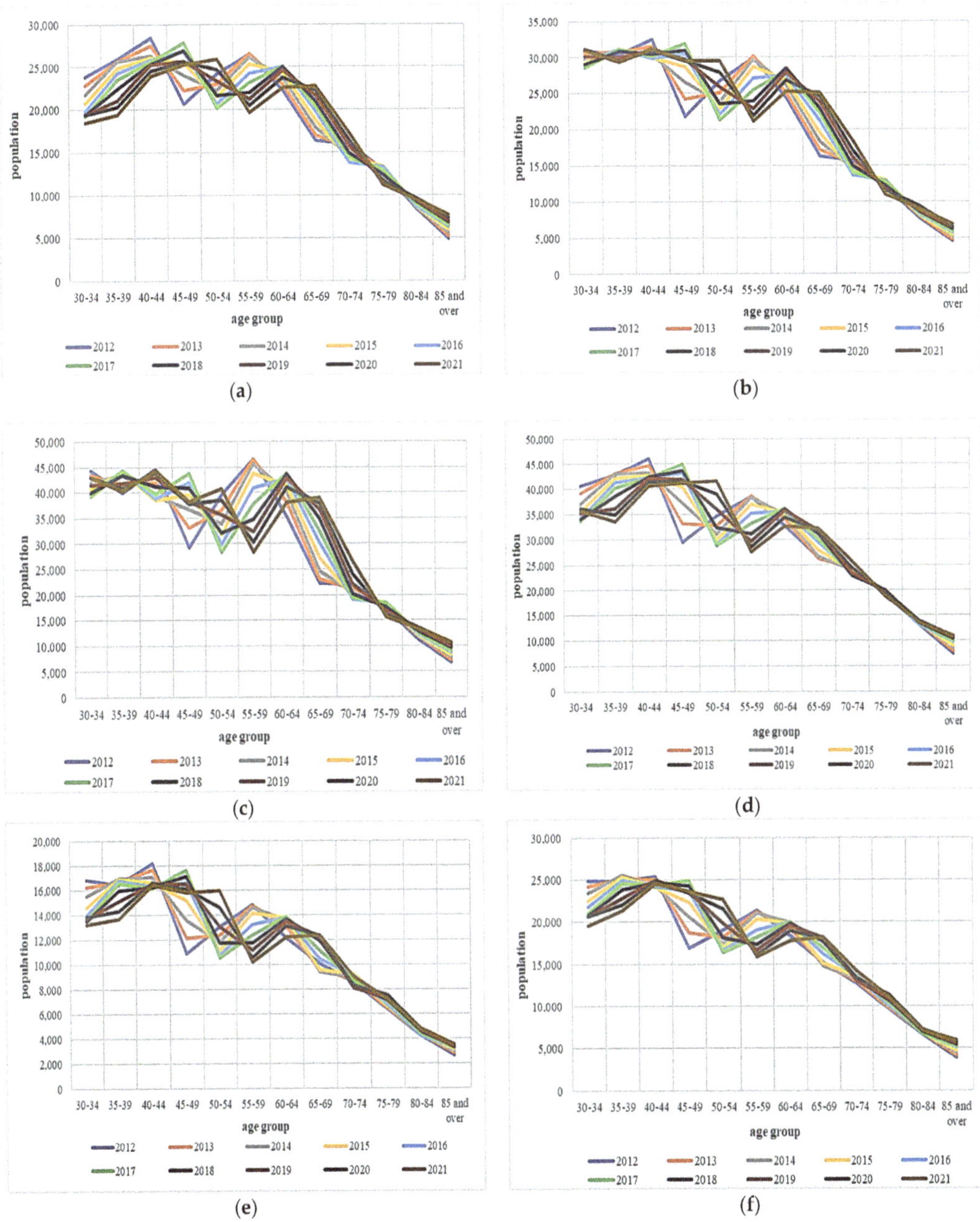

Figure 3. The population reported by age group for counties (**a**): Alba, (**b**) Sibiu, (**c**) Brasov, (**d**) Mures, (**e**) Covasna, and (**f**) Harghita.

The national mortality rate is presented in Figure 4. For the study period (2011–2021), 53% of the total mortality was caused by cardiovascular diseases (CVD) [29]. A positive trend is observed in the total mortality rate (Figure 4) caused by fatalities registered in 2020 and 2021 (note: the time series data were reviewed by NIS—but are not final). The same behavior is seen for the mortality rate due to CVD (Figure 5). The pandemic's indirect impact on the management of cardiovascular disease could partially be responsible for the higher-than-expected mortality toll associated with COVID-19 [30]. The total mortality rate by county is presented in Figure 5a. As can be seen, Mures County has the highest rate of total mortality (1925 per 100,000 inhabitants), followed by Brasov (1723 per 100,000 inhabitants). The lowest rate is registered in Covasna County. The mortality rate caused by CVD is presented in Figure 5b. As can be noticed, the highest mortality rate caused by CVD is registered in Mures County, while the lowest is in Covasna.

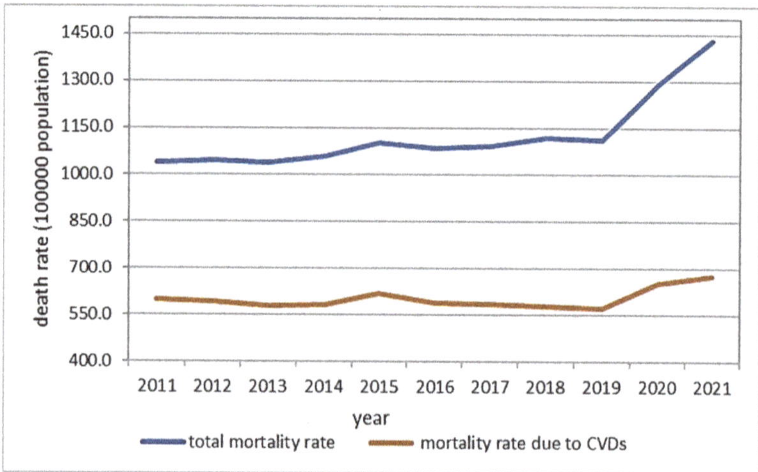

Figure 4. National mortality rate (2011–2021) (total mortality rate and rate due to CVDs).

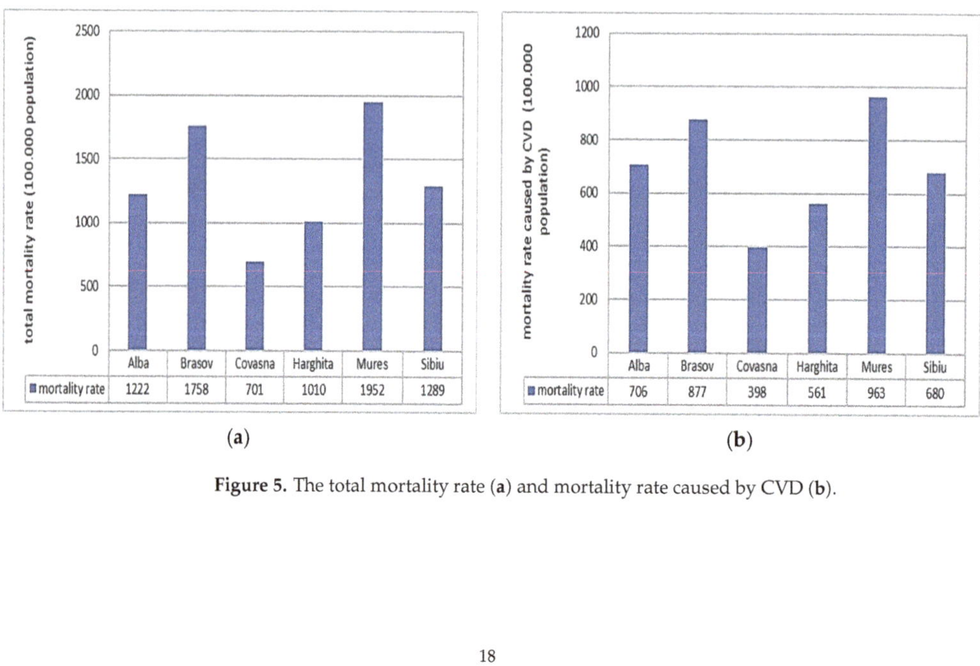

(a) (b)

Figure 5. The total mortality rate (a) and mortality rate caused by CVD (b).

The relative risk (RR) for PM_{10} and $PM_{2.5}$ was estimated for all-cause mortality, and the results are presented in Figures 6 and 7. Related to PM_{10}, RR for HR2 is not calculated due to missing data.

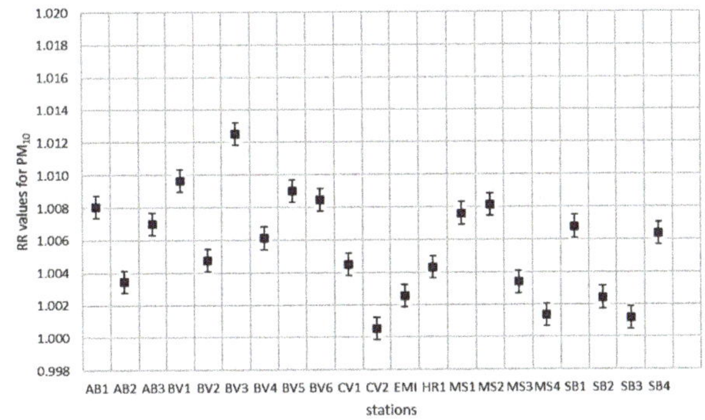

Figure 6. *RR* values for PM_{10}.

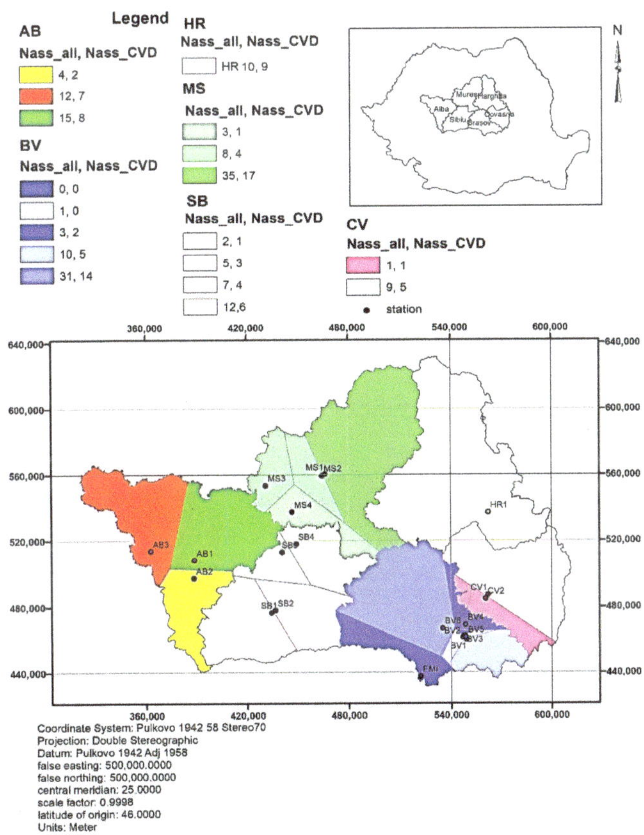

Figure 7. Spatial distribution of number of deaths attributed to all causes of mortality. (N_{ass} all) represents the number of deaths due to PM_{10} and (N_{ass}_CVD) represents the number of CVD-related deaths due to $PM_{2.5}$.

Summary statistics related to PM_{10} *RR* values are presented in the following table (Table 5). The average value of *RR* is 1.006 (+/−0.0014), which is in agreement with the research literature, as mentioned in the Section 1. The minimum value is obtained for CV2 (1.001), which is situated in an urban (residential) area near Elisabeta Park. Research on the effect of urban greenery on air pollution conducted by Cohen P. (2014) demonstrates that the urban park could influence the PM_{10} and NOx concentrations [31].

Table 5. Summary statistics of *RR* values related to PM_{10} for all monitoring stations.

Average	1.006	Maximum	1.0125
Standard Error	0.0007	Count	21
Median	1.006	CI (95.0%)	0.0014
Standard Deviation	0.00317	Upper Limit	1.0074
Minimum	1.00052	Lower Limit	1.0042

In Brasov's urban region, an average relative risk of 1.009 was observed in connection with PM_{10} and all-cause mortality. The highest recorded value was at BV3, situated near the train station—an area characterized by heavy road and rail traffic (1.013). The values obtained for urban areas in major cities (Sibiu, Târgu Mureș) are similar to those obtained in Brasov. Concerning $PM_{2.5}$, there are only five stations that do not cover all the region. For a short-term analysis, the values of *RR* are similar due to β coefficient, which is very close to the coefficient used for PM_{10}. For long-term exposure to $PM_{2.5}$, the values of *RR* are shown in Table 6.

Table 6. *RR* values for the five stations investigated.

Station	*RR*		$N_{assigned}$	$N_{assigned}$ CVD
	All Causes	CVD		
BV2	1.044	1.088	272	259
HR1 and HR2	1.054	1.127	312	100
MS1	1.072	1.148	616	568
SB1	1.084	1.171	138	130

The spatial distribution of the number of deaths attributed to air pollution with PM_{10} (total non-external causes mortality)—$N_{assigned}$ is presented in the following figure (Figure 7). The number of total deaths attributed to air pollution with PM_{10} varied from 53 (MS-Mures) to 10 (CV-Covasna) and the same parameter related to CVD varied from 26 to (MS-Mures) to 6 (CV-Covasna).

The number of deaths associated with exposure to $PM_{2.5}$ (all causes of mortality) and $PM_{2.5}$ (CVD mortality) for long-term analysis are presented in Table 7. The values represent the average for the period of study. As we mentioned in the methodology, to calculate the health impact assessment of $PM_{2.5}$ for the long term, the methodology proposed in the "Aphekom" project was used. Within this method, $PM_{2.5}$ is computed from PM_{10} using a correction factor of 0.7. In this case, we could use the incomplete time series data for the BV6 and HR2 stations. An example of such results, by this proposed method, is presented in Table 7.

Using the ratio $PM_{2.5}/PM_{10}$ of 0.7, the annual number of deaths avoided (for both all causes and CVD mortality) for the first scenario (viz., decrease by 5 $\mu g/m^3$) remains the same for the period of this study. For the second scenario, it has been found that there is a decrease of 6 to 97.6%. The annual number of fatalities avoided due to the reduction of $PM_{2.5}$ by 5 $\mu g/m^3$ and to 10 $\mu g/m^3$ (for both all causes and CVD mortality) is presented in Figure 8 for the study period. Decreasing air pollution levels (due to the reduction in $PM_{2.5}$ by 5 $\mu g/m^3$) to the updated WHO limits can save 161.5 lives (on average) in the case of total mortality (viz., 132.4—HR1 and 186.7—MS1) and an average of 147.5 (viz. 0.0—HR2 to 188.9—BV6) in the case of cardiovascular mortality.

Table 7. Potential benefits of reducing annual $PM_{2.5}$ levels on total non-external mortality and total cardiovascular mortality for Brasov.

Station	Scenarios	Total Non-External Mortality			Total Cardiovascular Mortality	
		Annual Number of Deaths Avoided	Annual Number of Deaths Avoided per 100,000	Gain in Life Expectancy (Months)	Annual Number of Deaths Avoided	Annual Number of Deaths Avoided per 100,000
BV2	Decrease by 5 µg/m³	172.4	50.6	4.74	188.9	55.4
	Decrease to 10 µg/m³	173.4	50.9	4.77	190.0	55.7

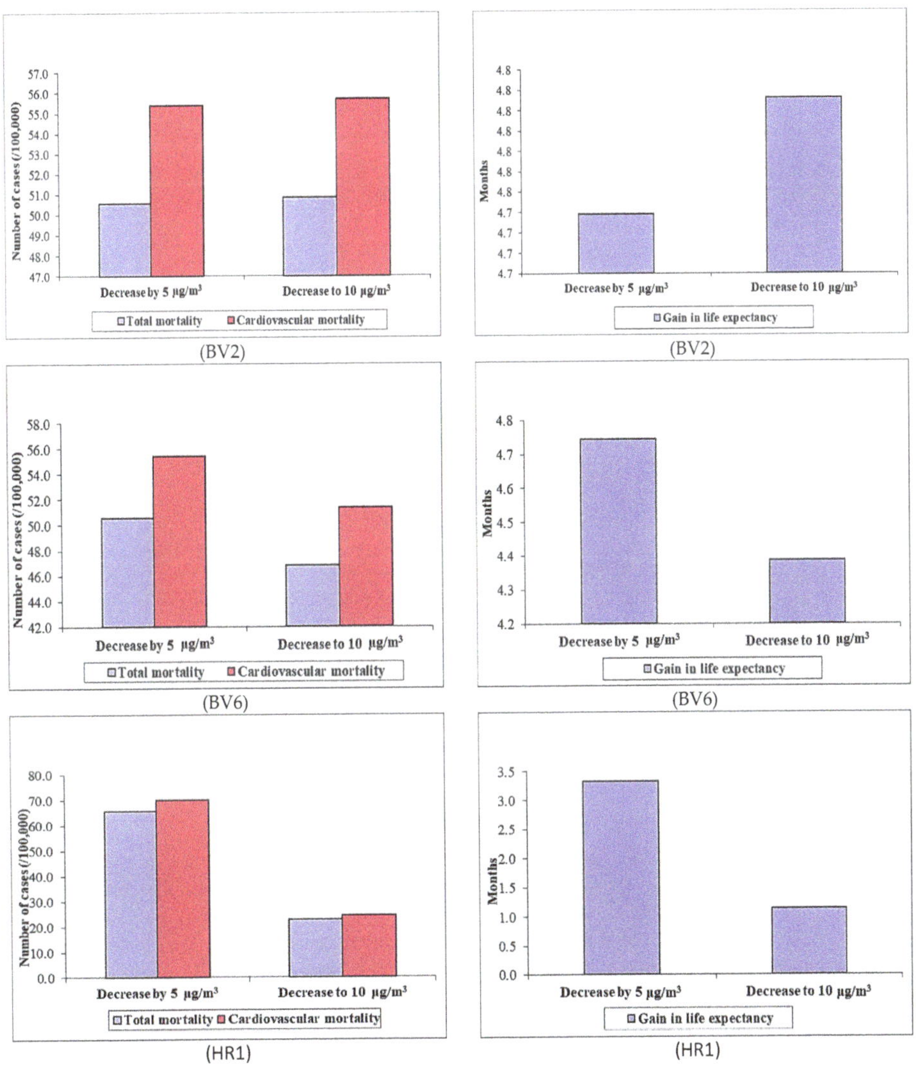

Figure 8. *Cont.*

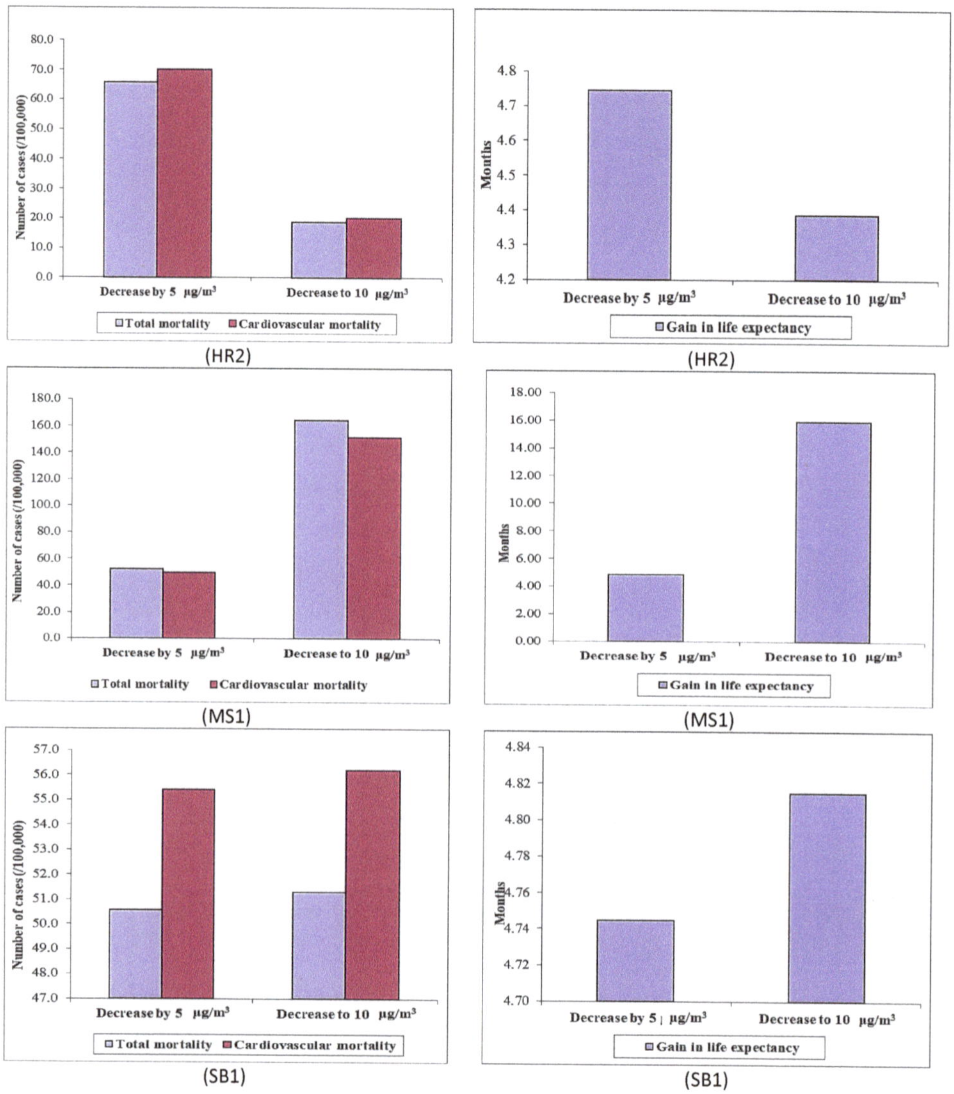

Figure 8. Potential benefits of reducing annual $PM_{2.5}$ levels on total non-external mortality (**left**) and on total cardiovascular mortality (**right**).

For the second scenario (with a decrease to 10 µg/m^3), a total of 196.1 deaths are avoided in the case of total mortality, and approximately 191 deaths are avoided for cardiovascular mortality. The gain in life expectancy is on average 4.3 months for the Central Region in the first scenario, both for all causes and CVD mortality. In the second scenario, the gain in life expectancy is 5.3 for total mortality and 3.5 for cardiovascular mortality.

Health benefits that are related to an improvement of ambient air quality in the Central Region of Romania are similar to previous estimates obtained from different other studies conducted worldwide [25,28,32–37]. According to our study, a reduction in short-term exposure to $PM_{2.5}$ by 5 µg/m^3 results in an annual avoidance of non-external deaths ranging from 50.6 to 65.7 per 100,000 inhabitants. If cities situated in the Central Region of Romania could lower the mean of $PM_{2.5}$ levels to 10 µg/m^3, approximately 196 annual deaths (total non-external mortality) would be delayed, and the population would gain more than 5 months in life expectancy. Those values are in concordance with the results published on the ISGlobal Ranking of Cities website (https://isglobalranking.org/, accessed on 12 September 2023). Based on our investigations, if the cities in the Central Region of Romania managed to meet the WHO limits for $PM_{2.5}$, approximately 198 deaths due to cardiovascular diseases could be avoided annually.

5. Conclusions, Limitations, and Future Directions

This article presents the health impact of an average of PM_{10} and $PM_{2.5}$ levels above the average limits recommended by Romanian legislation and WHO in the Central Region of Romania. During this study period, the average of PM_{10} was observed to be 1.09 times higher than the annual acceptable limit of 20 $\mu g/m^3$, which is 1.46 times higher than the annual acceptable limit of 15 $\mu g/m^3$ (limits recommended by WHO 2006 and 2021, respectively). The maximum values of PM_{10}, reaching 30.95 $\mu g/m^3$, were registered in Alba Iulia County at the AB2 station, which is situated in Sebes town. This town is located near the intersection of two major highways (A0 and A1) and has a well-developed industrial complex nearby (especially in the wood industry). The second place is the BV3 station located in an urban city area (Brasov County) in proximity to the central railway station, with a significant amount of traffic nearby. The minimum value was registered at EMI-Brasov at 9.2 $\mu g/m^3$; however, this station is a reference point for air quality assessment, being positioned at ~1350m elevation in a mountainous region of Brasov County. The multi-annual value of 10.21 $\mu g/m^3$ is registered in Mures County at MS4. The location of this station is Tarnaveni town. The industry developed here is based on methane gas resources, albeit significantly diminished after the 1989 Romanian revolution. The 10 $\mu g/m^3$ yearly limit of $PM_{2.5}$ recommended by WHO (2006) [3] was exceeded by ~1.1 times at the SB1 station, located in a residential area in Sibiu city. The limit of 5 $\mu g/m^3$ (WHO 2021) was exceeded by ~2.2 times at the same station and at the BV2 station, which is situated in a residential area in Brasov City. The monthly variation of PM_{10} and $PM_{2.5}$ shows a strong seasonality in all six counties. The maximum level was registered in winter and autumn due to commercial and institutional activities, as well as household heating and transportation. It is instructive to note here that the Brasov area is a tourist destination due to the predominance of mountains. Several ski resorts have been developed here, which, along with the historical monuments, have led to the development of cultural tourism [25]. Despite tourism-related traffic, the minimum level was registered during the summer period.

Related to the calculated risk, two analyses were conducted. One refers to the short-term exposure of $PM_{10}/PM_{2.5}$, and the second to the long-term exposure effect of $PM_{2.5}$. The higher calculated risk for PM_{10} risk was found in Brasov at three stations (BV3, BV1, BV5, and BV6). The first five stations mentioned here are located in Brasov city, while BV6 is located in Codlea town. Also, it is important to note that DN1—a national road—is an important source of pollution. This analysis provides the short-term calculated risk for $PM_{2.5}$ (cardiovascular disease), and it is not significantly different from the PM_{10} risk, especially due to the CFR coefficient, which is practically the same. The higher calculated risk for $PM_{2.5}$ for total mortality is obtained for MS1 and the two stations situated in Harghita County. The results show that exposure to these pollutants could cause an increase in both total mortality and cardiovascular mortality in the Central Region of Romania. The higher number of deaths assigned to $PM_{2.5}$ is obtained in Mures—49 (on average), of which 25 are due to cardiovascular disease. As a result of long-term exposure to $PM_{2.5}$, a higher number of deaths assigned to $PM_{2.5}$ pollutants is obtained in Mures, ~616, of which 568 are due to cardiovascular disease. To conclude, Mures County is the most exposed county to $PM_{2.5}$, followed by Brasov and Alba Iulia. The number of deaths assigned to PM_{10} pollutants varies from 10 in Covasna County to 56 in Mures County in terms of total mortality, from which 6 and 26, respectively, are assigned to cardiovascular mortality.

The results of this study offer the strongest argument currently available that prolonged exposure to airborne fine particulates that are typical of many urban areas is a significant risk factor for cardiovascular death. Related to health impact assessment, the present study shows that, by adopting the new WHO [3] limits, the potential benefits of reducing annual $PM_{2.5}$ levels on total mortality are 196 (an average) in the Central Region of Romania. This would increase life expectancy by approximately 5.3 months. Related to cardiovascular mortality, the average number of deaths reduced is 190, which means an approximate 3.5-month increase in life expectancy. Hence, the results of this investigation indicate that there is a need to reduce the risk of various health concerns that could arise from exposure to particulate matter by introducing the new limits recommended by WHO in Romanian regulation.

Notwithstanding these observations in Central Romania, the study has several limitations, which include gaps in the data series, especially in the one related to $PM_{2.5}$, which can lead to the establishment of an imprecise $PM_{2.5}/PM_{10}$ ratio. However, the results related to this ratio are slightly over the value established by Bodor et al. 2022 [12]; for this reason, we used 0.7 as the $PM_{2.5}/PM_{10}$ ratio. The small set of observations and/or lack of a complete dataset of $PM_{2.5}$ measurement stations prevented the establishment of a realistic spatial distribution of this pollutant. In addition, there is a lack of studies on new-generation heating systems and initiatives due to green and economic policies, and several studies are incomplete. This is due to the inherent fact that the speed of technology far outpaces policies and regulations. Another limitation of the study is that since the data were analyzed as received from the stations, parametric confounding among variables could not be addressed due to the availability of limited data. To conduct a systematic statistical analysis, more datasets would be required. Thus, we believe that having a larger sensor density is crucial. Concerning the influence of climate parameters (precipitation, wind, temperature, etc.) on PM, we consider that modeling the dispersion of pollutants is one aspect that can add value to the studies.

As a future expansion of the scope of this study, it is important to consider that suspended aerosols now contain micro/nanoparticle and micro/nanoplastics, since such materials are used due to their unique properties, enabling a broad range of possible applications, including cosmetic, pharmaceutical, and medical utilization. Although a discussion concerning their emission mechanisms is beyond the scope of this article, these materials are emitted into the aquatic environment and air. The toxicological impacts of these new and emerging contaminants are largely unknown in terms of their health effects. Some preliminary studies, especially in aquatic environments [38–40], may serve as references to airborne contamination and its adverse health impacts. However, the overall study of the health impacts of $PM_{2.5}$ and PM_{10}, especially in the context of new and emerging contaminants, needs careful and extensive investigation. Furthermore, it will be very beneficial to quantify these data in terms of disability-adjusted life year (DALY)—as a measure of overall disease burden due to PM and new and emergent contaminants. For the aquatic environment, DALY indicators are estimated for such contaminants through the Universal Water Quality Index (UWQI) [41] for new and emergent contaminants.

Author Contributions: Conceptualization, C.M. and I.P.; methodology, C.M., I.P. and A.V.; software, C.M. and I.P.; validation, C.M., I.P. and A.V.; formal analysis, C.M.; investigation, C.M. and I.P.; resources, C.M.; data curation, C.M. and I.P.; writing—original draft preparation, C.M. and A.V.; writing—review and editing, C.M. and A.V.; visualization, C.M.; supervision, C.M.; project administration, C.M. All authors have read and agreed to the published version of the manuscript.

Funding: This research received no external funding.

Informed Consent Statement: Not applicable.

Data Availability Statement: Most of the data are contained in the article. Any additional data, including links to publicly archived datasets analyzed or generated during the study, can be requested from the corresponding author, upon reasonable request.

Conflicts of Interest: The authors declare no conflicts of interest—financial or otherwise.

References

1. Thurston, G.D.; Burnett, R.T.; Turner, M.C.; Shi, Y.; Krewski, D.; Lall, R.; Ito, K.; Jerrett, M.; Gapstur, S.M.; Diver, W.R.; et al. Ischemic Heart Disease Mortality and Long-Term Exposure to Source-Related Components of U.S. Fine Particle Air Pollution. *Environ. Health Perspect.* **2016**, *124*, 785–794. [CrossRef]
2. Maftei, C.; Muntean, R.; Poinareanu, I. The Impact of Air Pollution on Pulmonary Diseases: A Case Study from Brasov County, Romania. *Atmosphere* **2022**, *13*, 902. [CrossRef]
3. World Health Organization (Ed.) *Air Quality Guidelines: Global Update 2005: Particulate Matter, Ozone, Nitrogen Dioxide, and Sulfur Dioxide*; World Health Organization: Copenhagen, Denmark, 2006; ISBN 978-92-890-2192-0.
4. Combes, A.; Franchineau, G. Fine Particle Environmental Pollution and Cardiovascular Diseases. *Metabolism* **2019**, *100*, 153944. [CrossRef] [PubMed]
5. Lu, Y.; Lin, S.; Fatmi, Z.; Malashock, D.; Hussain, M.M.; Siddique, A.; Carpenter, D.O.; Lin, Z.; Khwaja, H.A. Assessing the Association between Fine Particulate Matter ($PM_{2.5}$) Constituents and Cardiovascular Diseases in a Mega-City of Pakistan. *Environ. Pollut.* **2019**, *252*, 1412–1422. [CrossRef] [PubMed]
6. Zhang, W.; Lin, S.; Hopke, P.K.; Thurston, S.W.; van Wijngaarden, E.; Croft, D.; Squizzato, S.; Masiol, M.; Rich, D.Q. Triggering of Cardiovascular Hospital Admissions by Fine Particle Concentrations in New York State: Before, during, and after Implementation of Multiple Environmental Policies and a Recession. *Environ. Pollut.* **2018**, *242*, 1404–1416. [CrossRef] [PubMed]
7. Kim, S.Y.; Kim, S.H.; Wee, J.H.; Min, C.; Han, S.-M.; Kim, S.; Choi, H.G. Short- and Long-Term Exposure to Air Pollution Increases the Risk of Ischemic Heart Disease. *Sci. Rep.* **2021**, *11*, 5108. [CrossRef] [PubMed]
8. Pope, C.A.; Muhlestein, J.B.; May, H.T.; Renlund, D.G.; Anderson, J.L.; Horne, B.D. Ischemic Heart Disease Events Triggered by Short-Term Exposure to Fine Particulate Air Pollution. *Circulation* **2006**, *114*, 2443–2448. [CrossRef] [PubMed]
9. Orru, H.; Maasikmets, M.; Lai, T.; Tamm, T.; Kaasik, M.; Kimmel, V.; Orru, K.; Merisalu, E.; Forsberg, B. Health Impacts of Particulate Matter in Five Major Estonian Towns: Main Sources of Exposure and Local Differences. *Air Qual. Atmos. Health* **2011**, *4*, 247–258. [CrossRef]
10. Anenberg, S.C.; Horowitz, L.W.; Tong, D.Q.; West, J.J. An Estimate of the Global Burden of Anthropogenic Ozone and Fine Particulate Matter on Premature Human Mortality Using Atmospheric Modeling. *Environ. Health Perspect.* **2010**, *118*, 1189–1195. [CrossRef]
11. Pascal, M.; Corso, M.; Chanel, O.; Declercq, C.; Badaloni, C.; Cesaroni, G.; Henschel, S.; Meister, K.; Haluza, D.; Martin-Olmedo, P.; et al. Assessing the Public Health Impacts of Urban Air Pollution in 25 European Cities: Results of the Aphekom Project. *Sci. Total Environ.* **2013**, *449*, 390–400. [CrossRef]
12. Bodor, K.; Szép, R.; Bodor, Z. The Human Health Risk Assessment of Particulate Air Pollution ($PM_{2.5}$ and PM_{10}) in Romania. *Toxicol. Rep.* **2022**, *9*, 556–562. [CrossRef] [PubMed]
13. Burnett, R.; Ma, R.; Jerrett, M.; Goldberg, M.S.; Cakmak, S.; Pope, C.A.; Krewski, D. The Spatial Association between Community Air Pollution and Mortality: A New Method of Analyzing Correlated Geographic Cohort Data. *Environ. Health Perspect.* **2001**, *109*, 375–380. [PubMed]
14. Ostro, B. Outdoor Air Pollution: Assessing the Environmental Burden of Disease at National and Local Levels. In *Environmental Burden of Disease Series*; World Health Organization: Geneva, Switzerland, 2004; Volume 5, ISBN 9241591463.
15. Pope, C.A.; Burnett, R.T.; Thun, M.J.; Calle, E.E.; Krewski, D.; Ito, K.; Thurston, G.D. Lung Cancer, Cardiopulmonary Mortality, and Long-Term Exposure to Fine Particulate Air Pollution. *JAMA* **2002**, *287*, 1132–1141. [CrossRef] [PubMed]
16. Pope, C.A.; Burnett, R.T.; Thurston, G.D.; Thun, M.J.; Calle, E.E.; Krewski, D.; Godleski, J.J. Cardiovascular Mortality and Long-Term Exposure to Particulate Air Pollution. *Circulation* **2004**, *109*, 71–77. [CrossRef] [PubMed]
17. Anderson, H.R.; Atkinson, R.W.; Peacock, J.L.; Marston, L.; Konstantinou, K.; World Health Organization. *Meta-Analysis of Time-Series Studies and Panel Studies of Particulate Matter (PM) and Ozone (O3)*; Report of a WHO Task Group; World Health Organization: Copenhagen, Denmark, 2004.
18. Aphekom Group; Chanel, O.; Perez, L.; Künzli, N.; Medina, S. The Hidden Economic Burden of Air Pollution-Related Morbidity: Evidence from the Aphekom Project. *Eur. J. Health Econ.* **2016**, *17*, 1101–1115. [CrossRef]
19. Vaseashta, A. (Ed.) *Life Cycle Analysis of Nanoparticles: Risk, Assessment, and Sustainability*; Destech Publications, Inc: Lancaster, PA, USA, 2015; ISBN 978-1-60595-023-5.
20. CalitateAer. Available online: https://www.calitateaer.ro/ (accessed on 8 November 2023).
21. European Parliament Directive 2008/50/CE 2008. Available online: https://www.eea.europa.eu/policy-documents/directive-2008-50-ec-of (accessed on 11 November 2023).
22. Directive 2004/107/EC—European Environment Agency. Available online: https://www.eea.europa.eu/policy-documents/directive-2004-107-ec (accessed on 9 November 2023).
23. World Health Organization. *Review of Evidence on Health Aspects of Air Pollution: REVIHAAP Project: Technical Report*; World Health Organization. Regional Office for Europe: Copenhagen, Denmark, 2021.
24. Aphekom.Org. Available online: http://aphekom.org/web/aphekom.org/home (accessed on 9 November 2023).
25. Levei, L.; Hoaghia, M.-A.; Roman, M.; Marmureanu, L.; Moisa, C.; Levei, E.A.; Ozunu, A.; Cadar, O. Temporal Trend of PM_{10} and Associated Human Health Risk over the Past Decade in Cluj-Napoca City, Romania. *Appl. Sci.* **2020**, *10*, 5331. [CrossRef]

26. Stoian, I.M.; Parvu, S.; Neamtu, A.; Calota, V.; Voinoiu, A.; Pistol, A.; Cucuiu, R.; Minca, D.G.; Davila, C. PM_{10} and NO_2 Air Pollution and Evolution of COVID-19 Cases in Romania. *Maedica* **2022**, *17*, 777–784.
27. Park, S.K. Seasonal Variations of Fine Particulate Matter and Mortality Rate in Seoul, Korea with a Focus on the Short-Term Impact of Meteorological Extremes on Human Health. *Atmosphere* **2021**, *12*, 151. [CrossRef]
28. Popescu, L.L.; Popescu, R.S.; Catalina, T. Indoor Particle's Pollution in Bucharest, Romania. *Toxics* **2022**, *10*, 757. [CrossRef] [PubMed]
29. COVID-19 Excess Mortality Collaborators. Estimating excess mortality due to the COVID-19 pandemic: A systematic analysis of COVID-19-related mortality, 2020–2021. *Lancet* **2022**, *399*, 1513–1536. [CrossRef]
30. Vidal-Perez, R.; Brandão, M.; Pazdernik, M.; Kresoja, K.-P.; Carpenito, M.; Maeda, S.; Casado-Arroyo, R.; Muscoli, S.; Pöss, J.; Fontes-Carvalho, R.; et al. Cardiovascular disease, and COVID-19, a deadly combination: A review about direct and indirect impact of a pandemic. *World J. Clin. Cases* **2022**, *10*, 9556–9572. [CrossRef]
31. Cohen, P.; Potchter, O.; Schnell, I. The impact of an urban park on air pollution and noise levels in the Mediterranean city of Tel-Aviv, Israel. *Environ. Pollut.* **2014**, *195*, 73–83. [CrossRef]
32. Vaseashta, A.; Maftei, C.; Radu, D. Exploring Disruption Trajectories From COVID-19 on Education and the Impact of Policies: Lessons Learned and Path Forward. In *Transformation and Efficiency Enhancement of Public Utilities Systems: Multidimensional Aspects and Perspectives*; Gjorchev, J., Malcheski, S., Rađenović, T., Vasović, D., Živković, S., Eds.; IGI Global: Hershey, PA, USA, 2023; pp. 314–340. [CrossRef]
33. Baldwin, J.; Noorali, S.; Vaseashta, A. Biology and Behavior of Severe Acute Respiratory Syndrome Coronavirus Contagion with Emphasis on Treatment Strategies, Risk Assessment, and Resilience. *COVID* **2023**, *3*, 1259–1303. [CrossRef]
34. Abe, K.C.; Rodrigues, M.A.; Miraglia, S.G.E.K. Health impact assessment of air pollution in Lisbon, Portugal. *J. Air Waste Manag. Assoc.* **2022**, *72*, 1307–1315. [CrossRef] [PubMed]
35. Kowalski, M.; Kowalska, K.; Kowalska, M. Health benefits related to the reduction of PM concentration in ambient air, Silesian Voivodeship, Poland. *Int. J. Occup. Med. Environ. Health* **2015**, *29*, 209–217. [CrossRef] [PubMed]
36. Egerstrom, N.; Rojas-Rueda, D.; Martuzzi, M.; Jalaludin, B.; Nieuwenhuijsen, M.; So, R.; Lim, Y.-H.; Loft, S.; Andersen, Z.J.; Cole-Hunter, T. Health and economic benefits of meeting WHO air quality guidelines, Western Pacific Region. *Bull. World Health Organ.* **2023**, *101*, 130–139. [CrossRef] [PubMed]
37. Boboc, C. Cultural Tourism in Central Region of Romania. Available online: https://www.proquest.com/openview/ce3e67fa4c6b5c90e8592f005f7215d1/1?pq-origsite=gscholar&cbl=2036059 (accessed on 23 September 2023).
38. Mitrano, D.M.; Wick, P.; Nowack, B. Placing Nanoplastics in the Context of Global Plastic Pollution. *Nat. Nanotechnol.* **2021**, *16*, 491–500. [CrossRef] [PubMed]
39. Stabnikova, O.; Stabnikov, V.; Marinin, A.; Klavins, M.; Klavins, L.; Vaseashta, A. Microbial Life on the Surface of Microplastics in Natural Waters. *Appl. Sci.* **2021**, *11*, 11692. [CrossRef]
40. Vaseashta, A.; Ivanov, V.; Stabnikov, V.; Marinin, A. Environmental Safety and Security Investigations of Neustonic Microplastic Aggregates Near Water-Air Interphase. *Pol. J. Environ. Stud.* **2021**, *30*, 3457–3469. [CrossRef]
41. Vaseashta, A.; Gevorgyan, G.; Kavaz, D.; Ivanov, O.; Jawaid, M.; Vasović, D. Exposome, Biomonitoring, Assessment, and Data Analytics to Quantify Universal Water Quality. In *Water Safety, Security and Sustainability*; Advanced Sciences and Technologies for Security Applications; Vaseashta, A., Maftei, C., Eds.; Springer: Cham, Switzerland, 2021. [CrossRef]

toxics

MDPI

Article

Application of Machine Learning in Modeling the Relationship between Catchment Attributes and Instream Water Quality in Data-Scarce Regions

Miljan Kovačević [1,*], Bahman Jabbarian Amiri [2], Silva Lozančić [3], Marijana Hadzima-Nyarko [3], Dorin Radu [4] and Emmanuel Karlo Nyarko [5]

1 Faculty of Technical Sciences, University of Pristina, Knjaza Milosa 7, 38220 Kosovska Mitrovica, Serbia
2 Faculty of Economics and Sociology, Department of Regional Economics and the Environment, 3/5 P.O.W. Street, 90-255 Lodz, Poland; bahman.amiri@uni.lodz.pl
3 Faculty of Civil Engineering and Architecture Osijek, Josip Juraj Strossmayer University of Osijek, Vladimira Preloga 3, 31000 Osijek, Croatia; lozancic@gfos.hr (S.L.); mhadzima@gfos.hr (M.H.-N.)
4 Faculty of Civil Engineering, Department of Civil Engineering, Transilvania University of Brașov, 500152 Brașov, Romania; dorin.radu@unitbv.ro
5 Faculty of Electrical Engineering, Computer Science and Information Technology Osijek, Josip Juraj Strossmayer University of Osijek, Kneza Trpimira 2B, 31000 Osijek, Croatia; karlo.nyarko@ferit.hr
* Correspondence: miljan.kovacevic@pr.ac.rs; Tel.: +381-606173801

Abstract: This research delves into the efficacy of machine learning models in predicting water quality parameters within a catchment area, focusing on unraveling the significance of individual input variables. In order to manage water quality, it is necessary to determine the relationship between the physical attributes of the catchment, such as geological permeability and hydrologic soil groups, and in-stream water quality parameters. Water quality data were acquired from the Iran Water Resource Management Company (WRMC) through monthly sampling. For statistical analysis, the study utilized 5-year means (1998–2002) of water quality data. A total of 88 final stations were included in the analysis. Using machine learning methods, the paper gives relations for 11 in-stream water quality parameters: Sodium Adsorption Ratio (SAR), Na^+, Mg^{2+}, Ca^{2+}, $SO_4{}^{2-}$, Cl^-, HCO^{3-}, K^+, pH, conductivity (EC), and Total Dissolved Solids (TDS). To comprehensively evaluate model performance, the study employs diverse metrics, including Pearson's Linear Correlation Coefficient (R) and the mean absolute percentage error (MAPE). Notably, the Random Forest (RF) model emerges as the standout model across various water parameters. Integrating research outcomes enables targeted strategies for fostering environmental sustainability, contributing to the broader goal of cultivating resilient water ecosystems. As a practical pathway toward achieving a delicate balance between human activities and environmental preservation, this research actively contributes to sustainable water ecosystems.

Keywords: machine learning; water quality; land use; land cover; hydrologic soil groups; geological permeability

check for updates

Citation: Kovačević, M.; Jabbarian Amiri, B.; Lozančić, S.; Hadzima-Nyarko, M.; Radu, D.; Nyarko, E.K. Application of Machine Learning in Modeling the Relationship between Catchment Attributes and Instream Water Quality in Data-Scarce Regions. *Toxics* **2023**, *11*, 996. https://doi.org/10.3390/toxics11120996

Academic Editors: Alina Barbulescu and Lucica Barbes

Received: 30 October 2023
Revised: 30 November 2023
Accepted: 4 December 2023
Published: 7 December 2023

1. Introduction

River water quality plays a crucial role in ensuring the sustainability and health of freshwater ecosystems. Traditional monitoring methods often have spatial and temporal coverage limitations, leading to difficulties in effectively assessing and managing water quality [1]. However, recent developments in machine learning techniques have indicated the potential to predict water quality accurately based on catchment characteristics [1,2]. One of the underlying reasons for the growing interest in applying machine learning techniques to predict river water quality is the ability to simultaneously consider a wide range of catchment characteristics. These characteristics include various features that affect water quality, including land use, soil properties, climate data, topography, and hydrological characteristics [3]. These variables interact in complex ways, and their associations may not easily be distinguished using prevalent analytical methods. In practice, for the modeling of water quality parameters, different mathematical models can be used that show satisfactory accuracy, as in papers [4,5]. A more comprehensive understanding of the factors influencing water quality can be achieved by applying machine learning algorithms, which can disclose hidden patterns and capture nonlinear relationships in large and diverse datasets [6].

The performance of supervised machine learning algorithms has been proved by recent studies in predicting river water quality [7]. These algorithms are trained on historical water quality data, in line with associating catchment characteristics, to explore the patterns and relationships between them [8]. Researchers have been able to develop accurate predictive models by applying algorithms such as decision trees, random forests, support vector machines, and neural networks [9–11]. Water quality parameters, such as nutrient concentrations, pollutant levels, and biological indicators, can then be applied by these models to make predictions using catchment characteristics [12]. Such predictions can help determine potential pollution points, prioritize management actions, and support decision-making processes for water resource management.

In addition to catchment characteristics, integrating remote sensing data has emerged as a valuable tool by which the accuracy of water quality predictions can be significantly enhanced [13,14], because the spatially explicit information about land cover/land use, vegetation, and surface characteristics can be provided by remote sensing techniques, which include satellite imagery and aerial photographs. Researchers can improve the predictive performance of machine learning models by combining remote sensing data with catchment characteristics [15]. For example, satellite data can provide insight into vegetation dynamics, land use/land cover changes, and the extent of impervious surfaces, which encompass urban and semi-urban areas, all of which can influence water quality. Integrating these additional spatial data sources into the machine learning models can result in more accurate and spatially explicit predictions, allowing a more comprehensive assessment of water quality dynamics across large river basins.

Furthermore, applying machine learning techniques to predict river water quality has led to collaborative efforts between researchers and stakeholders [16,17], which can be crucial to facilitate achieving the objectives of water resources management. These efforts aim to develop standardized frameworks and models that can be applied across different catchments and regions. Data sharing, methodologies, and the best practices can collectively improve the accuracy and reliability of predictive models that researchers develop. Collaborative initiatives also facilitate the identification of common challenges, such as data availability and quality issues, and foster the development of innovative solutions.

Advances in machine learning applications to predict river water quality underscore the growing importance of this field. Researchers are moving toward a more sustainable and informed water resource management by integrating advanced machine learning algorithms with catchment characteristics and remote sensing data. These predictive models can support policymakers, water resource managers, and environmental authorities in making evidence-based decisions, implementing targeted pollution control measures, and maintaining the ecological integrity of river ecosystems.

Integrating machine learning techniques with catchment characteristics gives researchers, water resource engineers, planners, and managers immense potential to predict river water quality. These models can provide accurate and timely information to support water resource management, pollution mitigation efforts, and the preservation of freshwater ecosystems by leveraging the power of advanced algorithms and incorporating diverse environmental data sources [18]. The advancements made in this field highlight the growing significance of machine learning in addressing the challenges associated with water quality prediction and paving the way for a more sustainable and informed management of our precious water resources.

Although machine learning applications in predicting river water quality based on catchment characteristics have shown promising findings, several gaps in this research field warrant further investigation. They include but are not limited to (1) data availability and quality, (2) the incorporation of temporal dynamics, (3) uncertainty estimation, (4) across-catchments transferability, (5) the integration of socio-economic factors, and (6) interpretability and transparency, which are briefly addressed as follows:

Data availability and quality, particularly historical water quality data and comprehensive catchment characteristics data remain challenges in many regions. Limited data may lead to biased or incomplete models, limiting the accuracy and the generality of predictions. Efforts should be made to improve data collection methods, establish standardized data protocols, and improve data sharing among researchers and stakeholders.

Current machine learning models often ignore the temporal dimension and assume static relationships between catchment characteristics and water quality parameters [19]. While integrating temporal dynamics into machine learning models could increase their predictive capacity and allow for more accurate short-term and long-term water quality forecasts, various temporal factors, such as seasonal variations, climate change, and short-term events such as precipitation events or pollution incidents, affect river water quality [20].

Machine learning models typically provide point predictions, but quantifying and communicating the uncertainties associated with these predictions is crucial for decision-making and risk assessment [21]. Developing methods to quantify and propagate uncertainties through the modeling process, considering the sources of uncertainty such as data quality, model structure, and parameter estimation, would enhance the reliability and applicability of predictive models.

The use of deep learning models is considered for modeling changes in water reservoirs. The methodology of Long Short-Term Memory (LSTM) networks was applied in the work, and a number of criteria including the Coefficient of Determination (R^2), Root Mean Square Error (RMSE), mean absolute percentage error (MAPE), Mean Absolute Deviation (MAD), and Nash–Sutcliffe Efficiency (NSE) were used to assess the accuracy. Satisfactory accuracy of the model was achieved on a series of samples that covered the period from 2003 to 2025 for five basins in Saudi Arabia [22].

Research regarding the application of artificial intelligence techniques—artificial neural networks (ANN), a group method of data handling (GMDH), and support vector machines (SVM)—for predicting water quality components in Tireh River, southwest Iran showed that the application of ANN and SVM models, using tansig and RBF functions, respectively, showed satisfactory performance. The database included samples collected over a period of 55 years [23].

In addition, models developed for a particular catchment cannot be applied directly to another due to variations in catchment characteristics, land use/cover, soil, geology, and climatic conditions. The development of transferable models that can account for specific variations in catchment while taking general patterns would be valuable for the management of water resources on a larger scale [24]. On the other hand, capturing the characteristics related to human activities, such as agricultural practices, urbanization, and industrial activities, has a significant impact on water quality. Incorporating socio-economic factors into machine learning models can improve their predictive power and enable more comprehensive water quality assessments. However, the integration of socio-economic data and understanding the complex interactions between human activity and water quality present challenges that must be addressed.

It should be noted that the need for greater ambiguity and transparency in machine learning models can limit the adoption and acceptance of these models by policymakers and stakeholders [25]. The development of logical machine learning techniques that provide insights into the model decision-making process and highlight the most influential catch characteristics would improve the reliability and usability of predictive models.

Addressing these gaps requires interdisciplinary collaborations among hydrologists, ecologists, data scientists, policymakers, and other relevant stakeholders. Furthermore, focusing on data-driven approaches, data-sharing initiatives, and advances in computational methods will be critical to advancing the field and harnessing the full potential of machine learning in predicting river water quality based on catchment characteristics. In the study, we intend to investigate how we can address spatial variations in the characteristics of the catchment to explain river water quality using machine learning techniques.

This paper explores the relationship between catchment attributes and in-stream water quality parameters using machine learning methods. It evaluates model accuracy with RMSE, MAE, R, and MAPE, identifies optimal models for 11 parameters, and determines significant influencing variables.

The predictive models developed for each water parameter demonstrate strong performance in most cases. The significant variables identified provide insights into the key factors influencing water quality in the studied catchment. This research, therefore, serves as a catalyst for fostering a nuanced and effective approach to water resource management, underpinned by the empirical foundation laid by the predictive models and the discerned influential variables. As a result, the integration of these findings into decision-making processes holds the potential to optimize resource allocation, mitigate environmental impacts, and ultimately contribute to the overarching goal of achieving sustainable and resilient water ecosystems. The study highlights the potential of artificial intelligence for quick and accurate water quality assessment, tailored to watershed attributes.

2. Materials and Methods

To establish relationships between the catchment attributes and water quality parameters, machine learning methods were employed. Multiple algorithms, such as regression trees, TreeBagger, Random Forests, and Gaussian process regression (GPR) models, were applied to construct predictive models for each water quality parameter.

2.1. Regression Tree Models

The fundamental concept behind regression trees is to partition the input space into distinct regions and assign predictive values to these regions. This segmentation enables the model to make predictions based on the most relevant conditions and characteristics of the data. A regression tree (RT) is a simple and comprehensible machine learning model applicable to both regression and classification problems. It follows a tree-like structure composed of nodes and branches (Figure 1).

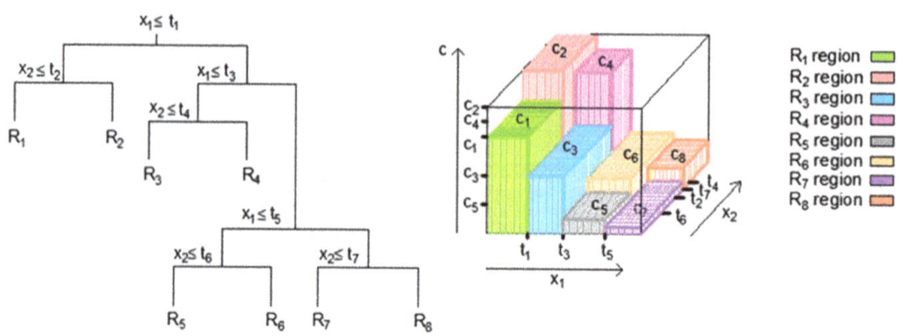

Figure 1. The partitioning of an input space into distinct regions and the representation of a 3D regression surface within a regression tree [26].

Each node corresponds to a specific condition related to the input data, and this condition is evaluated at each node as the data progresses through the tree.

To predict an outcome for a given input, the starting point is the root node of the tree (Figure 1). Here, the initial condition associated with the input feature(s) is considered. Depending on whether this condition is deemed true or false, the branches are followed to reach the next node. This process is repeated recursively until a leaf node is arrived at. At the leaf node, a value is found, which serves as the predicted result for the input instance. For regression tasks, this value is typically a numeric prediction.

As the tree is traversed, the input space undergoes changes. Initially, all instances are part of a single set represented by the root node. However, as the algorithm progresses, the input space is gradually divided into smaller subsets. These divisions are based on conditions that help in tailoring predictions to different regions within the input space.

The process of constructing regression trees involves determining the optimal split variable ("j") and split point ("s") to partition the input space effectively. These variables are chosen by minimizing a specific expression (Equation (1)) that considers all input features. The goal is to minimize the sum of squared differences between observed values and predicted values in resulting regions [27–29].

$$\min_{j,s}\left[\min_{c_1}\sum_{x_i\in R_1(j,s)}(y_i-c_1)^2+\min_{c_2}\sum_{x_i\in R_2(j,s)}(y_i-c_2)^2\right] \quad (1)$$

Once "j" and "s" are identified, the tree-building process continues by iteratively dividing regions. This process is referred to as a "greedy approach" because it prioritizes local optimality at each step. The binary recursive segmentation approach divides the input space into non-overlapping regions characterized by their mean values.

The depth of a regression tree serves as a critical factor in preventing overfitting (too much detail) or underfitting (too simplistic).

2.2. Ensembles of Regression Trees: Bagging, Random Forest, and Boosted Trees

Bagging is another ensemble method that involves creating multiple subsets of the training dataset through random sampling with replacement (bootstrap samples).

The process begins with the creation of multiple bootstrap samples from the original dataset. Bootstrap sampling involves randomly selecting data points from the dataset with replacement. This means that the same data point can be selected multiple times, while others may not be selected at all. In this way, subsets of the same size as the original data set are formed and are used to train the model.

Each subset is used to train a separate regression tree model, and their predictions are aggregated to make the final prediction (Figure 2). Bagging helps reduce variance by averaging predictions from multiple models, making it particularly effective when the base models are unstable or prone to overfitting [27–29].

In bagging, multiple training sets are generated by repeatedly selecting samples from the original dataset, and this process involves sampling with replacement. This technique is utilized to create diverse subsets of data. The primary goal is to reduce the variance in the model's predictions by aggregating the results from these different subsets. Consequently, each subset contributes to the final prediction, and the averaging of multiple models enhances the model's robustness and predictive accuracy.

Random forests, a variant of bagging, stand out by introducing diversity among the constituent models within the ensemble. This diversity is achieved through the creation of multiple regression trees, each trained on a distinct bootstrap sample from the data. Moreover, before making decisions at each split within these trees, only a randomly selected subset of available features is considered. This approach helps in decorrelating the individual trees within the ensemble, thereby further reducing variance. The ensemble's final prediction is generated by aggregating the predictions from these decorrelated trees, resulting in a robust and high-performing model [26–29].

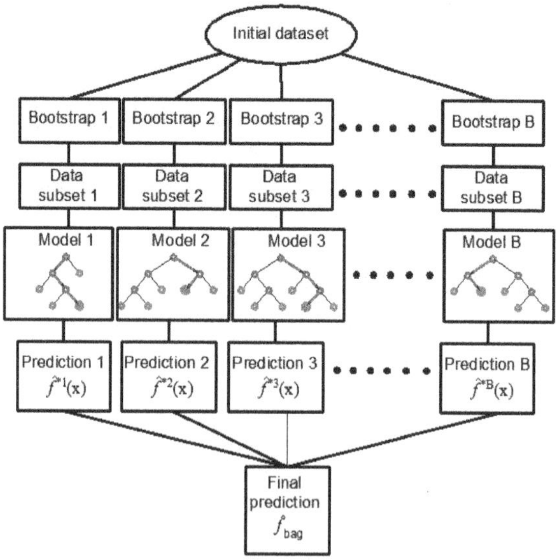

Figure 2. Creating regression tree ensembles using the bagging approach [30].

The boosting tree method is a sequential training method, and within this paradigm, gradient boosting stands out as a widely employed technique for enhancing overall model performance (Figure 3). In gradient boosting, submodels are introduced iteratively, with each new model selected based on its capacity to effectively estimate the residuals or errors of the preceding model in the sequence. The distinctive feature of gradient boosting lies in its commitment to minimizing these residuals during the iterative process.

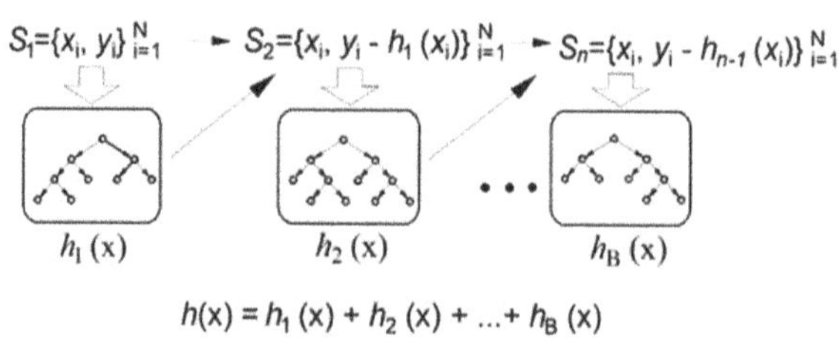

Figure 3. The application of gradient boosting within regression tree ensembles [26].

By focusing on minimizing residuals, gradient boosting ensures that each new submodel added to the ensemble is adept at correcting the errors left by its predecessors. This emphasis on addressing the shortcomings of prior models leads to the creation of a robust and adaptive ensemble model. The iterative nature of gradient boosting allows it to systematically refine its predictions, making the final ensemble proficient in capturing intricate patterns and nuances within the data. The result is a powerful model capable of delivering highly accurate predictions by continuously learning and adapting to the complexities present in the dataset.

The fundamental concept is rooted in gradient-based optimization techniques, which involve refining the current solution to an optimization problem by incorporating a vector that is directly linked to the negative gradient of the function under minimization, as referenced in previous works [31–33]. This approach is logical because a negative gradient signifies the direction in which the function decreases. When it is applied a quadratic error function, each subsequent model aims to correct the discrepancies left by its predecessors, essentially reinforcing and improving the model with a focus on the residual errors from earlier stages.

In the context of gradient-boosting trees, the learning rate is a crucial hyperparameter that controls the contribution of each tree in the ensemble to the final prediction. It is often denoted as "lambda" (λ). The learning rate determines how quickly or slowly the model adapts to the errors from the previous trees during the boosting process. A lower learning rate means that the model adjusts more gradually and may require a larger number of trees to achieve the same level of accuracy, while a larger learning rate leads to faster adaptation but may risk overfitting with too few trees.

BT models are significantly more complex regarding computational complexity because they are trained sequentially compared to TR and RF models that can be trained in parallel.

2.3. Gaussian Process Regression (GPR)

Gaussian processes provide a probabilistic framework for modeling functions, capturing uncertainties, and making predictions in regression tasks. The choice of covariance functions and hyperparameters allows for flexibility in modeling relationships among variables [34].

Gaussian process modeling involves estimating an unknown function $f(\cdot)$ in nonlinear regression problems. It assumes that this function follows a Gaussian distribution characterized by a mean function $\mu(\cdot)$ and a covariance function $k(\cdot,\cdot)$. The covariance matrix K is a fundamental component of GPR and is determined by the kernel function (k).

The kernel function (k) plays a pivotal role in capturing the relationships between input data points (x and x′). This function is essential for quantifying the covariance or similarity between random values f(x) and f(x′). One of the most widely used kernels is defined by the following expression:

$$k(xx') = \sigma^2 \exp\left(-\frac{(x-x')^2}{2l^2}\right) \quad (2)$$

In this expression, several elements are critical:

σ^2 represents the signal variance, a model parameter that quantifies the overall variability or magnitude of the function values.

The exponential function "exp" is used to model the similarity between x and x′. It decreases as the difference between x and x′ increases, capturing the idea that values close to each other are more strongly correlated.

The parameter l, known as the length scale, is another model parameter. It controls the smoothness and spatial extent of the correlation. A smaller l results in more rapid changes in the function, while a larger l leads to smoother variations.

The observations in a dataset $\mathbf{y} = \{y_1, \ldots, y_n\}$ can be viewed as a sample from a multivariate Gaussian distribution.

$$(y_1, \ldots, y_n)^T \sim N(\boldsymbol{\mu}, K), \quad (3)$$

Gaussian processes are employed to model the relationship between input variables x and target variables y, considering the presence of additive noise $\varepsilon \sim N(0, \sigma^2)$. The goal is to estimate the unknown function $f(\cdot)$. The observations y are treated as a sample from a multivariate Gaussian distribution with mean vector μ and covariance matrix K. This distribution captures the relationships among the data points. The conditional distribution of a test point's response value y^*, given the observed data $\mathbf{y} = (y_1, \ldots, y_n)^T$, is represented as $N(\hat{\mathbf{y}}^*, \hat{\boldsymbol{\sigma}}^{*2})$ with the following:

$$\hat{y}^* = \mu(x^*) + K^{*T} K^{-1}(y - \mu), \quad (4)$$

$$\hat{\sigma}^{*2} = K^{**} + \sigma^2 - K^{*T} K^{-1} K^{*}. \tag{5}$$

In traditional GPR, a single length-scale parameter (l) and signal variance (σ^2) are used for all input dimensions. In contrast, the Automatic Relevance Determination (ARD) approach employs a separate length-scale parameter (l_i) for each input dimension, where 'i' represents a specific dimension. This means that for a dataset with 'm' input dimensions, you have 'm' individual length-scale parameters [34].

The key advantage of ARD is that it automatically determines the relevance of each input dimension in the modeling process. By allowing each dimension to have its own length scale parameter, the model can assign different degrees of importance to each dimension. This means that the model can adapt and focus more on dimensions that are more relevant to the target variable and be less influenced by less relevant dimensions.

GPR involves matrix operations, and the computational complexity can become an issue for large datasets. Techniques such as sparse approximations or using specialized kernels can be employed to address these computational challenges. GPR is frequently used in Bayesian optimization problems where the goal is to optimize an unknown objective function that is expensive to evaluate.

3. Case Study of the Caspian Sea Basin

This study took place in the Caspian Sea catchment area (Figure 4) in Northern Iran, covering approximately 618 m^2 with coordinates ranging from 49°48′ to 54°41′ longitude and from 35°36′ to 37°19′ latitude (Figure 4).

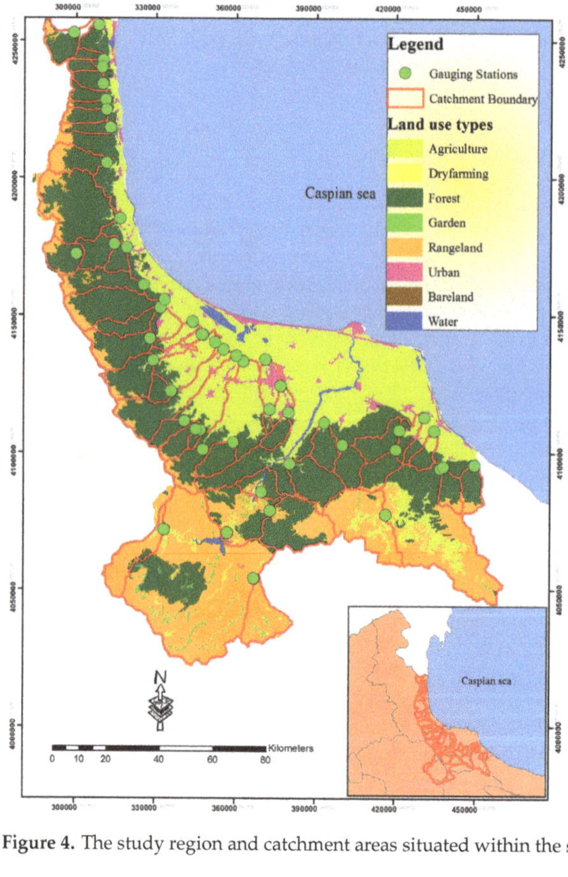

Figure 4. The study region and catchment areas situated within the southern Caspian Sea basin.

The majority of this area, approximately 65.10%, is forested, while the rest consists of rangelands (24.41%), agricultural land (9.41%), urban areas (0.88%), water bodies (0.0126%), and bare land (0.186%) [35].

Initially, 108 water quality monitoring stations scattered across the southern basin of the Caspian Sea were selected for analysis (Figure 4). To define the upstream catchment boundaries, digital elevation models (DEMs) with a resolution of 30 m by 30 m from the USGS database were used, with boundary refinement achieved through a user digitizing technique. Macro-sized catchments, those exceeding 1000 square kilometers, totaling 18 catchments, were excluded from the modeling process due to their significant impact on hydrological dynamics.

Water quality data, including parameters like SAR, Na^+, Mg^{2+}, Ca^{2+}, $SO_4{}^{2-}$, Cl^-, HCO^{3-}, K^+, pH, EC, and TDS, were obtained from the Iran Water Resource Management Company (WRMC) through monthly sampling. Collection adhered to the WRMC Guidelines for Surface Water Quality Monitoring (2009) and EPA-841-B-97-003 standards [36]. For statistical analysis, the 5-year means (1998–2002) of water quality data were calculated. After scrutinizing for normality and identifying outliers, 88 final stations were used in the study. The geographic scope of the study area is illustrated in Figure 4.

A land cover dataset was created using a 2002 digital land cover map (Scale 1:250,000) from the Forest, Ranges, and Watershed Management Organization of Iran. The original land cover categories were consolidated into six classes: bare land, water bodies, urban areas, agriculture, rangeland, and forests, following [37] land use and land cover classification systems. Furthermore, digital geological and soil feature maps (1:250,000 scale) were obtained from the Geological Survey of Iran (www.gsi.ir, accessed on 24 April 2021). Detailed information about the characteristics of the catchments and their statistical attributes can be found in Tables 1 and 2.

In this study, hydrologic soil groups and geological permeability classes were developed and applied in conjunction with land use/land cover types within the modeling process. Hydrologic soil groups are influenced by runoff potential and can be used to determine runoff curve numbers. They consist of four classes (A, B, C, and D), with A having the highest runoff potential and D the lowest. Notably, soil profiles can undergo significant alterations due to changes in land use/land cover. In such cases, the soil textures of the new surface soil can be employed to determine the hydrologic soil groups as described in Table 1 [38]. Furthermore, the study incorporates the application of geological permeability attributes related to catchments, with the development of three geological permeability classes: Low, Medium, and High. These classes are associated with various characteristics of geological formations, such as effective porosity, cavity type and size, their connectivity, rock density, pressure gradient, and fluid properties like viscosity.

The range and statistical properties of training and test data play a fundamental role in the development and evaluation of machine learning models. They impact the model's generalization, robustness, fairness, and ability to perform effectively in diverse real-world scenarios. Statistical properties of input and output data are given in Tables 1 and 2.

The machine learning methods used in this paper were assessed using five-fold cross-validation. In this approach, the dataset was randomly divided into five subsets, with four of them dedicated to training the model and the remaining subset utilized for model validation (testing). This five-fold cross-validation process was repeated five times, ensuring that each subset was used exactly once for validation. Subsequently, the results from these five repetitions were averaged to produce a single estimation.

All models were trained and tested under identical conditions, ensuring a fair and consistent evaluation of their performance. This practice is essential in machine learning to provide a level playing field for comparing different algorithms and models.

When machine learning models are trained and tested under equal conditions, it means that they are exposed to the same datasets, preprocessing steps, and evaluation metrics.

Table 1. Statistical properties of the input variables used for modeling.

Input Parameter (Acronym)	Min	Max	Average	Std
Hydrometric Station Elevation (HSE)	−23.0000	2360.0000	163.3333	363.3326
Catchment Area (CA)	22.0300	6328.2800	422.7320	1085.4846
Stream Order (SO)	1.0000	4.0000	2.6275	1.3261
Percentage of Land Use or Land Cover Types:				
Barren Land (BL)	0.0000	3.1825	0.1246	0.6083
Forest (F)	1.1805	100.0000	70.0401	29.5955
Rangeland (RL)	0.0000	90.3170	17.0144	23.9666
Urban Area (UA)	0.0000	20.2095	1.1367	3.5475
Water Body (WB)	0.0000	0.3567	0.0074	0.0499
Agricultural Area (AA)	0.0000	84.3857	11.6768	20.3896
Hydrological Soil Group:				
A—Sand, loamy sand, or sandy loam (HSGA)	0.0000	79.3654	8.0039	16.2217
B—Silt loam or loam (HSGB)	0.0000	48.4653	3.0354	8.4226
C—Sandy clay loam (HSGC)	12.9196	100.0000	80.4068	27.0174
D—Clay loam, silty clay loam, sandy clay, silty clay, or clay (HSGD)	0.0000	56.4129	8.5539	15.9743
Geological Permeability:				
Low (Geological hydrological group M—GHGM)	0.0143	100.0000	69.2656	28.0000
Average (Geological hydrological group N—GHGN)	0.0000	96.9436	23.4102	24.3102
High (Geological hydrological group T—GHGT)	0.0000	90.9015	7.3243	15.3979

Table 2. Statistical properties of the output variables used for modeling.

Parameter	Min	Max	Average	Std
SAR	7.1500	9.0900	7.5318	0.3976
Na^+	0.1200	15.8400	0.9978	2.3957
Mg^{2+}	0.4100	4.4100	1.0331	0.6790
Ca^{2+}	1.0600	5.8800	2.3584	0.9521
$SO_4{}^{2-}$	0.2100	4.4500	0.6643	0.8449
Cl^-	0.1900	18.2000	1.1861	2.7131
HCO^{3-}	1.3500	4.0900	2.5978	0.7729
pH	172.0500	2879.9700	453.7716	428.3365
EC	108.8900	3892.8200	375.2543	579.1442
TDS	0.1000	3.4400	0.4750	0.5971
K^+	0.0200	0.1400	0.0447	0.0316

The quality of the model was assessed using several evaluation and performance measures, which include RMSE, MAE, Pearson's Linear Correlation Coefficient (R), and MAPE.

The RMSE criterion, expressed in the same units as the target values, serves as a measure of the model's general accuracy. It is calculated as the square root of the average squared differences between the actual values (d_k) and the model's predictions (o_k) across the training samples (N).

$$\text{RMSE} = \sqrt{\frac{1}{N}\sum_{k=1}^{N}(d_k - o_k)^2}, \tag{6}$$

The MAE criterion represents the mean absolute error of the model, emphasizing the absolute accuracy. It calculates the average absolute differences between the actual values and the model's predictions.

$$\text{MAE} = \frac{1}{N}\sum_{k=1}^{N}|d_k - o_k|. \tag{7}$$

Pearson's Linear Correlation Coefficient (R) provides a relative measure of accuracy assessment. It considers the correlation between the actual values (d_k) and the model's predictions (o_k) relative to their respective means ($\overline{d}$ and $\overline{o}$). Values of R greater than 0.75 indicate a strong correlation between the variables.

$$\text{R} = \frac{\sum_{k=1}^{N}\left(d_k - \overline{d}\right)(o_k - \overline{o})}{\sqrt{\left[\sum_{k=1}^{N}\left(d_k - \overline{d}\right)^2 (o_k - \overline{o})^2\right]}}. \tag{8}$$

The MAPE is a relative criterion that evaluates accuracy by calculating the average percentage differences between the actual values and the model's predictions.

$$\text{MAPE} = \frac{100}{N}\sum_{k=1}^{N}\left|\frac{d_k - o_k}{d_k}\right|. \tag{9}$$

This research deals with a limited dataset, and in this case, there is a higher risk of overfitting, where a model performs well on the training data but needs to generalize to new, unseen data. Five-fold cross-validation helps mitigate overfitting by partitioning the dataset into five subsets, using four for training and one for testing in each iteration. This process allows for a more robust evaluation of the model's performance.

Five-fold cross-validation efficiently utilizes the available data by rotating through different subsets for training and testing, ensuring that each data point contributes to training and evaluation.

Moreover, cross-validation provides a more robust estimate of the model's performance by averaging the evaluation metrics across multiple folds. This helps ensure that our results are not overly dependent on the particular random split of the data. Additionally, cross-validation allows us to iteratively train and evaluate the model on different subsets, aiding in the fine-tuning of hyperparameters and ensuring the model's performance is consistently reliable (Figure 5).

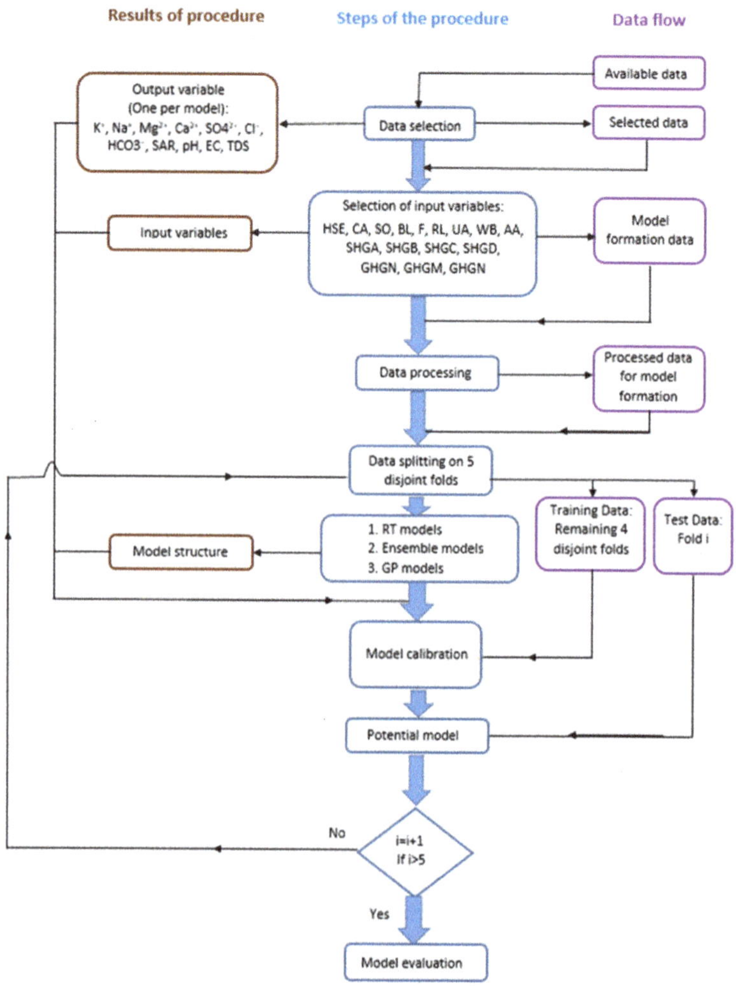

Figure 5. Applied methodology for creating prediction models.

4. Results

The paper analyzes the application of regression trees, bagging, RF, gradient boosting, and Gaussian process regression models using a systemic approach (Figure 5). For each of the models, the hyperparameters of the model were varied in the appropriate range, and optimal values were determined using a grid-search method. The following values were analyzed:

- Regression trees (RT) model

The depth of a regression tree is a crucial factor in preventing overfitting, which occurs when the tree becomes too detailed and fits the training data too closely, as well as underfitting, which happens when the tree is too simplistic and fails to capture the underlying patterns. Table 3 illustrates the impact of the minimum leaf size on the accuracy of the regression tree (RT) model. It shows how changing the minimum leaf size affects key performance metrics such as RMSE, MAE, MAPE, and R. Accordingly, the leaf size is almost positively associated with the RMSE but inversely correlated with the R values.

Table 3. Influence of the minimum leaf size on regression tree (RT) model accuracy.

Min Leaf Size	RMSE	MAE	MAPE	R
1	0.5646	0.2753	0.5948	0.4363
2	0.6077	0.2923	0.6322	0.3226
3	0.6096	0.2875	0.6311	0.2909
4	0.5984	0.2914	0.6188	0.3186
5	0.5894	0.2931	0.6728	0.3621
6	0.5833	0.2925	0.6616	0.3593
7	0.5813	0.2949	0.6968	0.3542
8	0.5841	0.3095	0.7603	0.3226
9	0.5821	0.2990	0.7078	0.3432
10	0.5969	0.3083	0.7369	0.2906

- TreeBagger (TB) model
    1. Number of generated trees (B): Investigated values up to a maximum of 500 trees, with 100 as the standard setting in Matlab. The bootstrap aggregation method was employed, generating a specific number of samples in each iteration. The study considered an upper limit of 500 trees.
    2. Minimum amount of data/samples per tree leaf: Analyzed values ranging from 2 to 15 samples per leaf, with a step size of 1 sample. The standard setting in Matlab is 5 samples per leaf for regression, but here, a broader range was examined to assess its impact on model generalization.
- Random Forest (RF) model:
    1. Number of generated trees (B): Analyzed within a range of 100–500 trees, with 100 as the standard setting in Matlab. Cumulative MSE values for all base models in the ensemble were presented. Bootstrap aggregation was used to create trees, generating 181 samples per iteration. The study explored an extended ensemble of up to 500 regression trees, aligning with recommended Random Forest practices.
    2. Number of variables used for tree splitting: Based on guidance, the study selected a subset of approximately $\sqrt{p}$ predictors for branching, where p is the number of input variables. With 16 predictors, this translated to a subset of 4 variables, but in those research, a wider number of variables, ranging from 1 to 16, is investigated.
    3. Minimum number of samples per leaf: the study considered values from 2 to 10 samples per tree leaf, with a 1-sample increment.
- Boosted tree model:
    1. Number of generated trees (B): analyzed within a range of 1–100 trees.
    2. Learning rate (λ): explored a range, including 0.001, 0.01, 0.1, 0.5, 0.75, and 1.0.
    3. Tree splitting levels (d): analyzed from 1 (a decision stump) to $2^7 = 128$ in an exponential manner.
- Gaussian process regression (GPR) model With the GPR method, the application of different kernel functions were explored:
    1. Exponential, quadratic exponential, Mattern 3/2, Mattern 5/2, rational quadratic.
    2. ARD Exponential, ARD quadratic exponential, ARD Mattern 3/2, ARD Mattern 5/2, ARD rational quadratic.

All models were evaluated in terms of optimality in terms of the mean square error, and then the optimal model obtained from all the analyzed ones was evaluated on the test data using the RMSE, MAE, MAPE, and R criteria.

In the paper, a detailed procedure is illustrated for determining the optimal model for the prediction of the SAR parameter. In contrast, for all other models for the prediction, it is given in a more concise form. Accompanying results for other models, except the SAR parameter, can be found in the Appendix A (Tables A1–A10) of the paper.

4.1. Prediction of SAR Parameter Values

- Regression tree models

Table 3 illustrates the impact of the minimum leaf size on the accuracy of the regression tree (RT) model. It shows how changing the minimum leaf size affects key performance metrics such as RMSE, MAE, MAPE, and R.

In this particular case, it was found that models with less complexity, i.e., the amount of data per terminal sheet is 10, have higher accuracy (Figure 6).

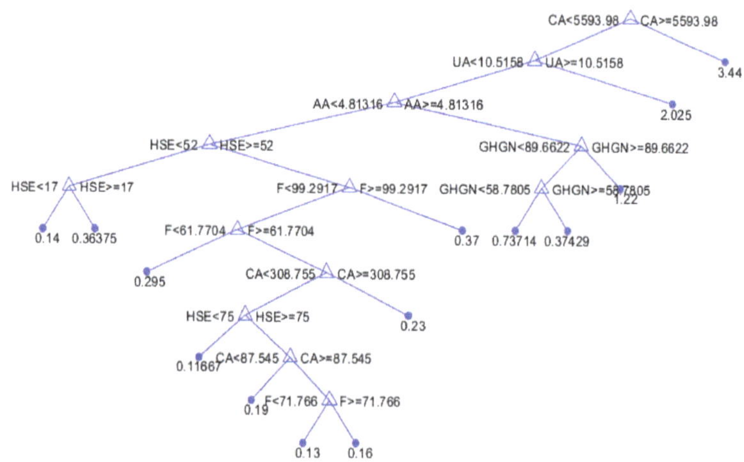

Figure 6. An optimal individual model for SAR parameter prediction based on a regression tree.

- TreeBagger models and Random Forest models

The application of TB and RF models was analyzed simultaneously (Figure 7). The figure shows the dependence of the achieved accuracy of the model on the hyperparameter value. The TB model represents the borderline case of the RF model when all variables are taken into account for potential calculations.

Among the optimal models in this group, the RF model with 500 generated trees proved to be the best. In contrast, the model that uses a subset of eight variables and has a minimum amount of data per terminal leaf equal to one has a higher accuracy according to the RMSE and R criteria, while the model that uses a subset with six variables and has a minimum amount of data per terminal sheet equal to one and has a higher accuracy according to MAE and MAPE criteria (Table 4). Optimal values according to different accuracy criteria are marked with bold numbers in Table 4.

With the BT model (Figure 8), it was shown that the highest accuracy is obtained by applying complex models with a large number of branches.

The optimal obtained model had a structure of 32 branches and a Learning Rate value equal to 0.01.

- **GPR models**

The optimal values for the parameters of the applied models with different kernel functions were obtained using marine probability (Tables 5 and 6).

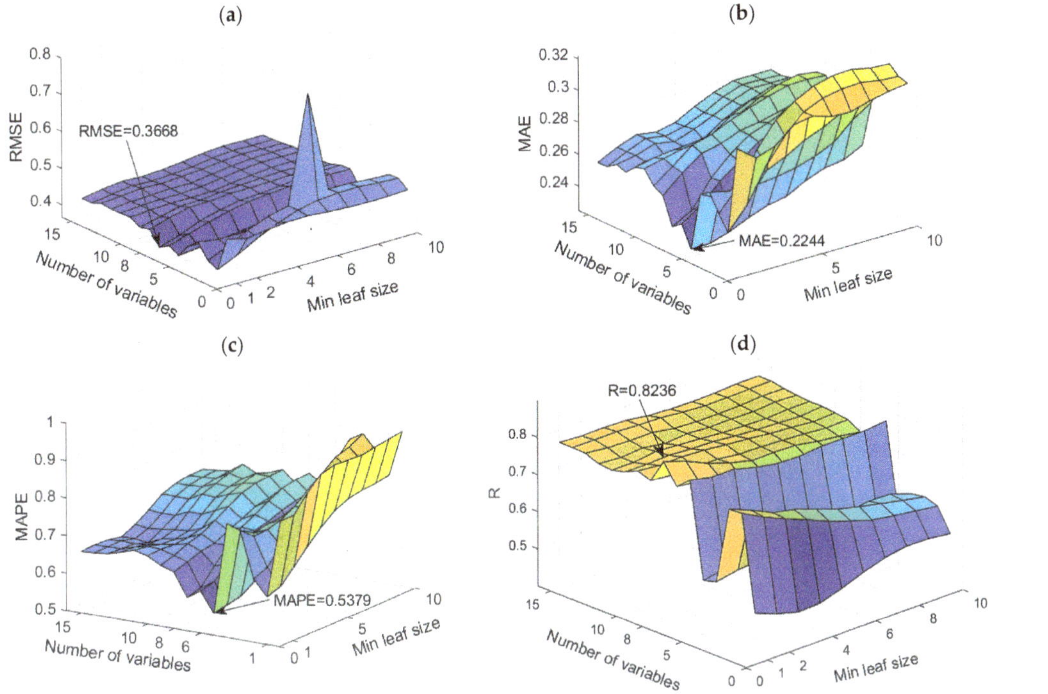

Figure 7. Comparison of different accuracy criteria for the RF model for the SAR parameter as a function of the number of randomly selected splitting variables and minimum leaf size: (**a**) RMSE, (**b**) MAE, (**c**) MAPE, (**d**) R.

Table 4. Accuracy of obtained models for SAR parameter prediction according to defined criteria.

Criteria	RMSE	MAE	MAPE	R
RF 1 (var 8, leaf 1)	**0.3668**	0.2328	0.5679	**0.8236**
RF 2 (var 6, leaf 1)	0.3696	**0.2244**	0.5379	0.8012

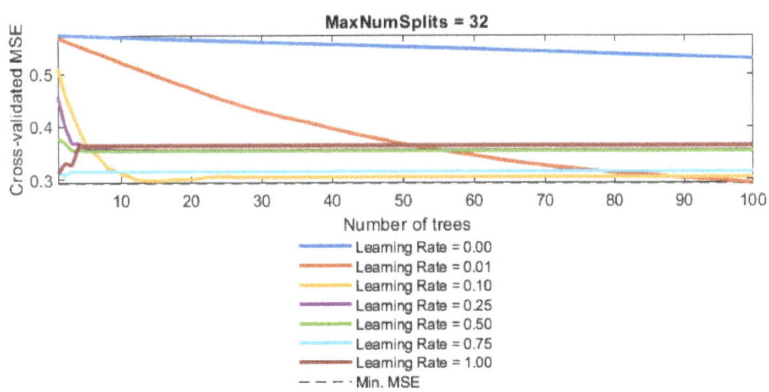

Figure 8. Dependence of the MSE value on the reduction parameter λ and the number of trees (base models) in the boosted tree model for the SAR parameter.

Table 5. Values of optimal parameters in GPR models with different covariance functions.

GP Model Covariance Function	Covariance Function Parameters		
Exponential	$k\left(\left(x_i, x_j \middle| \Theta\right)\right) = \sigma_f^2 exp\left[-\frac{1}{2}\frac{r}{\sigma_l^2}\right]$		
	$\sigma_l = 111.9371$		$\sigma_f = 1.8001$
Squared Exponential	$k\left(\left(x_i, x_j \middle| \Theta\right)\right) = \sigma_f^2 exp\left[-\frac{1}{2}\frac{(x_i - x_j)^T (x_i - x_j)}{\sigma_l^2}\right]$		
	$\sigma_l = 8.3178$		$\sigma_f = 1.2040$
Matern 3/2	$k\left(\left(x_i, x_j \middle| \Theta\right)\right) = \sigma_f^2 \left(1 + \frac{\sqrt{3}r}{\sigma_l}\right) exp\left[-\frac{\sqrt{3}r}{\sigma_l}\right]$		
	σ_l =14.7080		σ_f =1.4023
Matern 5/2	$k\left(\left(x_i, x_j \middle| \Theta\right)\right) = \sigma_f^2 \left(1 + \frac{\sqrt{5}r}{\sigma_l} + \frac{5r^2}{3\sigma_l^2}\right) exp\left[-\frac{\sqrt{5}r}{\sigma_l}\right]$		
	$\sigma_l = 9.9890$		$\sigma_f = 1.1947$
Rational Quadratic	$k\left(\left(x_i, x_j \middle| \Theta\right)\right) = \sigma_f^2 \left(1 + \frac{r^2}{2a\sigma_l^2}\right)^{-\alpha}; r = 0$		
	$\sigma_l = 8.3178$	$a = 3{,}156{,}603.8854$	$\sigma_f = 1.2040$

where $r = \sqrt{(x_i - x_j)^T (x_i - x_j)}$.

Table 6. Values of optimal parameters in GPR ARD models with different covariance functions.

Covariance Function Parameters															
σ_1	σ_2	σ_3	σ_4	σ_5	σ_6	σ_7	σ_8	σ_9	σ_{10}	σ_{11}	σ_{12}	σ_{13}	σ_{14}	σ_{15}	σ_{16}
ARD Exponential: $k\left(\left(x_i, x_j \middle| \Theta\right)\right) = \sigma_f^2 \exp(-r); \sigma_F = 1.; r = \sqrt{\sum_{m=1}^{d} \frac{(x_{im} - x_{jm})^2}{\sigma_m^2}}$															
119.8634	29.7051	300.3391	7.6459×10^5	158.5921	1.0859×10^7	15.3385	6.9058	89.5438	109.1181	121.9050	3.5665×10^5	3.6154×10^6	114.0696	5.0314×10^5	245.9937
ARD Squared exponential: $k\left(\left(x_i, x_j \middle| \Theta\right)\right) = \sigma_f^2 exp\left[-\frac{1}{2}\sum_{m=1}^{d} \frac{(x_{im} - x_{jm})^2}{\sigma_m^2}\right]; \sigma_f = 1.0577$															
0.9131	4.9648	3.2490×10^{11}	166.9588	3.6087×10^6	4.9117×10^6	2.3307	1.1875	11.1584	14.6207	15.1518	1.4141×10^6	32.6135	3.4365×10^9	1.5821×10^5	2.2696×10^4
ARD Matern 3/2: $k\left(\left(x_i, x_j \middle| \Theta\right)\right) = \sigma_f^2 \left(1 + \sqrt{3}r\right) exp\left[-\sqrt{3}r\right]; \sigma_f = 1.3295$															
2.3961×10^5	9.2294	1.3527×10^9	9.4195×10^{16}	32.1227	9.4054×10^8	9.6482	4.7958	1.1130×10^5	28.2837	4.3381	7.8940×10^4	131.4068	62.5967	1.7599×10^6	7.2779×10^{10}

Table 6. *Cont.*

Covariance Function Parameters															
σ_1	σ_2	σ_3	σ_4	σ_5	σ_6	σ_7	σ_8	σ_9	σ_{10}	σ_{11}	σ_{12}	σ_{13}	σ_{14}	σ_{15}	σ_{16}
ARD Matern 5/2: $k\left(\left(x_i, x_j \middle| \Theta\right)\right) = \sigma_f^2\left(1 + \sqrt{5}r + \frac{5r^2}{3}\right)exp\left[-\sqrt{5}r\right]; \sigma_f = 1.0452$															
0.2981	3.4788×10^4	27.1402	16.2408	6.4079×10^4	16.8653	3.3999	3.4057	9.1495×10^4	11.2644	6.3690×10^5	1.4039×10^6	30.9904	7.7608	23.1035	33.8984
ARD Rational quadratic: $k\left(\left(x_i, x_j \middle| \Theta\right)\right) = \sigma_f^2\left(1 + \frac{1}{2\alpha}\sum_{m=1}^{d}\frac{\left(x_{im} - x_{jm}\right)^2}{\sigma_m^2}\right)^{-\alpha}; \alpha = 0.7452; \sigma_f = 1.07$															
2.4546×10^6	3.7479	2.3486×10^7	5.3012×10^7	3.9435×10^3	10.1652	2.2446	0.7733	7.0275	18.4361	13.7803	1.3226×10^7	30.2548	5.2024×10^5	41.8960	7.2334×10^7

where $r = \sqrt{\sum_{m=1}^{d}\frac{\left(x_{im} - x_{jm}\right)^2}{\sigma_m^2}}$.

The marginal likelihood is a function influenced by the observed data (y(X)) and model parameters $\{l, \sigma^2, \eta^2\}$. The determination of the model parameters is achieved through the maximization of this function.

Importantly, when the marginal likelihood is transformed by taking the logarithm, identical results are achieved as when optimizing the original likelihood. Therefore, model parameter optimization is typically carried out by employing gradient-based procedures on the log marginal probability expression, simplifying the optimization process without altering the final outcomes. The comparative results of the implemented ML models are presented in Table 7. Optimal values according to different accuracy criteria are marked with bold numbers in Table 7.

The values of all accuracy criteria according to the adopted accuracy criteria on the test data set are shown in Table 7. According to the RMSE and R criteria, the RF model had the highest accuracy (it uses a subset of eight variables for calculation, and the amount of data per terminal sheet is equal to one), while according to the MAE and MAPE criteria, the GP model with an exponential kernel function stood out as the most accurate. On the optimal RF model, the significance of each of the input variables was determined such that the values of the considered variable are permuted within the training data, and the out-of-bag error for such permuted data is recalculated. The significance of the variable (Figure 9) is then determined by calculating the mean value of the difference before and after a permutation. This value is then divided by the standard deviation of these differences. The variable for which a higher value was obtained in relation to the others is ranked as more significant in relation to the variables for which smaller values were obtained.

Table 7. Comparative analysis of the results of different machine learning models for the SAR prediction.

Model	RMSE	MAE	MAPE/100	R
Decision Tree	0.5646	0.2753	0.5948	0.4363
TreeBagger	0.4021	0.2513	0.6413	0.7652
RF 1 (var 8, leaf 1)	**0.3668**	0.2328	0.5679	**0.8236**
RF 2 (var 6, leaf 1)	0.3696	0.2244	0.5379	0.8012
Boosted Trees	0.5592	0.3348	0.6047	0.5867
GP exponential	0.4625	**0.2104**	**0.4998**	0.6317
GP Sq. exponential	0.4868	0.2393	0.5810	0.5733
GP matern 3/2	0.4757	0.2293	0.5406	0.5992
GP matern 5/2	0.4779	0.2307	0.5520	0.5941
GP Rat. quadratic	0.4868	0.2393	0.5810	0.5733
GP ARD exponential	0.5917	0.2873	0.6991	0.3302
GP ARD Sq. exponential	0.5669	0.2788	0.7736	0.3568
GP ARD matern 3/2	0.5276	0.2707	0.7206	0.4702
GP ARD matern 5/2	0.5464	0.2875	0.8794	0.4223
GP ARD Rat. quadratic	0.6573	0.3285	0.9059	0.2349

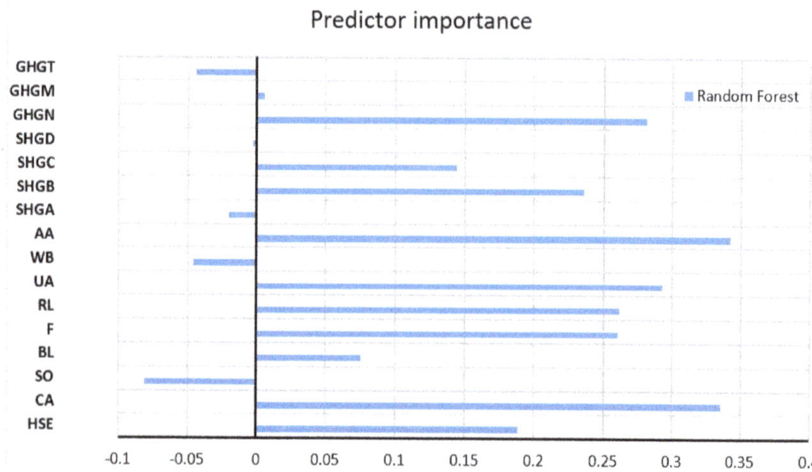

Figure 9. Significance of individual variables for SAR parameter prediction in an optimal RF model.

4.2. Prediction of Na^+ Parameter Values

RF models proved to be the optimal models for predicting sodium ion (Na^+) concentrations, while the analysis of all models in terms of accuracy is given in Appendix A (Table A1). The dependence of the adopted accuracy criteria on the model parameters is shown in Figure 10. Based on the defined accuracy criteria, four models with the following criteria values were selected (Table 8).

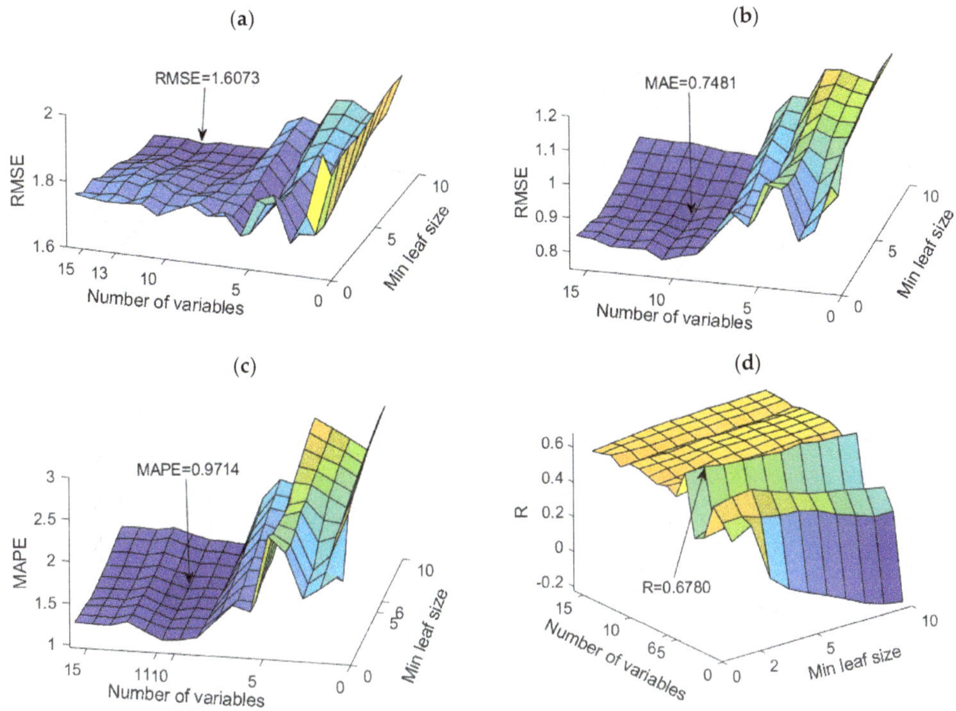

Figure 10. Comparison of different accuracy criteria for the RF model for the Na^+ parameter as a function of the number of randomly selected splitting variables and minimum leaf size: (**a**) RMSE, (**b**) MAE, (**c**) MAPE, (**d**) R.

Table 8. Accuracy of obtained models for Na^+ parameter prediction according to defined criteria.

Criteria	RMSE	MAE	MAPE	R
RF 1 (var 13, leaf 10)	1.6073	0.8086	1.1651	0.5817
RF 2 (var 11, leaf 5)	1.6755	0.7481	0.9734	0.5919
RF 3 (var 11, leaf 6)	1.6595	0.7516	0.9714	0.5923
RF 4 (var 6, leaf 2)	1.6385	0.8772	1.3929	0.6780

The "Weighted Sum Model" or "Simple Multi-Criteria Ranking" method was used to select the optimal model. For the minimization objectives (RMSE, MAE, MAPE), Min-Max normalization is applied, and for the maximization objective (R), Max-Min normalization is applied to ensure that all metrics are on the same scale. Equal weights are assigned to the normalized evaluation metrics to indicate their relative importance in the decision-making process. The weighted sum method calculated an aggregated value for each model, which considers all four normalized metrics. All models are ranked based on their aggregated values, with the lower aggregated value indicating better overall performance (Table 9).

Table 9. Determining the optimal prediction model for the Na^+ parameter using Simple Multi-Criteria Ranking.

Weighted Criteria	w.RMSE	w.MAE	w.MAPE	w.R	Agg. Value
RF 1 (var 13, leaf 10)	0.2500	0.1328	0.1351	0.0000	0.5180
RF 2 (var 11, leaf 5)	0.0000	0.2500	0.2488	0.0265	0.5253
RF 3 (var 11, leaf 6)	0.0587	0.2432	0.2500	0.0275	0.5794
RF 4 (var 6, leaf 2)	0.1356	0.0000	0.0000	0.2500	0.3856

As the optimal model, the RF model with 500 trees was obtained, which uses a subset of 11 variables, where the minimum amount of data per sheet is six. The assessment of the significance of individual input variables for the accuracy of the prediction was performed precisely on the obtained model with the highest accuracy (Figure 11).

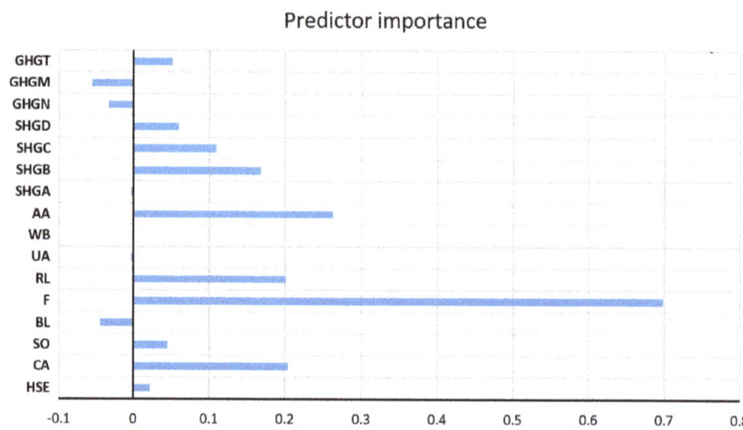

Figure 11. The significance of individual variables for Na^+ parameter prediction in an optimal RF model.

4.3. Prediction of Magnesium (Mg^{2+}) Parameter Values

RF models proved to be the optimal models for predicting sodium ion (Mg^{2+}) concentrations. An analysis of all models in terms of accuracy is given in Appendix A (Table A2). The dependence of the adopted accuracy criteria on the model parameters is shown in Figure 12. Based on the defined accuracy criteria, three models with the following values were selected (Table 10). Optimal values according to different accuracy criteria are marked with bold numbers in Table 10.

Table 10. The accuracy of the obtained models for Mg^{2+} prediction according to defined criteria.

Criteria	RMSE	MAE	MAPE	R
RF 1 (var 13, leaf 1)	**0.3988**	0.2640	**0.2662**	0.7377
RF 2 (var 12, leaf 1)	0.4014	**0.2608**	0.2706	0.7516
RF 3 (var 10, leaf 1)	0.4020	0.2631	0.2717	**0.7567**

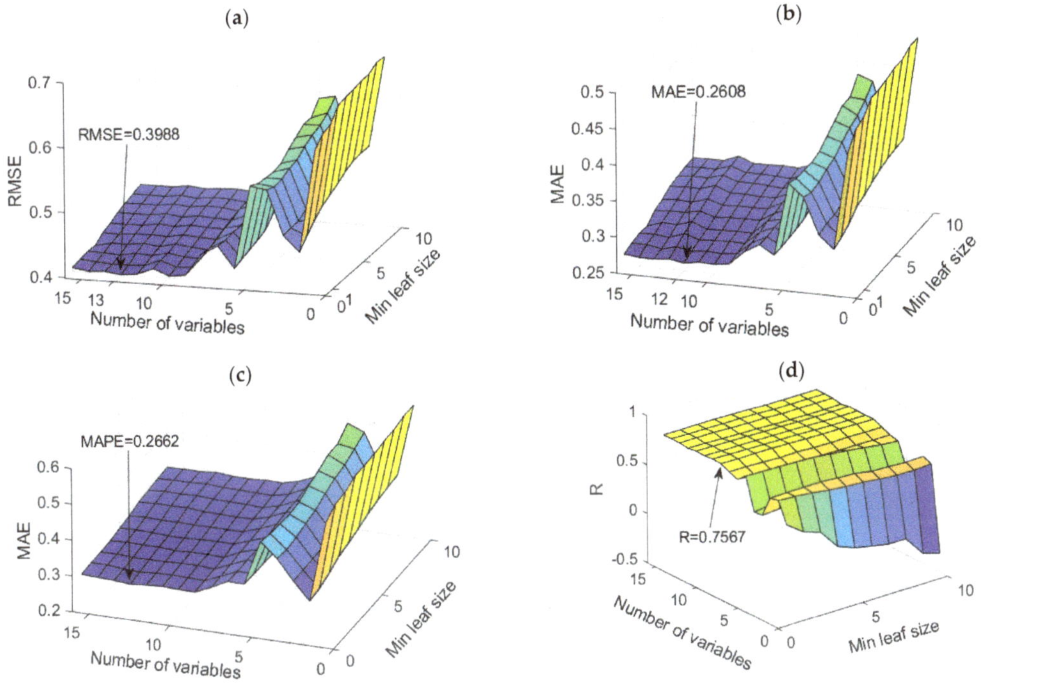

Figure 12. Comparison of different accuracy criteria for the RF model for the Mg^{2+} parameter as a function of the number of randomly selected splitting variables and minimum leaf size: (**a**) RMSE, (**b**) MAE, (**c**) MAPE, (**d**) R.

"Simple Multi-Criteria Ranking" was applied again when extracting the optimal model (Table 11).

Table 11. Determining the optimal prediction model for the Mg^{2+} parameter using Simple Multi-Criteria Ranking.

Weighted Criteria	w.RMSE	w.MAE	w.MAPE	w.R	Agg. Value
RF 1 (var 13, leaf 1)	0.2500	0.0000	0.2500	0.0000	0.5000
RF 2 (var 12, leaf 1)	0.0469	0.2500	0.0500	0.1829	0.5298
RF 3 (var 10, leaf 1)	0.0000	0.0703	0.0000	0.2500	0.3203

As the optimal model, the RF model with 500 trees was obtained, which uses a subset of 12 variables, where the minimum amount of data per sheet is one. The assessment of the importance of individual input variables on the accuracy of the prediction was performed precisely on the obtained model with the highest accuracy (Figure 13).

4.4. Prediction of Ca^{2+} Parameter Values

RF models proved to be the optimal models for Ca^{2+} (calcium ion concentration). An analysis of all models in terms of accuracy is given in Appendix A (Table A3). The dependence of the adopted accuracy criteria on the model parameters is shown in Figure 14. According to all the defined accuracy criteria, only one model stood out with values for RMSE, MAE, MAPE, and R of 0.5847, 0.4500, 0.2007, and 0.7496, respectively.

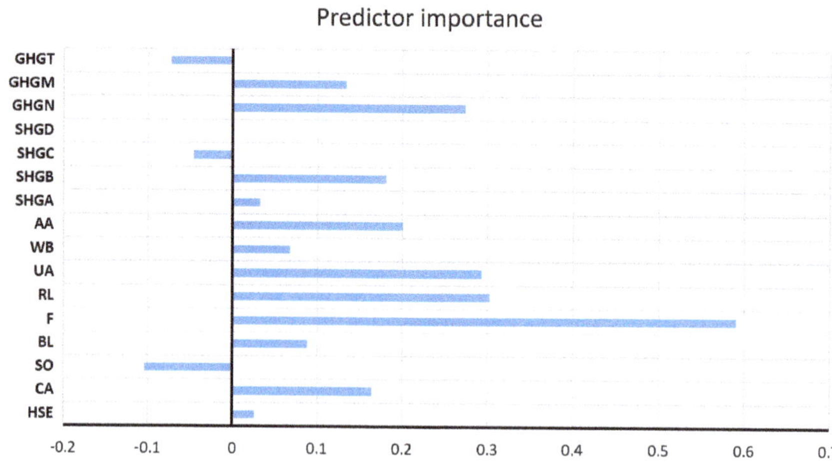

Figure 13. Significance of individual variables for Mg^{2+} parameter prediction in an optimal RF model.

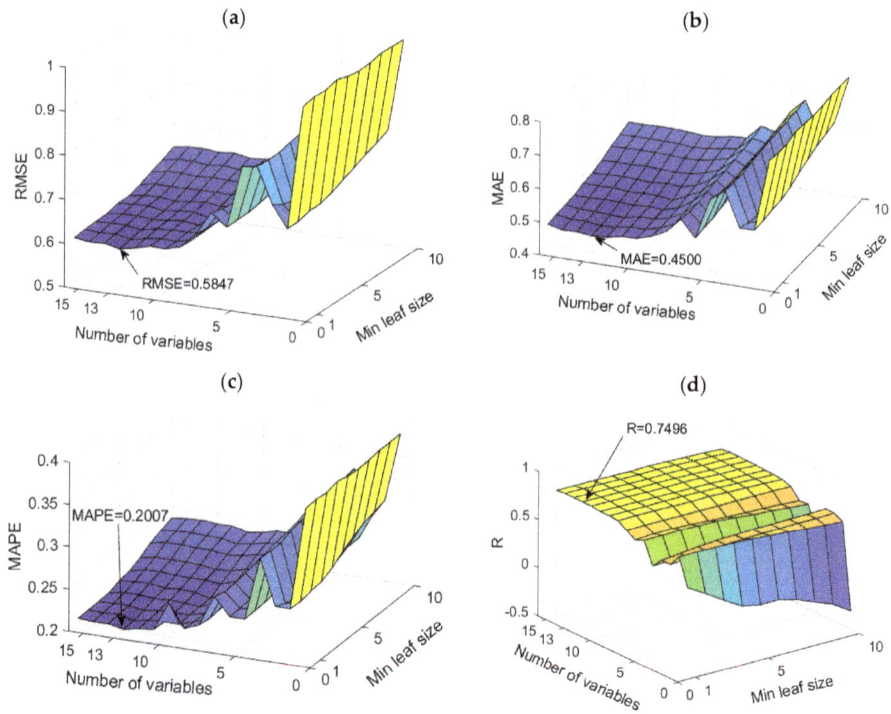

Figure 14. Comparison of different accuracy criteria for the RF model for the Ca^{2+} parameter as a function of the number of randomly selected splitting variables and minimum leaf size: (**a**) RMSE, (**b**) MAE, (**c**) MAPE, (**d**) R.

The assessment of the significance of individual input variables on the accuracy of the prediction was performed precisely on the obtained model with the highest accuracy (Figure 15).

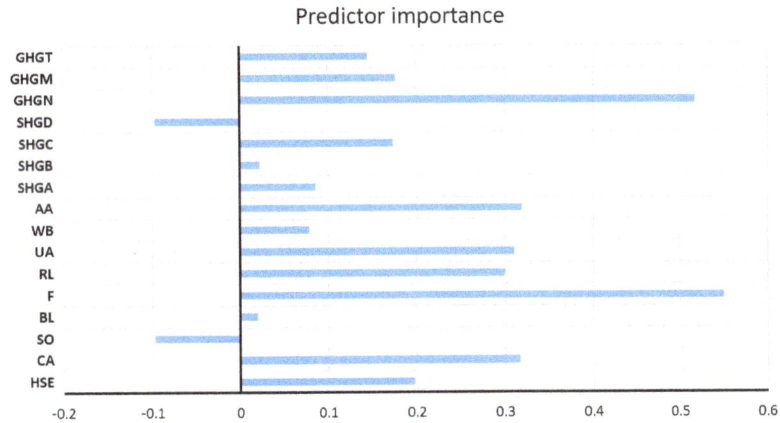

Figure 15. Significance of individual variables for Ca^{2+} parameter prediction in an optimal RF model.

4.5. Prediction of SO_4^{2-} Parameter Values

The RF models proved to be the optimal models for predicting SO_4^{2-} levels. An analysis of all models in terms of accuracy is given in Appendix A (Table A4). The dependence of the adopted accuracy criteria on the model parameters is shown in Figure 16.

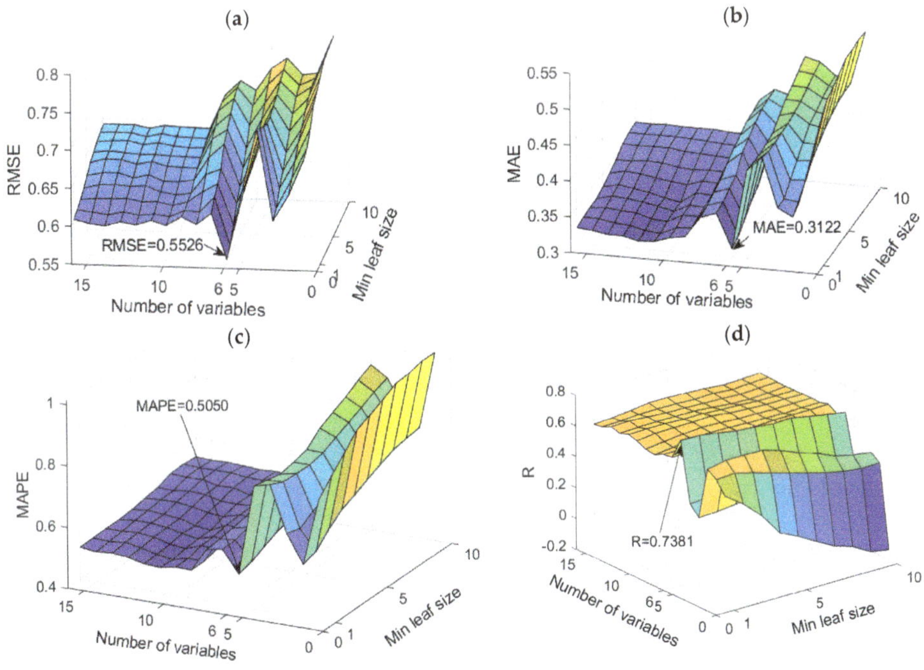

Figure 16. Comparison of different accuracy criteria for the RF model for the SO_4^{2-} parameter as a function of the number of randomly selected splitting variables and minimum leaf size: (**a**) RMSE, (**b**) MAE, (**c**) MAPE, (**d**) R.

According to all defined accuracy criteria, only one model was singled out with values for RMSE, MAE, MAPE, and R of 0.5526, 0.3122, 0.5050, and 0.7381, respectively.

The assessment of the significance of the individual input variables for the accuracy of the prediction was performed directly on the obtained model with the highest accuracy (Figure 17).

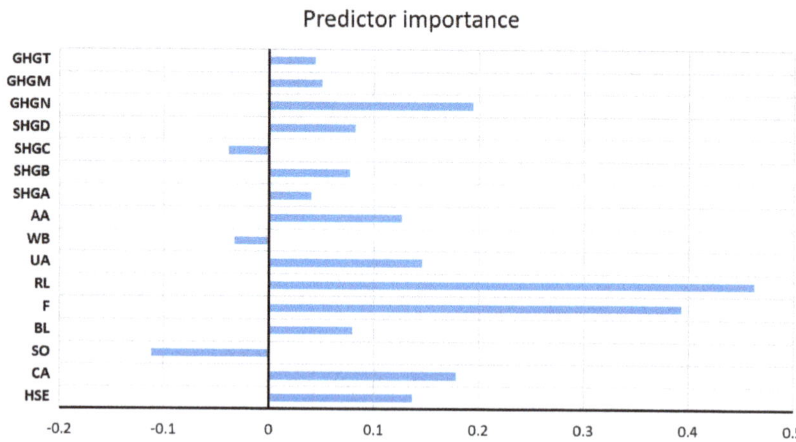

Figure 17. Significance of the individual variables for SO_4^{2-} parameter prediction in an optimal RF model.

4.6. Prediction of Cl^- Parameter Values

RF models proved to be the optimal models for predicting Cl^- concentrations. An analysis of all models in terms of accuracy is given in Appendix A (Table A5). The dependence of the adopted accuracy criteria on the model parameters is shown in Figure 18. Based on the defined accuracy criteria, three models were selected (Table 12). Optimal values according to different accuracy criteria are marked with bold numbers in Table 12.

Table 12. Accuracy of the obtained models for Cl^- prediction according to defined criteria.

RF Model	RMSE	MAE	MAPE	R
RF 1 (var 13, leaf 10)	**1.7999**	0.9111	1.1120	0.5691
RF 2 (var 11, leaf 5)	1.8831	**0.8316**	0.8589	0.5964
RF 3 (var 11, leaf 4)	1.8904	0.8323	**0.8557**	0.5933
RF 4 (var 6, leaf 2)	1.8473	0.9370	1.0288	**0.6940**

As the optimal model, the RF model with 500 trees was obtained, which uses a subset of 11 variables, where the minimum amount of data per leaf is five (Table 13).

Table 13. Determining the optimal prediction model for the Cl^- parameter using Simple Multi-Criteria Ranking.

RF Model	w.RMSE	w.MAE	w.MAPE	w.R	Agg. Value
RF 1 (var 13, leaf 10)	0.2500	0.0614	0.0000	0.0000	0.3114
RF 2 (var 11, leaf 5)	0.0202	0.2500	0.2469	0.0546	0.5717
RF 3 (var 11, leaf 4)	0.0000	0.2483	0.2500	0.0484	0.5468
RF 4 (var 6, leaf 2)	0.1191	0.0000	0.0812	0.2500	0.4502

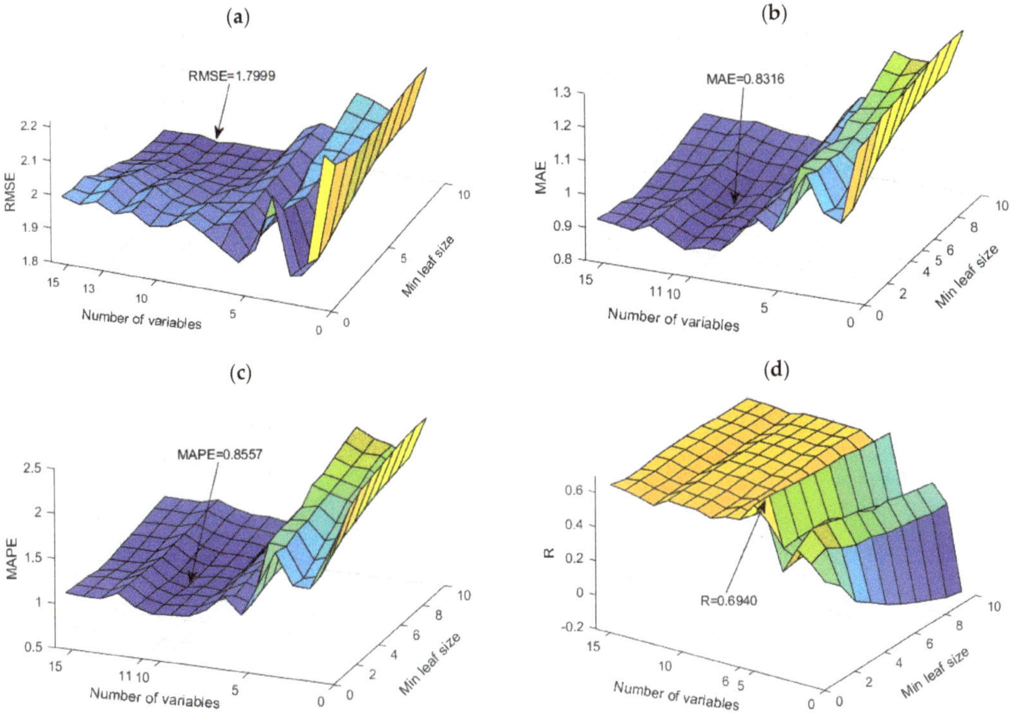

Figure 18. Comparison of different accuracy criteria for the RF model for the Cl^- parameter as a function of the number of randomly selected splitting variables and minimum leaf size: (**a**) RMSE, (**b**) MAE, (**c**) MAPE, (**d**) R.

The assessment of the importance of individual input variables on the accuracy of the prediction was performed precisely on the obtained model with the highest accuracy (Figure 19).

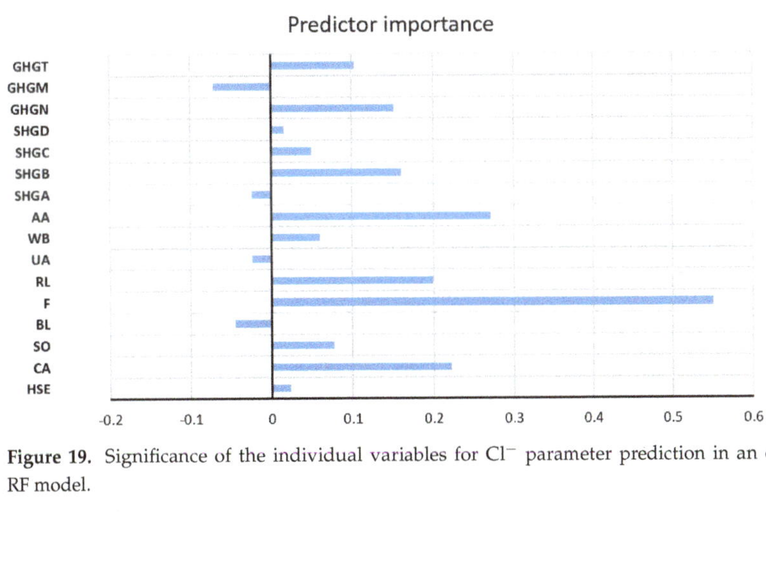

Figure 19. Significance of the individual variables for Cl^- parameter prediction in an optimal RF model.

4.7. Prediction of HCO^{3-} Parameter Values

GPR models proved to be the optimal models for predicting HCO3− concentrations. Very similar values in terms of accuracy were also given by the RF models. However, since the difference between the GPR model and the RF model is practically negligible, and since it is not possible to obtain the significance value for individual input variables on the obtained GPR model because it has the same length scale parameter for all variables, RF models were used for the analysis. An analysis of all models in terms of accuracy is given in Appendix A (Table A6).

The dependence of the adopted accuracy criteria on the parameters of the RF model is shown in Figure 20.

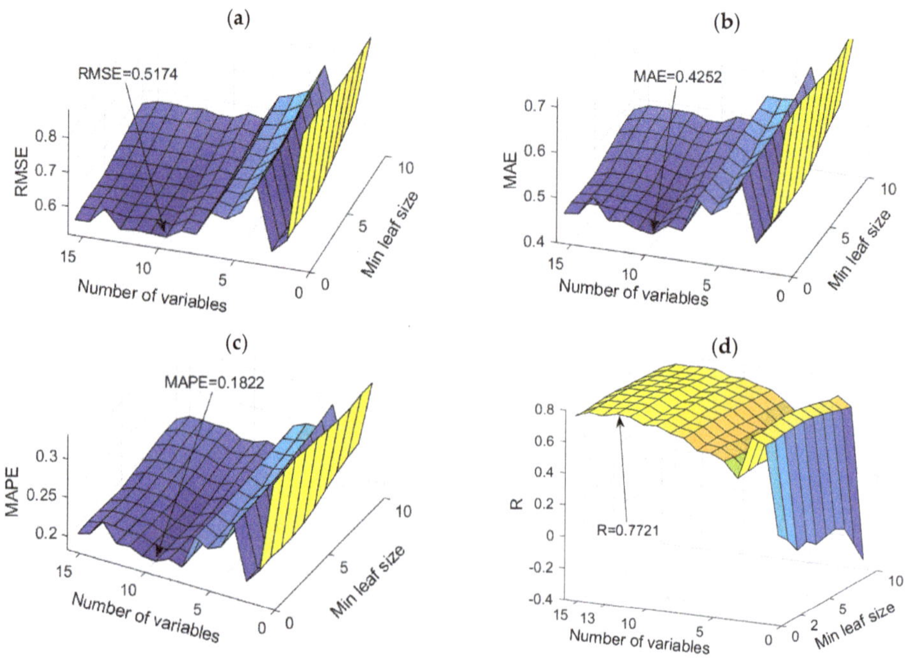

Figure 20. Comparison of different accuracy criteria for the RF model for the HCO^{3-} parameter as a function of the number of randomly selected splitting variables and minimum leaf size: (**a**) RMSE, (**b**) MAE, (**c**) MAPE, (**d**) R.

In the specific case of applying the RF model, two models were distinguished, namely the RF model that uses ten variables as a subset for analysis and where the amount of data per terminal sheet is equal to one, which is optimal according to the RMSE, MAE, and MAPE criteria and the model that uses 13 variables as a subset for analysis and where the amount of data per terminal sheet is equal to two, which is optimal according to the R criterion. Since the first-mentioned model is optimal according to the three adopted accuracy criteria, RMSE, MAE, and MAPE, and the difference compared to the R criterion is practically negligible, the first model can be considered optimal.

The optimal model has the following criterion values for RMSE, MAE, MAPE, and R of 0.5174, 0.4252, 0.1822, and 0.7721, respectively.

The assessment of the importance of the individual input variables on the accuracy of the prediction was performed precisely on the obtained model with the highest accuracy (Figure 21).

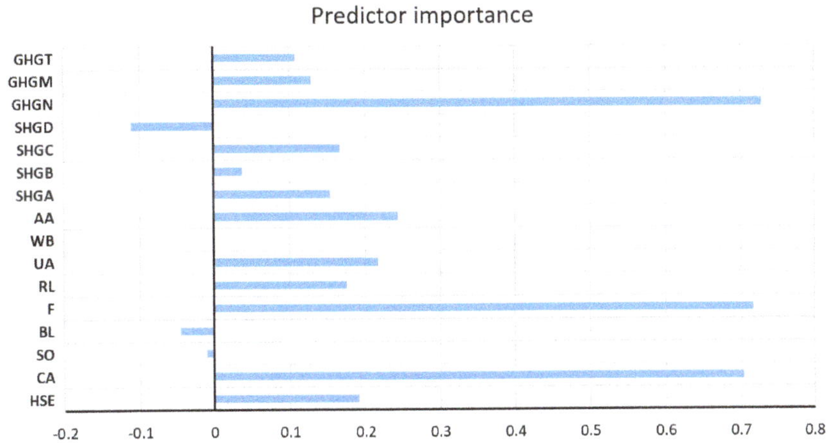

Figure 21. Significance of individual variables for HCO^{3-} parameter prediction in an optimal RF model.

4.8. Prediction of K^+ Parameter Values

The RF models proved to be the optimal models for predicting K^+ levels. An analysis of all models in terms of accuracy is given in Appendix A (Table A7). The dependence of the adopted accuracy criteria on the model parameters is shown in Figure 22.

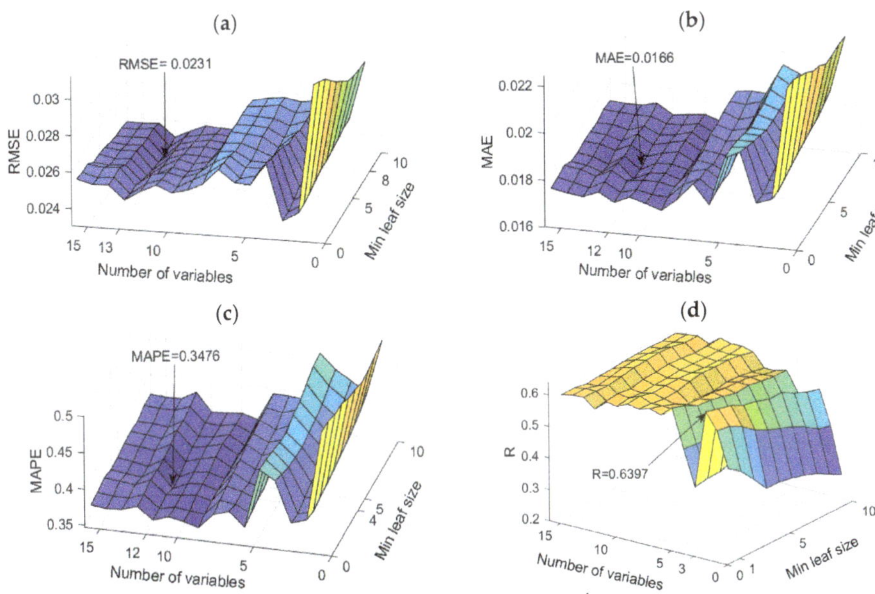

Figure 22. Comparison of different accuracy criteria for the RF model for the K+ parameter as a function of the number of randomly selected splitting variables and minimum leaf size: (**a**) RMSE, (**b**) MAE, (**c**) MAPE, (**d**) R.

In terms of accuracy, three models were singled out, and the optimal model was obtained by applying the Simple Multi-Criteria Ranking method (Tables 14 and 15). Optimal values according to different accuracy criteria are marked with bold numbers in Table 14.

Table 14. Accuracy of obtained models for K^+ parameter prediction according to defined criteria.

Criteria	RMSE	MAE	MAPE	R
Var 13, leaf 8	**0.0231**	0.0172	0.3755	0.5689
Var 3, leaf 1	0.0236	0.0174	0.3700	**0.6397**
Var 12, leaf 4	0.0241	**0.0166**	**0.3476**	0.6024

Table 15. Determining the optimal prediction model for the K^+ parameter using Simple Multi-Criteria Ranking.

Weighted Criteria	w.RMSE	w.MAE	w.MAPE	w.R	Agg. Value
Var 13, leaf 8	0.2500	0.0625	0.0000	0.0000	0.3125
Var 3, leaf 1	0.1250	0.0000	0.0493	0.2500	0.4243
Var 12, leaf 4	0.0000	0.2500	0.2500	0.1183	0.6183

The analysis of the significance of the individual input variables of the model was performed on the optimal RF model with hyperparameter values for the value of the number of trees, a subset of variables for splitting, and the amount of data per terminal leaf, which are 500, 12, and 4, respectively (Table 15). The assessment of the importance of the individual input variables was performed precisely on the obtained model with the highest accuracy (Figure 23).

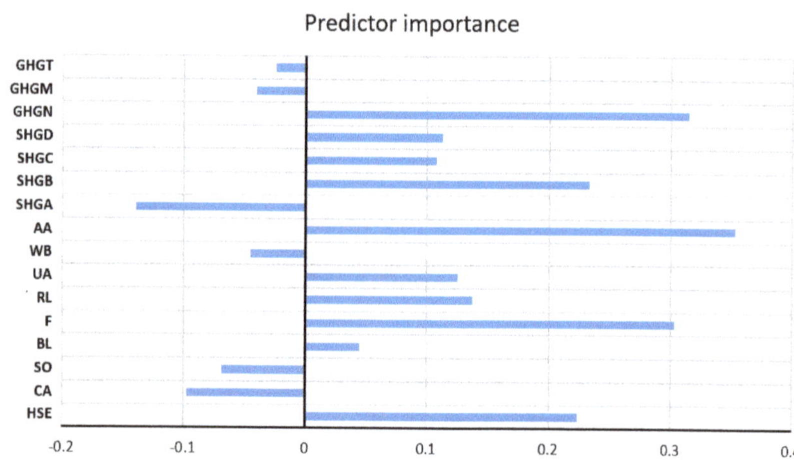

Figure 23. Significance of individual variables for K^+ parameter prediction in an optimal RF model.

4.9. Prediction of pH Parameter Values

The RF models proved to be the optimal models for predicting SO_4 levels. An analysis of all models in terms of accuracy is given in Appendix A (Table A8). The dependence of the adopted accuracy criteria on the model parameters is shown in Figure 24.

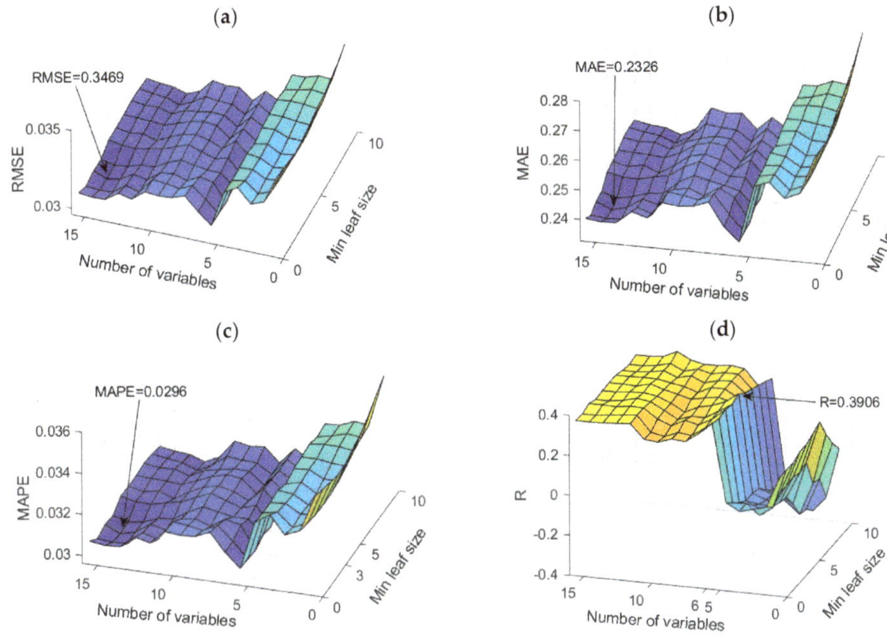

Figure 24. Comparison of different accuracy criteria for the RF model for the pH parameter as a function of the number of randomly selected splitting variables and minimum leaf size: (**a**) RMSE, (**b**) MAE, (**c**) MAPE, (**d**) R.

In terms of accuracy, three models were singled out, and the optimal model was obtained by applying the Simple Multi-Criteria Ranking method (Tables 16 and 17). Optimal values according to different accuracy criteria are marked with bold numbers in Table 16.

Table 16. Accuracy of the obtained models for pH parameter prediction according to defined criteria.

Criteria	RMSE	MAE	MAPE	R
RF 1 (var 7, leaf 1)	**0.3469**	0.2383	0.0306	0.3331
RF 2 (var 15, leaf 3)	0.3554	**0.2326**	**0.0296**	0.3476
RF 3 (var 6, leaf 5)	0.3531	0.2338	0.0298	**0.3906**

Table 17. Determining the optimal prediction model for the PH parameter using Simple Multi-Criteria Ranking.

Weighted Criteria	w.RMSE	w.MAE	w.MAPE	w.R	Agg. Value
RF 1 (var 7, leaf 1)	0.2500	0.0000	0.0000	0.0000	0.2500
RF 2 (var 15, leaf 3)	0.0000	0.2500	0.2500	0.0630	0.5630
RF 3 (var 6, leaf 5)	0.0676	0.1974	0.2000	0.2500	0.7150

Using the weighted sum method, an aggregated value for each model is calculated, which takes into account all four normalized metrics.

The analysis of the significance of the individual input variables of the model was performed on the optimal RF model and shown in Figure 25.

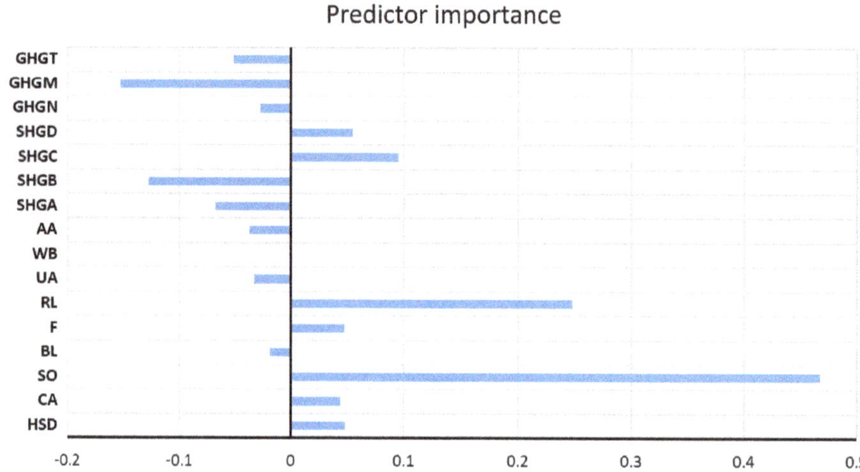

Figure 25. Significance of the individual variables for PH parameter prediction in an optimal RF model.

4.10. Prediction of EC Parameter Values

RF models proved to be optimal models for EC parameter prediction. An analysis of all models in terms of accuracy is given in Appendix A (Table A9). The dependence of the adopted accuracy criteria on the model parameters is shown in Figure 26.

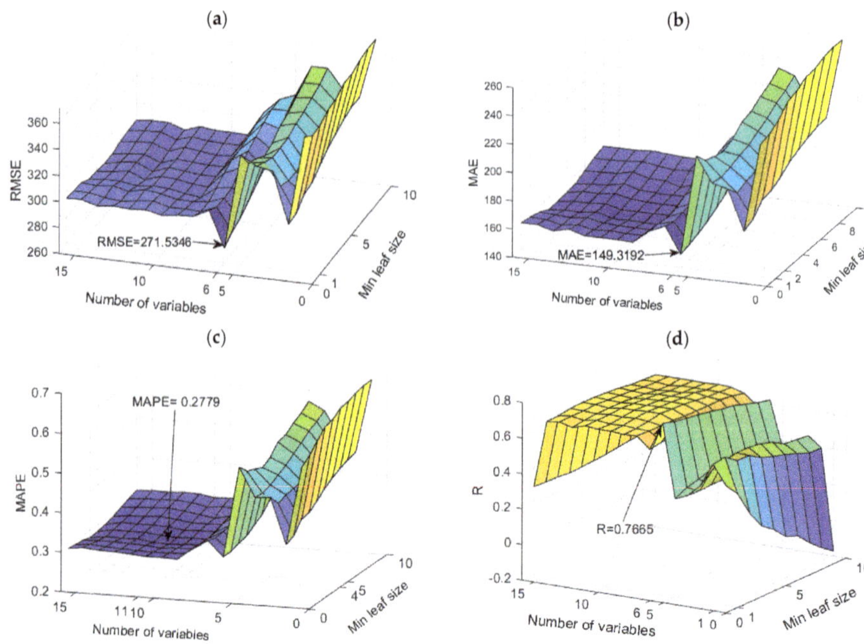

Figure 26. Comparison of different accuracy criteria for the RF model for the EC parameter as a function of the number of randomly selected splitting variables and minimum leaf size: (**a**) RMSE, (**b**) MAE, (**c**) MAPE, (**d**) R.

According to all defined accuracy criteria, only one model was singled out with values for RMSE, MAE, MAPE, and R of 271.5346, 149.3192, 0.2779, and 0.7665, respectively.

The obtained hyperparameter values for the number of trees, the subset of splitting variables, and the minimum amount of data per leaf are 500, 6, and 1, respectively. The analysis of the significance of the individual input variables of the model was performed on the optimal RF model and shown in Figure 27.

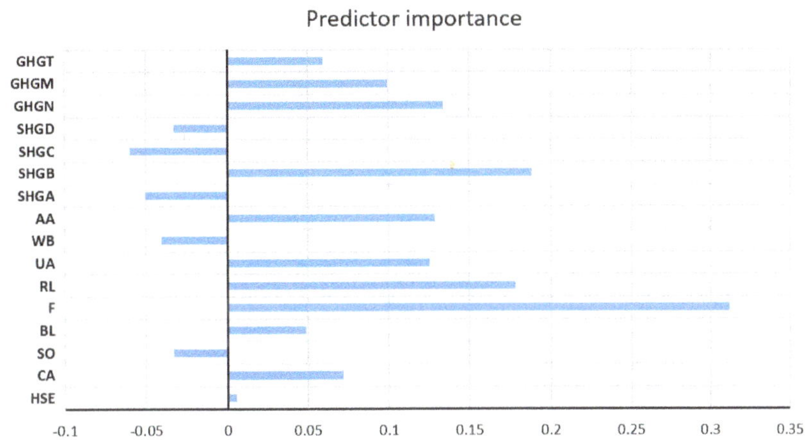

Figure 27. Significance of the individual variables for EC parameter prediction in an optimal RF model.

4.11. Prediction of TDS Parameter Values

The RF models proved to be the optimal models for predicting SO_4^{2-} levels. An analysis of all models in terms of accuracy is given in Appendix A (Table A10). The dependence of the adopted accuracy criteria on the model parameters is shown in Figure 28.

In terms of accuracy, three models were singled out, and the optimal model was obtained by applying the Simple Multi-Criteria Ranking method (Tables 18 and 19). Optimal values according to different accuracy criteria are marked with bold numbers in Table 18.

Table 18. Accuracy of the obtained models for TDS parameter prediction according to defined criteria.

Criteria	RMSE	MAE	MAPE	R
RF 1 (Var 9, leaf 10)	**417.2155**	201.8572	0.4863	0.5467
RF 1 (Var 12, leaf 7)	422.6822	**196.7117**	0.4578	**0.5521**
RF 1 (Var 12, leaf 5)	435.3533	198.3639	**0.4562**	0.5502

Table 19. Determining the optimal prediction model for the TDS parameter using Simple Multi-Criteria Ranking.

Weighted Criteria	w.RMSE	w.MAE	w.MAPE	w.R	Agg. Value
RF 1 (Var 9, leaf 10)	0.2500	0.0000	0.0000	0.0000	0.2500
RF 1 (Var 12, leaf 7)	0.1747	0.2500	0.2367	0.2500	0.9114
RF 1 (Var 12, leaf 5)	0.0000	0.1697	0.2500	0.1620	0.5818

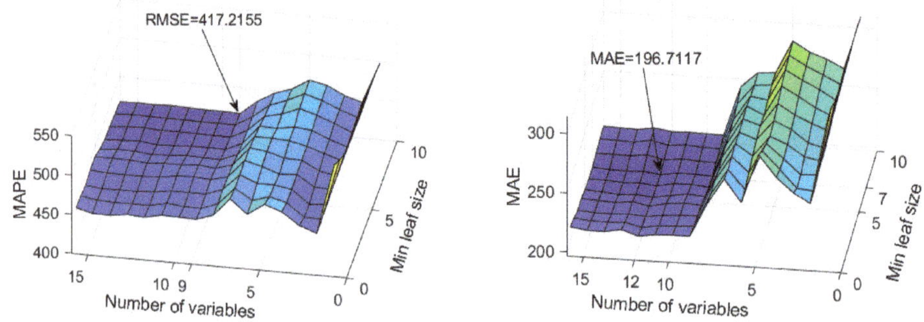

(**c**) (**d**)

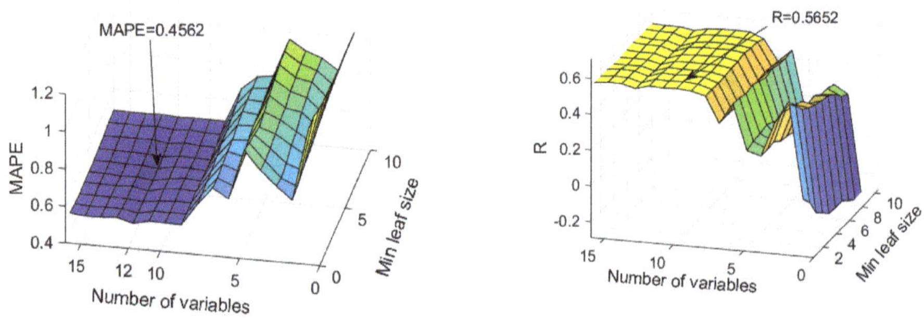

Figure 28. Comparison of different accuracy criteria for the RF model for the TDS parameter as a function of the number of randomly selected splitting variables and minimum leaf size: (**a**) RMSE, (**b**) MAE, (**c**) MAPE, (**d**) R.

The analysis of the significance of the individual input variables of the model was performed on the optimal RF model and shown in Figure 29.

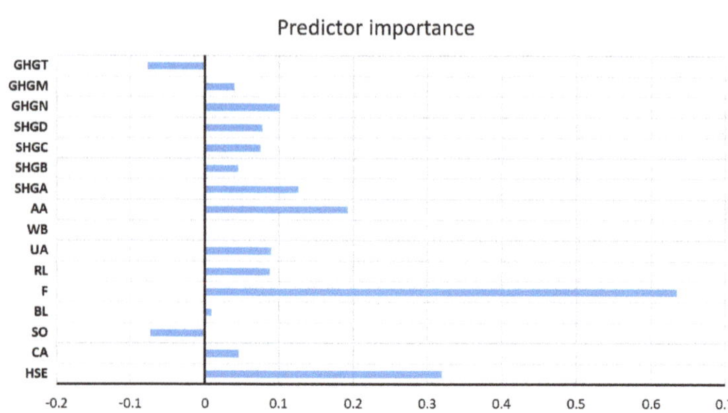

Figure 29. Significance of the individual variables for TDS parameter prediction in an optimal RF model.

5. Discussion

In our research, most models demonstrated satisfactory accuracy, meeting the predefined criteria. However, a subset of models exhibited shortcomings in specific criteria. To gauge accuracy effectively, we leaned on relative metrics, notably accuracy (R) and mean absolute percentage error (MAPE), as they offer more insightful perspectives compared to absolute criteria such as RMSE and MAE (Table 20).

Table 20. Accuracy of the ML model in predicting individual water parameters.

Output Parameter	Best Model	RMSE	MAE	MAPE	R
SAR	RF	0.3668	0.2328	0.5679	0.8236
Na^+	RF	16.385	0.8772	13.929	0.678
Mg^{2+}	RF	0.402	0.2631	0.2717	0.7567
Ca^{2+}	RF	0.5847	0.45	0.2007	0.7496
$SO_4{}^{2-}$	RF	0.5526	0.3122	0.505	0.6148
Cl^-	RF	18.831	0.8316	0.8589	0.5964
HCO^{3-}	GP	0.5056	0.4144	0.1782	0.7668
K^+	RF	0.0241	0.0166	0.3476	0.6024
pH	RF	0.3531	0.2338	0.0298	0.3906
EC	RF	271.5346	149.3192	0.3013	0.7665
TDS	RF	422.6822	196.7117	0.4578	0.5521

Table 20 highlights the accuracy of the machine learning models in predicting individual water parameters. Notably, the RF model emerged as the best performer across various parameters, underscoring its efficacy.

Analyzing the R values reveals the overall satisfactory performance of most models, except for the pH prediction model. Examining MAPE values identified five models—SAR, Na+, SO_4, Cl, and TDS—where this metric is relatively higher than other ones. Despite these nuances, our primary research focus was unraveling the significance of individual input variables within the constraints of limited data.

When we delve into the significance of the individual input variables, our conclusions (Table 21) unveil the following crucial insights:

Table 21. The most influential input variables for predicting water parameters.

	Input Variable															
Output	**HSE**	**CA**	**SO**	**BL**	**F**	**RL**	**UA**	**WB**	**AA**	**HSGA**	**HSGB**	**HSGC**	**HSGD**	**GPGM**	**GPGN**	**GPGT**
SAR		3			5	1	4		2							
Na^+		3			1	4			2		5					
Mg^{2+}					1	2	3		5					4		
Ca^{2+}		4			1		5		3					2		
$SO_4{}^{2-}$		4			2	1	5							3		
Cl^-	5				1		3				2					4
$HCO_3{}^-$	5	3			2		4							1		
K^+	5				3				1		4			2		
pH			1			2					4	5			3	
EC					1	2	5		4						3	
TDS	2				1				3	4				5		

Forest Cover ('F'): Forest areas significantly influence diverse water quality parameters. Trees and vegetation in forests contribute organic matter to water bodies, influencing ion concentrations. The root systems of trees can affect the uptake of certain ions. Forests strongly impact the concentrations of sodium, magnesium, calcium, chloride, sulfate, bicarbonate, and potassium ions. Also, forests act as natural filters, reducing the transport of sediments and pollutants into water bodies. Cleaner water, with fewer suspended solids, tends to have lower TDS and EC. Additionally, forest areas often have minimal human activities compared to urban or agricultural areas.

Rangeland (RL) is essential for predicting water sulfate ion concentrations. This suggests that the characteristics associated with the rangeland, such as land cover and land use patterns, significantly influence sulfate levels. Additionally, rangeland strongly affects SAR by influencing sodium concentrations, vital for evaluating water's suitability for irrigation and soil health. Also, the notable impact on magnesium levels showcases rangeland's role in shaping water quality. Rangeland's influence on pH highlights its role in determining water acidity or alkalinity, which is crucial for aquatic ecosystems and nutrient availability. Additionally, rangeland significantly influences electrical conductivity, providing insights into water quality and dissolved ion content, essential for understanding overall water composition. While having a somewhat lesser impact, rangeland still plays a discernible role in shaping sodium concentrations, contributing to insights into water salinity and its ecological implications.

Urban Area ('UA'): Urban areas have a moderate impact on ion levels, magnesium, chloride, bicarbonate, and SAR parameters, owing to urbanization and land use changes, introducing contaminants and altering water chemistry. Calcium, sulfate, and EC parameters have less impact.

The Agricultural Area (AA) substantially impacts potassium, SAR, and sodium, with a moderate impact on calcium, TDS, and magnesium. The influence of AA on these parameters can be explained by the agricultural areas' use of potassium-containing fertilizers, leading to elevated potassium concentrations in water. Cultivation practices and nutrient management contribute to increased potassium levels. Additionally, agricultural activities often involve irrigation, and water with high sodium content can increase SAR. Sodium in the soil can be introduced through irrigation water, affecting sodium levels in the water. Moreover, agricultural runoff can introduce calcium, magnesium, and other dissolved solids into water sources.

Catchment Area ('CA'): The size of catchment areas plays a moderate role in ion transport, particularly affecting SAR, sodium, bicarbonate, calcium, and sulfate levels. The size of the catchment area could moderately impact SAR, as larger areas may interact with more diverse geological and soil features, affecting sodium adsorption ratios.

Considering different soil types (HSGA, HSGB, HSGC, HSGD) and geological permeability (GHGM, GHGN, GHGT) underscores their impact on ion retention and release. Sandy soils facilitate easier ion movement, while clayey soils retain ions. Geological permeability influences potassium, magnesium, calcium, and bicarbonate levels, showcasing the interconnectedness of soil and geological characteristics with water parameters.

6. Conclusions

Our study demonstrates the effectiveness of machine learning methods in predicting and assessing water quality parameters within a catchment area. With the Random Forest (RF) model as the standout performer, the model provides a robust tool for efficient and accurate water quality evaluation.

While certain models may fall short on specific criteria, a nuanced evaluation leveraging relative criteria like accuracy (R) and mean absolute percentage error (MAPE) underscores the overall robustness of the predictive models. Table 20 encapsulates the detailed results, highlighting the efficacy of the RF model across various water parameters.

Evaluation of R values showcases all models' satisfactory performance except for pH prediction. Despite marginally elevated MAPE values in five models (SAR, Na+, SO_4, Cl,

TDS), the core research objective—unraveling the importance of individual input variables within data constraints—was largely achieved.

This accomplishment paves the way for selecting and implementing optimal models from a broader ML spectrum. To further elevate model accuracy, future research will focus on dataset expansion, a strategic initiative to address current limitations and achieve heightened accuracy, particularly in parameters exhibiting slight deviations.

The significance of individual input variables, as outlined in Table 21, provides crucial insights for understanding their roles in influencing water parameters. Forest cover, catchment area characteristics, stream order, barren land, and urban areas are pivotal factors shaping water quality.

Incorporating these research insights into decision-making processes presents transformative opportunities for strategic resource allocation and environmental impact mitigation. Furthermore, integrating these outcomes empowers decision-makers to adopt targeted strategies for fostering environmental sustainability, contributing to the broader goal of cultivating resilient water ecosystems. This integration signifies a practical pathway toward achieving a delicate balance between human activities and environmental preservation, actively contributing to sustainable water ecosystems.

Author Contributions: Conceptualization, M.K., B.J.A., S.L., E.K.N., M.H.-N. and D.R.; methodology, M.K.; software, M.K. and E.K.N.; validation, M.K., S.L., M.H.-N. and E.K.N.; formal analysis, M.K. and E.K.N.; investigation, M.K., B.J.A. and M.H.-N.; resources, B.J.A., S.L., M.H.-N. and D.R.; data curation, M.K., S.L. and E.K.N.; writing—original draft preparation, M.K., S.L., E.K.N. and M.H.-N.; writing—review and editing, M.K., S.L., E.K.N. and M.H.-N.; funding acquisition, D.R. All authors have read and agreed to the published version of the manuscript.

Funding: This research received no external funding.

Institutional Review Board Statement: Not applicable.

Informed Consent Statement: Not applicable.

Data Availability Statement: The data presented in this study are available on request from the corresponding author.

Conflicts of Interest: The authors declare no conflict of interest.

Appendix A

- **Na parameter**

Table A1. Comparative analysis of results of different machine learning models for Na parameter prediction.

Model	RMSE	MAE	MAPE/100	R
Decision Tree	2.1568	**0.7220**	**0.8754**	0.4381
TreeBagger	1.6809	0.7675	1.0412	0.5813
Random Forest	**1.6385**	0.8772	1.3929	**0.6780**
Boosted Trees	1.8603	0.9529	1.5436	0.5024
GP exponential	2.5655	1.0096	1.8659	0.0642
GP Sq.exponential	2.8037	1.1203	2.1615	0.0133
GP matern 3/2	2.7865	1.0860	2.0261	0.0314
GP matern 5/2	2.8302	1.1018	2.0680	0.0240
GP Rat. quadratic	2.8037	1.1203	2.1615	0.0133
GP ARD exponential	3.3385	1.3350	2.8318	−0.0282
GP ARD Sq. exponential	3.6399	1.4212	2.5305	−0.0099
GP ARD matern 3/2	4.2629	1.5668	2.8746	0.0350
GP ARD matern 5/2	4.4450	1.6865	3.0962	0.0170
GP ARD Rat. quadratic	4.2855	1.5028	2.6716	0.0366

- **Mg parameter**

Table A2. Comparative analysis of results of different machine learning models for Mg parameter prediction.

Model	RMSE	MAE	MAPE/100	R
Decision Tree	0.6043	0.3313	0.3001	0.4911
TreeBagger	0.4048	0.2641	0.2735	0.7472
Random Forest	**0.4020**	**0.2631**	**0.2717**	**0.7567**
GP exponential	0.6173	0.3524	0.3443	0.4355
GP Sq.exponential	0.6711	0.4076	0.4150	0.3733
GP matern 3/2	0.6532	0.3786	0.3731	0.3915
GP matern 5/2	0.6633	0.3907	0.3875	0.3802
GP Rat. quadratic	0.6711	0.4076	0.4150	0.3733
GP ARD exponential	0.6925	0.4114	0.3927	0.3953
GP ARD Sq. exponential	0.7831	0.4364	0.4129	0.3459
GP ARD matern 3/2	0.6877	0.4295	0.4213	0.4068
GP ARD matern 5/2	0.7180	0.4323	0.4207	0.3371
GP ARD Rat. quadratic	0.7291	0.4207	0.4208	0.3987

- **Ca parameter**

Table A3. Comparative analysis of results of different machine learning models for Ca parameter prediction.

Model	RMSE	MAE	MAPE/100	R
Decision Tree	0.7057	0.5504	0.2511	0.6804
TreeBagger	0.5949	0.4642	0.2054	0.7496
Random Forest	**0.5847**	**0.4500**	**0.2007**	**0.7496**
Boosted Trees	0.7730	0.6435	0.2808	0.5093
GP exponential	0.9379	0.5540	0.2225	0.3910
GP Sq.exponential	0.8241	0.5888	0.2566	0.5166
GP matern 3/2	0.7853	0.5538	0.2374	0.5662
GP matern 5/2	0.7989	0.5687	0.2447	0.5505
GP Rat. quadratic	0.8093	0.5755	0.2498	0.5364
GP ARD exponential	0.8347	0.6113	0.2617	0.5626
GP ARD Sq. exponential	0.8156	0.5878	0.2515	0.5497
GP ARD matern 3/2	0.7873	0.5772	0.2391	0.5873
GP ARD matern 5/2	0.7825	0.5809	0.2409	0.5948
GP ARD Rat. quadratic	0.9471	0.6581	0.2822	0.4985

- **SO_4 parameter**

Table A4. Comparative analysis of results of different machine learning models for SO4 parameter prediction.

Model	RMSE	MAE	MAPE/100	R
Decision Tree	0.6585	0.3319	**0.4713**	**0.6249**
TreeBagger	0.5997	0.3228	0.5064	0.5535
Random Forest	**0.5526**	**0.3122**	0.5050	0.6148
Boosted Trees	0.6421	0.4283	0.8503	0.5900
GP exponential	0.8183	0.4002	0.6450	0.3296
GP Sq.exponential	0.9090	0.4751	0.8683	0.1869
GP matern 3/2	0.8811	0.4453	0.7749	0.2489
GP matern 5/2	0.8930	0.4592	0.8144	0.2260
GP Rat. quadratic	0.9060	0.4713	0.8580	0.1918

Table A4. *Cont.*

Model	RMSE	MAE	MAPE/100	R
GP ARD exponential	0.8579	0.4061	0.5984	0.2182
GP ARD Sq. exponential	1.0228	0.5025	0.8036	0.1891
GP ARD matern 3/2	0.9006	0.4476	0.8242	0.3506
GP ARD matern 5/2	0.8340	0.4127	0.7664	0.3954
GP ARD Rat. quadratic	0.8370	0.4631	0.8300	0.3486

- **Cl parameter**

Table A5. Comparative analysis of results of different machine learning models for Cl parameter prediction.

Model	RMSE	MAE	MAPE/100	R
Decision Tree	2.4687	0.8348	0.9213	0.4090
TreeBagger	1.9022	0.8556	0.6413	0.5878
Random Forest	**1.8831**	**0.8316**	**0.8589**	**0.5964**
Bosted Trees	2.2544	1.0919	1.3900	0.4431
GP exponential	2.9457	1.1626	1.8895	0.0196
GP Sq.exponential	3.2492	1.2872	2.1291	−0.0253
GP matern 3/2	3.2183	1.2545	2.0769	−0.0135
GP matern 5/2	3.2735	1.2793	2.1214	−0.0177
GP Rat. quadratic	3.2492	1.2871	2.1291	−0.0253
GP ARD exponential	3.8178	1.5359	2.7443	−0.0370
GP ARD Sq. exponential	4.1299	1.5817	2.4557	−0.0557
GP ARD matern 3/2	4.9069	1.8010	2.7176	−0.0523
GP ARD matern 5/2	5.3636	2.0316	3.6072	−0.0620
GP ARD Rat. quadratic	4.1299	1.5817	2.4557	−0.0557

- **HCO_3 parameter**

Table A6. Comparative analysis of results of different machine learning models for HCO3 parameter prediction.

Model	RMSE	MAE	MAPE/100	R
Decision Tree	0.6782	0.5473	0.2228	0.5477
TreeBagger	0.5231	0.4400	0.1875	0.7386
Random Forest	0.5174	0.4252	0.1822	0.7280
GP exponential	**0.5056**	**0.4144**	**0.1782**	**0.7668**
GP Sq.exponential	0.5803	0.4791	0.2006	0.6541
GP matern 3/2	0.5312	0.4309	0.1827	0.7287
GP matern 5/2	0.5437	0.4404	0.1859	0.7109
GP Rat. quadratic	0.5516	0.4473	0.1872	0.6994
GP ARD exponential	0.6389	0.5309	0.2225	0.5902
GP ARD Sq. exponential	0.5596	0.4529	0.1860	0.6829
GP ARD matern 3/2	0.5692	0.4750	0.2001	0.6773
GP ARD matern 5/2	0.5986	0.4949	0.2026	0.6417
GP ARD Rat. quadratic	0.5951	0.4967	0.2070	0.6340

- **K parameter**

Table A7. Comparative analysis of results of different machine learning models for K parameter prediction.

Model	RMSE	MAE	MAPE/100	R
Decision Tree	0.0308	0.0201	0.4365	0.3385
TreeBagger	**0.0238**	0.0172	0.3722	0.5919
Random Forest	0.0241	**0.0166**	**0.3476**	**0.6024**
GP exponential	0.0275	0.0181	0.4008	0.4880
GP Sq.exponential	0.0306	0.0205	0.4568	0.3393
GP matern 3/2	0.0292	0.0196	0.4344	0.4168
GP matern 5/2	0.0297	0.0199	0.4434	0.3924
GP Rat. quadratic	0.0299	0.0201	0.4486	0.3769
GP ARD exponential	0.0293	0.0192	0.4225	0.4182
GP ARD Sq. exponential	0.0301	0.0189	0.3988	0.3328
GP ARD matern 3/2	0.0311	0.0207	0.4621	0.3086
GP ARD matern 5/2	0.0307	0.0206	0.4682	0.3449
GP ARD Rat. quadratic	0.0315	0.0209	0.4728	0.3148

- **Ph parameter**

Table A8. Comparative analysis of results of different machine learning models for Ph parameter prediction.

Model	RMSE	MAE	MAPE/100	R
Decision Tree	0.3719	0.2457	0.0316	**0.4241**
TreeBagger	0.3558	0.2376	0.0303	0.3380
Random Forest	**0.3531**	**0.2338**	**0.0298**	**0.3906**
Boosted Trees	0.3817	0.2576	0.0330	0.3187
GP exponential	0.4155	0.2586	0.0330	−0.0197
GP Sq.exponential	0.4201	0.2622	0.0335	0.0009
GP matern 3/2	0.4183	0.2573	0.0328	−0.0002
GP matern 5/2	0.4172	0.2560	0.0326	0.0114
GP Rat. quadratic	0.4192	0.2613	0.0334	−0.0247
GP ARD exponential	0.4972	0.2970	0.0381	−0.0636
GP ARD Sq. exponential	0.5655	0.3499	0.0453	−0.0511
GP ARD matern 3/2	0.5025	0.3098	0.0400	0.0272
GP ARD matern 5/2	0.5096	0.3127	0.0403	−0.0118
GP ARD Rat. quadratic	0.4154	0.2497	0.0318	0.1367

- **EC parameter**

Table A9. Comparative analysis of results of different machine learning models for EC parameter prediction.

Model	RMSE	MAE	MAPE/100	R
Decision Tree	352.6501	168.7962	0.3289	0.5627
TreeBagger	286.5049	151.5407	**0.2797**	0.7664
Random Forest	**271.5346**	**149.3192**	0.3013	**0.7665**
Bosted Trees	297.9335	170.2860	0.3620	0.6393

Table A9. *Cont.*

Model	RMSE	MAE	MAPE/100	R
GP exponential	432.0779	200.1383	0.4037	0.2465
GP Sq.exponential	474.4095	227.5836	0.4762	0.1730
GP matern 3/2	467.6136	217.1865	0.4402	0.1915
GP matern 5/2	472.1707	221.6132	0.4518	0.1856
GP Rat. quadratic	474.4095	227.5836	0.4762	0.1730
GP ARD exponential	461.6100	216.0258	0.4514	0.2684
GP ARD Sq. exponential	674.4802	269.8017	0.5526	0.1942
GP ARD matern 3/2	631.0788	275.7947	0.5870	0.1756
GP ARD matern 5/2	674.8450	287.7216	0.5782	0.1310
GP ARD Rat. quadratic	470.4831	237.1822	0.4714	0.2560

- **TDS parameter**

Table A10. Comparative analysis of results of different machine learning models for TDS parameter prediction.

Model	RMSE	MAE	MAPE/100	R
Decision Tree	509.6578	212.4422	0.5367	0.4610
TreeBagger	422.7209	199.4986	0.4718	**0.5535**
Random Forest	**422.6822**	**196.7117**	**0.4578**	0.5521
Boosted Trees	457.8293	235.2232	0.6106	0.5367
GP exponential	617.8458	**274.5861**	**0.7459**	0.0383
GP Sq.exponential	663.3802	302.3803	0.8247	0.0276
GP matern 3/2	662.2055	297.1371	0.8050	0.0191
GP matern 5/2	666.2775	299.6533	0.8082	0.0207
GP Rat. quadratic	663.3802	302.3803	0.8247	0.0276
GP ARD exponential	765.3184	377.8162	1.1147	−0.0131
GP ARD Sq. exponential	818.4166	408.9820	1.1664	−0.0367
GP ARD matern 3/2	881.9864	460.2564	1.4128	−0.0318
GP ARD matern 5/2	828.6321	416.7358	1.3426	0.0524
GP ARD Rat. quadratic	785.2499	392.9596	1.2369	0.0238

References

1. Dyer, F.; Elsawah, S.; Croke, B.; Griffiths, R.; Harrison, E.; Lucena-Moya, P.; Jakeman, A. The Effects of Climate Change on Ecologically-relevant Flow Regime and Water Quality Attributes. *Stoch. Environ. Res. Risk. Assess.* **2013**, *1*, 67–82. [CrossRef]
2. Bisht, A.; Singh, R.; Bhutiani, R.; Bhatt, A. Application of Predictive Intelligence in Water Quality Forecasting of the River Ganga Using Support Vector Machines. In *Predictive Intelligence Using Big Data and the Internet of Things*; Gupta, P.K., Ören, T., Singh, M., Eds.; IGI Global: Hershey, PA, USA, 2019; pp. 206–218. [CrossRef]
3. Liu, S.; Ryu, D.; Webb, J.A.; Lintern, A.; Guo, D.; Waters, D.; Western, A.W. A multi-model approach to assessing the impacts of catchment characteristics on spatial water quality in the Great Barrier Reef catchments. *Environ. Pollut.* **2021**, *288*, 117337. [CrossRef]
4. Xia, M.; Craig, P.M.; Schaeffer, B.; Stoddard, A.; Liu, Z.; Peng, M.; Zhang, H.; Wallen, C.M.; Bailey, N.; Mandrup-Poulsen, J. Influence of physical forcing on bottom-water dissolved oxygen within Caloosahatchee River Estuary, Florida. *J. Environ. Eng.* **2010**, *136*, 1032–1044. [CrossRef]
5. Liu, Y.; Weisberg, R.H.; Zheng, L.; Heil, C.A.; Hubbard, K.A. Termination of the 2018 Florida red tide event: A tracer model perspective. *Estuar. Coast. Shelf Sci.* **2022**, *272*, 107901. [CrossRef]
6. Mazher, A. Visualization Framework for High-Dimensional Spatio-Temporal Hydrological Gridded Datasets using Machine-Learning Techniques. *Water* **2020**, *12*, 590. [CrossRef]
7. Nasir, N.; Kansal, A.; Alshaltone, O.; Barneih, F.; Sameer, M.; Shanableh, A.; Al-Shamma'a, A. Water quality classification using machine learning algorithms. *J. Water. Process. Eng.* **2022**, *48*, 102920. [CrossRef]
8. Wang, R.; Kim, J.H.; Li, M.H. Predicting stream water quality under different urban development pattern scenarios with an interpretable machine learning approach. *Sci. Total. Environ.* **2021**, *761*, 144057. [CrossRef]
9. Rodriguez-Galiano, V.; Sanchez-Castillo, M.; Chica-Olmo, M.; Chica-Rivas, M.J.O.G.R. Machine learning predictive models for mineral prospectivity: An evaluation of neural networks, random forest, regression trees and support vector machines. *Ore. Geol. Rev.* **2015**, *71*, 804–818. [CrossRef]

10. Koranga, M.; Pant, P.; Kumar, T.; Pant, D.; Bhatt, A.K.; Pant, R.P. Efficient water quality prediction models based on machine learning algorithms for Nainital Lake, Uttarakhand. *Mater. Today Proc.* **2022**, *57*, 1706–1712. [CrossRef]
11. Kovačević, M.; Ivanišević, N.; Dašić, T.; Marković, L. Application of artificial neural networks for hydrological modelling in Karst. *Gradjevinar* **2018**, *70*, 1–10.
12. Zhu, M.; Wang, J.; Yang, X.; Zhang, Y.; Zhang, L.; Ren, H.; Ye, L. A review of the application of machine learning in water quality evaluation. *Eco. Environ. Health* **2022**, *1*, 107–116. [CrossRef]
13. Leggesse, E.S.; Zimale, F.A.; Sultan, D.; Enku, T.; Srinivasan, R.; Tilahun, S.A. Predicting Optical Water Quality Indicators from Remote Sensing Using Machine Learning Algorithms in Tropical Highlands of Ethiopia. *Hydrology* **2023**, *10*, 110. [CrossRef]
14. Zhu, Y.; Liu, K.; Liu, L.; Myint, S.W.; Wang, S.; Liu, H.; He, Z. Exploring the potential of worldview-2 red-edge band-based vegetation indices for estimation of mangrove leaf area index with machine learning algorithms. *Remote Sens.* **2017**, *9*, 1060. [CrossRef]
15. Cai, J.; Chen, J.; Dou, X.; Xing, Q. Using machine learning algorithms with in situ hyperspectral reflectance data to assess comprehensive water quality of urban rivers. *IEEE Geosci. Remote Sens. Lett.* **2022**, *60*, 5523113. [CrossRef]
16. Castrillo, M.; García, Á.L. Estimation of high frequency nutrient concentrations from water quality surrogates using machine learning methods. *Water Res.* **2020**, *172*, 115490. [CrossRef]
17. Gladju, J.; Kamalam, B.S.; Kanagaraj, A. Applications of data mining and machine learning framework in aquaculture and fisheries: A review. *Smart Agric. Technol.* **2022**, *2*, 100061. [CrossRef]
18. Chen, K.; Chen, H.; Zhou, C.; Huang, Y.; Qi, X.; Shen, R.; Ren, H. Comparative analysis of surface water quality prediction performance and identification of key water parameters using different machine learning models based on big data. *Water Res.* **2020**, *171*, 115454. [CrossRef]
19. Zhang, H.; Xue, B.; Wang, G.; Zhang, X.; Zhang, Q. Deep Learning-Based Water Quality Retrieval in an Impounded Lake Using Landsat 8 Imagery: An Application in Dongping Lake. *Remote Sens.* **2022**, *14*, 4505. [CrossRef]
20. Jin, T.; Cai, S.; Jiang, D.; Liu, J. A data-driven model for real-time water quality prediction and early warning by an integration method. *Environ. Sci. Pollut. Res.* **2019**, *26*, 30374–30385. [CrossRef]
21. Uddin, M.G.; Nash, S.; Rahman, A.; Olbert, A.I. A novel approach for estimating and predicting uncertainty in water quality index model using machine learning approaches. *Water Res.* **2023**, *229*, 119422. [CrossRef]
22. Haq, M.A.; Jilani, A.K.; Prabu, P. Deep Learning Based Modeling of Groundwater Storage Change. *Comput. Mater. Contin.* **2022**, *70*, 4599–4617. [CrossRef]
23. Haghiabi, A.H.; Ali Heidar Nasrolahi, A.H.; Parsaie, A. Water quality prediction using machine learning methods. *Water Qual. Res. J.* **2018**, *53*, 3–13. [CrossRef]
24. Cheng, S.; Cheng, L.; Qin, S.; Zhang, L.; Liu, P.; Liu, L.; Wang, Q. Improved understanding of how catchment properties control hydrological partitioning through machine learning. *Water Resour. Res.* **2022**, *58*, e2021WR031412. [CrossRef]
25. Krishnan, M. Against interpretability: A critical examination of the interpretability problem in machine learning. *Philos. Technol.* **2020**, *33*, 487–502. [CrossRef]
26. Kovačević, M.; Lozančić, S.; Nyarko, E.K.; Hadzima-Nyarko, M. Application of Artificial Intelligence Methods for Predicting the Compressive Strength of Self-Compacting Concrete with Class F Fly Ash. *Materials* **2022**, *15*, 4191. [CrossRef] [PubMed]
27. Hastie, T.; Tibsirani, R.; Friedman, J. *The Elements of Statistical Learning*; Springer: Berlin/Heidelberg, Germany, 2009.
28. Breiman, L. Bagging predictors. *Mach. Learn.* **1996**, *24*, 123–140. [CrossRef]
29. Breiman, L. Random forests. *Mach. Learn.* **2001**, *45*, 5–32. [CrossRef]
30. Kovačević, M.; Ivanišević, N.; Petronijević, P.; Despotović, V. Construction cost estimation of reinforced and prestressed concrete bridges using machine learning. *Građevinar* **2021**, *73*, 727.
31. Freund, Y.; Schapire, R.E. A decision-theoretic generalization of on-line learning and an application to boosting. *J. Comput. Syst. Sci.* **1997**, *55*, 119–139. [CrossRef]
32. Elith, J.; Leathwick, J.R.; Hastie, T. A working guide to boosted regression trees. *J. Anim. Ecol.* **2008**, *77*, 802–813. [CrossRef]
33. Friedman, J.H. Greedy function approximation: A gradient boosting machine. *Ann. Stat.* **2001**, *29*, 1189–1232. [CrossRef]
34. Rasmussen, C.E.; Williams, C.K. *Gaussian Processes for Machine Learning*; The MIT Press: Cambridge, MA, USA, 2006.
35. Fatehi, I.; Amiri, B.J.; Alizadeh, A. Modeling the Relationship between Catchment Attributes and In-stream Water Quality. *Water Resour. Manag.* **2015**, *29*, 5055–5072. [CrossRef]
36. Dohner, E.; Markowitz, A.; Barbour, M.; Simpson, J.; Byrne, J.; Dates, G. *Volunteer Stream Monitoring: A Methods Manual*; Office of Water (EPA 841-B-97-003); Environmental Protection Agency: Washington, DC, USA, 1997.
37. Anderson, J.R. *A Land Use and Land Cover Classification System for Use with Remote Sensor Data*; US Government Printing Office: Washington, DC, USA, 1996; Volume 964.
38. Brakensiek, D.; Rawls, W. Agricultural management effects on soil-water Processes 2: Green and Ampt Parameters for Crusting Soils. *Trans. ASAE* **1983**, *26*, 1753–1757. [CrossRef]

toxics MDPI

Article

Assessing the Efficiency of a Drinking Water Treatment Plant Using Statistical Methods and Quality Indices

Alina Bărbulescu [1] and Lucica Barbeş [2,3,*]

[1] Department of Civil Engineering, Transilvania University of Braşov, 5 Turnului Str., 500152 Braşov, Romania; alina.barbulescu@unitbv.ro
[2] Department of Chemistry and Chemical Engineering, Ovidius University of Constanța, 124 Mamaia Bd., 900112 Constanța, Romania
[3] Doctoral School of Biotechnical Systems Engineering, Politehnica University of Bucharest, 313, Splaiul Independenței, 060042 Bucharest, Romania
* Correspondence: lucille.barbes2020@gmail.com

Abstract: This study presents the efficiency of a drinking water treatment plant from Constanța, Romania. Individual and aggregated indices are proposed and built using nine water parameters for this aim. The analysis of individual indices permits the detection of the period of malfunctioning of the water treatment plant with respect to various parameters at various sampling points. In contrast, the cumulated indices indicate the overall performance of the treatment plant during the study period, considering all water parameters. It was shown that the outliers significantly impact the values of some indices. Comparisons between the simple average and weighted average indices (built taking into account the importance of each parameter) better reflect the impact on the water quality of some chemical elements that might harm people's health when improperly removed.

Keywords: water quality; treatment plant; water parameters; efficiency indices

Citation: Bărbulescu, A.; Barbeş, L. Assessing the Efficiency of a Drinking Water Treatment Plant Using Statistical Methods and Quality Indices. *Toxics* **2023**, *11*, 988. https://doi.org/10.3390/toxics11120988

Academic Editor: Xue Bai

Received: 14 November 2023
Revised: 25 November 2023
Accepted: 29 November 2023
Published: 5 December 2023

1. Introduction

Water is an essential resource for life. Human history shows that the primary freshwater sources have been rivers. They still play a significant role in socio-economic development [1]. In the last decades, water quality has been affected by environmental pollution produced by anthropic activities, becoming inappropriate for drinking, irrigation, and other uses [2]. Therefore, its consumption can harm organisms, especially humans, given that more than two-thirds of organisms are formed of water [3].

Unfortunately, people in some regions or countries lack sufficient access to clean water or use water from contaminated sources with disease-carrying organisms, pathogens, or unacceptable levels of toxic substances and suspended solids [4–6]. Olukanni et al. [7] show that over 2.2 million people in developing countries die annually from diseases provoked by contaminated water. Inefficient water treatment and the distribution of drinking water, as well as the consumption of contaminated water, can lead to the apparition of many diseases [8]. To avoid such effects, drinking water must be tasteless, odorless, and colorless, and free from physical, chemical, and biological contaminants.

An extended analysis of the factors affecting the spatial variation in stream water composition is presented in [9], emphasizing natural causes. The surface water quality and the pollutants' transport can be assessed utilizing statistical methods [10–14] and water quality indicators [15–17]. Modeling and forecasting water quality and the parameters that influence it has been performed recently by Artificial Intelligence methods (Fuzzy techniques, ANFIS, C&RT) and hybrid method [14,18–20]. Water quality simulation and forecast utilizing exponential models, differential equations, deep learning neural networks, and fuzzy clustering have been developed by some scientists [21–24].

Romania has abundant sources of drinking water. However, the demand for water resources is constantly rising due to population growth, intensified agricultural and industrial activities, and the recent years of low rainfall and adverse conditions, which impact the quality of drinking water sources [10]. The quality of drinking water is essential for EU residents [25]. The necessary treatments for producing drinking water, depending on the quality of water sources, are presented in Directive EC 2184/2020 [26]. Researchers' studies reflect the interest in the topic [27–32]. Romania is also tasked with finding cost-effective and innovative approaches that address environmental, regulatory, and public concerns for maintaining a clean environment [33,34].

Since ensuring good water quality is essential for the population's health, research has been developed to propose advanced technologies for drinking water treatment. Some of the most recent technologies are presented in the books and articles of Thomas and Burgess [35], Brar et al. [36], Vara Prasad [37], Caratar et al. [38], Brusseau et al. [39], and Farhaoui and Derraz [40].

Most studies written by Romanian scientists present wastewater analysis, proposing solutions for cleaning them [41–45]. Chirilă et al. [31] studied the water supply sources in Constanta town (Romania), the applied treatments based on their quality, and the performances of the water purification process. Some authors [46–48] addressed the disinfection by-products in drinking water, modeling the chlorine decay or proposing the analysis of the chlorine concentration in the distribution system.

This study aims to fill a gap in the knowledge related to the efficiency evaluation of a drinking water treatment plant utilizing a series of individual and aggregated indices introduced by the authors. The originality of this work consists of (1) proposing individual and composite efficiency indices for assessing the plant's efficiency, (2) building indices that are not restricted to a certain number of parameters or a determined period, and (3) introducing an objective evaluation method of the treatment plant's efficiency.

2. Materials and Methods

2.1. Studied Region and Data Series

Constanța city is situated in the Dobrogea region, in the southeastern part of Romania (Europe), with one of the biggest metropolitan areas in Romania. It has a circular drinking water distribution system with a length of about 575 km. The hydrographic region of Dobrogea contains two river basins: the Littoral basin and a portion of the Danube basin (341.5 km along the Danube River), covering an area of 11,809 km^2 (excluding the Danube Delta), with a network length of 1624 km and an average density of 0.13 km/km^2. Approximately 73% of this hydrographic network is affected by drying phenomena.

To ensure the best quality and circulation of drinking water, four treatment–storage and pumping stations operate—Constanța Nord, Constanța Sud, Călăraşi, and Palas complex. Groundwater and surface water sources are used for the city's water supply. The groundwater sources include Caragea Dermen, Cișmea I A, B, C, Cișmea II, Constanța Nord, and Medgidia. The surface water is extracted from the Priza Galeșu (44°15′0″ N and 28°25′60″ E) located on the Danube–Poarta Albă–Midia Năvodari channel. The Caragea Dermen source, situated between Constanța and Ovidiu, is the oldest groundwater source, consisting of 18 wells with depths from 35 to 90 m. It provides water to Ovidiu, Mihail Kogălniceanu, as well as the Palazu Mare neighborhood and the Călărași storage–pumping complex, with a flow rate of 3549 m^3/h. Cișmea I A, B, and C consist of three groups of 36 wells located in the northern part of Constanța, with a total captured flow rate of about 8500 m^3/h. The Cișmea I sources provide water to neighborhoods in the northern part of Constanța (also pumped to the Palas storage–pumping complex and the Călărași complex). Cișmea II is situated between the Caragea Dermen and Cișmea I sources and contains ten wells with a captured flow rate of approximately 1700 m^3/h. The water from this source is transported to the Palas storage–pumping complex. The Constanța Nord source, situated in the northern part of Constanța, south of the Siutghiol Lake, consists of 5 wells with a captured flow rate of about 2200 m^3/h. The water is pumped from here to the Constanța

Nord complex. The Medgidia source is located along the Danube–Black Sea Canal and comprises 11 wells with a captured flow rate of approximately 1500 m^3/h. The extracted water is pumped to the Constanța Sud complex. The surface source Galeșu captures water from the Poarta Albă–Midia Năvodari Channel at km 6 + 398 and pumps it to the Palas Constanța storage–treatment and pumping complex at 17.4 km. It provides a water supply of 13,050 m^3/h. This surface source was created to meet the high summer water demand and to supplement the water supply for Constanța city if necessary. The intake system uses five sorbs with a diameter of 1200 mm, equipped with metallic screens to retain suspended solids [48,49]. The Palas–Constanța water treatment plant (PCTP) provides drinking water to the city's 350,000 inhabitants through nine large-diameter pipelines. Details on the PCTP can be found in [48].

The geographical locations of the Constanta county and city in Romania are shown in Figure 1.

Figure 1. The map of Romania (with Constanța highlighted).

This study proposes the evaluation of the treatment efficiency of surface water from the Galeșu source and groundwater for the purification and distribution of drinking water to consumers. The experimental results were obtained from the analyses of the surface water and groundwater quality from four sampling points of the treatment plant, denoted by S1–S4 in Figure 2 and representing (1)—raw surface water, (2)—raw pre-chlorinated groundwater, (3)—treated surface water, and (4)—drinking water distributed to consumers from the treatment plant.

The monitored parameters include temperature—T (°C), pH [SR ISO 10523:2012] [50], electrical conductivity—EC (μS/cm) [SR EN 27888:1997] [51], turbidity—TUR (NTU) [SR EN ISO 7027-1:2016] [52], total hardness—TH (^{0}dH) [SR ISO 6059:2008] [53], permanganate index—PMI (mg O_2/L) [SR EN ISO 8467:2001] [54], free residual chlorine (mg/L) [SR EN ISO 7393-2:2018] [55], Cl^- (chlorides, mg/L) [SR ISO 9297:2001] [56], SO_4^{2-} (sulphates, mg/L) [Romanian standard: STAS 3069-87] [57], and nutrients—NH_4^+ (ammonium, mg/L) [SR ISO 7150-1:2001] [58], NO_2^- (nitrites, mg/L) [SR EN 26777:2006] [59], and NO_3^- (nitrates, mg/L) [SR ISO 7890-3:2000] [60]. PMI provides information on the quantity of oxidizable inorganic and organic substances in water. This index is utilized to assess the quality of the freshwater and treated potable waters in the European Union (EU), according to [52].

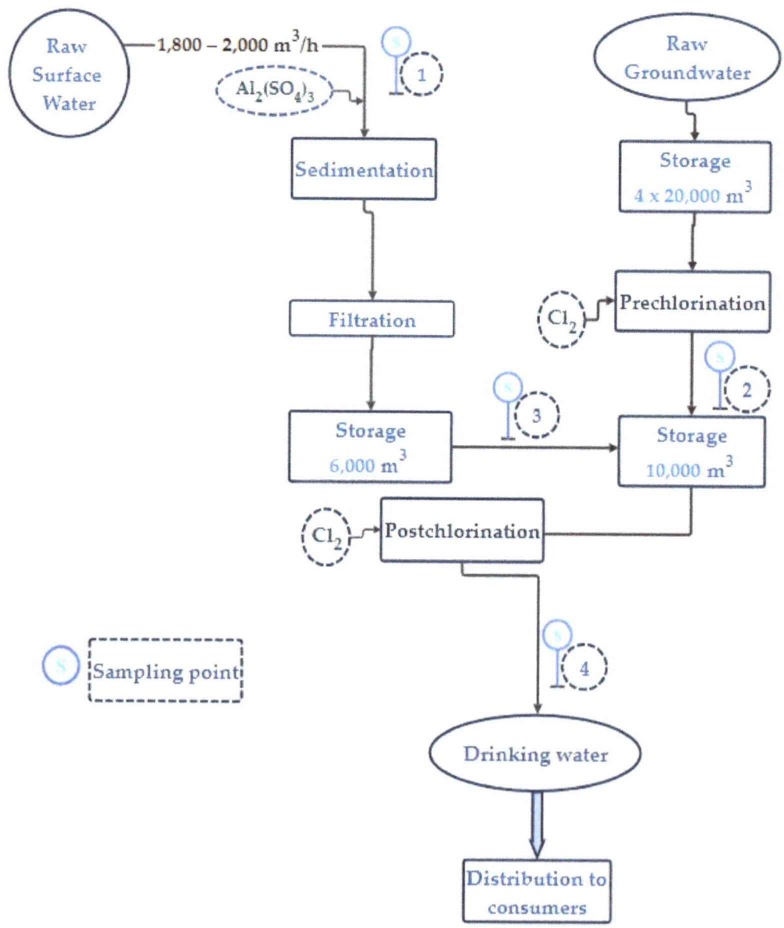

Figure 2. Palas–Constanţa water treatment plant (PCTP) process flow diagram.

The data series consists of the monthly average values of the mentioned parameters for 2016–2019. The obtained values were compared with the maximum allowable values (MAVs) from the Romanian legislation [61]. Given that the water temperature may significantly influence the efficiency indexes because there is a high variation between its value at the treatment plant's entrance and the distribution system's entrance, we shall not consider it when building the indices.

2.2. Statistical Analysis and Efficiency Indicators

The basic statistics of the recorded data series have been calculated, and the histograms and boxplots have been drawn to show the series characteristics and determine the outliers' existences.

2.3. Efficiency Indices

2.3.1. Efficiency Indices of the Treatment Process at a Given Moment t

1. *The individual efficiency at the moment t with respect to the k-th water parameter,* ef_{kt}, is defined by:

$$ef_{kt} = \left[1 - (C_{o,t})_k / (C_{in,t})_k\right] \times 100 \quad (1)$$

where $(C_{in,t})_k$ and $(C_{o,t})_k$ are the concentrations of the water parameter k, in the input and output at a certain point (2, 3, or 4), at the moment t.

2. *The mean cumulated efficiency with respect to n water parameters at the moment t,* MCE_t is defined by:

$$MCE_t = \frac{1}{n}\sum_{j=1}^{n} ef_{jt} \quad (2)$$

3. *The weighted cumulated efficiency with respect to n water parameters at the moment t,* WCE_t is defined by:

$$WCE_t = \left(\sum_{j=1}^{n} ef_{jt} \times w_j\right) / \left(\sum_{j=1}^{n} w_j\right) \quad (3)$$

where w_j is the j-th water parameter weight.

2.3.2. Efficiency Indices of the Treatment Process during the Study Period (T moments)

1. *The individual average efficiency with respect to the k-th water parameter,* AE_k, is defined by the Formula (4):

$$AE_k = \frac{1}{T}\sum_{t=1}^{T} ef_{kt} = \left(1 - \overline{(C_o/C_{in})_k}\right) \times 100. \quad (4)$$

or by JAE_k, whose formula is [62]:

$$JAE_k = (1 - \overline{C_{o,k}}/\overline{C_{in,k}}) \times 100 \quad (5)$$

where

$$\overline{(C_o/C_{in})_k} = \frac{1}{T}\sum_{t=1}^{T} (C_{0,t})_k / (C_{in,t})_k \quad (6)$$

and $\overline{C_{in,k}}$ and $\overline{C_{o,k}}$ are the averages of the k-th parameter concentrations as input and output of a treatment stage during the study period.

2. *The cumulated average efficiency with respect to n water parameter* is defined by one of the formulas:

$$\overline{CAE} = \frac{1}{T}\sum_{t=1}^{T} MCE_t = \frac{1}{n}\sum_{k=1}^{n} AE_k \quad (7)$$

$$\overline{JAE} = \frac{1}{n}\sum_{k=1}^{n} JAE_k \quad (8)$$

3. *The weighted cumulated efficiency with respect to n water parameters* is defined by one of the formulas:

$$\overline{WCE} = \frac{1}{T}\sum_{t=1}^{T} WCE_t = \left(\sum_{k=1}^{n} \overline{AE}_k \times w_k\right) / \left(\sum_{k=1}^{n} w_k\right) \quad (9)$$

$$WJAE = \left(\sum_{k=1}^{n} JAE_k \times w_k\right) / \left(\sum_{k=1}^{n} w_k\right) \quad (10)$$

The values assigned to the weights are from 1 to 5, considering the harmful potential of some chemicals to human health. The higher the harm potential, the higher the index is. In the present article, we took advantage of the scientific literature findings related to the water quality indices (WQIs) for drinking water. In the manuscript, we used the indices provided in [63] (that gives the most-used weights attached to different water parameters). The weights utilized here are 1 for pH, EC, free residual chlorine, and chlorides, 2 for permanganate index, sulfates, nitrates, and nitrites, and 3 for ammonia.

The highest value of all indices but pH is 100%, corresponding to a perfect working of the treatment plant. Any positive value indicates a certain degree of efficiency. The closer the value is to 100%, the better the station performance. Negative indices indicate the water treatment plant's incapability to remove certain elements from the water. The lower the indices are, the worse the treatment plant's performance is. In the case of pH, efficiency around zero means keeping the pH within almost constant limits (6.5–8.5 recommended).

3. Results

3.1. Results of the Statistical Analysis

Table 1 contains the basic statistics computed for the water parameters analyzed at the sampling points S1–S4.

Table 1. Basic statistics of the water parameters at the sampling points and MAVs [61].

	T (°C)	pH	TUR (NTU)	EC (μS/cm)	Cl^- (mg/L)	SO_4^{2-} (mg/L)	PMI (mg O_2/L)	NH_4^+ (mg/L)	NO_2^- (mg/L)	NO_3^- (mg/L)
Admissible Limit		6.5–8.5	5	2500	250	250	5	0.5	0.10	50
Sampling point S1										
min	2.40	7.60	0.00	366.00	20.50	29.60	1.07	0.00	0.00	1.70
mean	14.71	8.11	2.10	479.27	43.37	69.55	1.91	0.05	0.04	7.76
median	15.00	8.03	1.49	450.50	44.45	65.05	1.83	0.023	0.029	8.07
max	26.00	9.07	14.40	696.00	66.40	111.00	3.55	0.80	0.11	14.80
st.dev	7.64	0.29	2.32	86.84	11.35	20.17	0.54	0.12	0.02	3.78
skewness	−0.01	1.31	3.90	0.96	0.01	0.15	0.88	6.11	0.84	0.13
kurtosis	−1.51	2.46	18.16	−0.10	−0.59	−0.78	0.82	39.97	0.33	−1.06
Sampling point S2										
min	2.80	7.50	0.00	371.00	22.00	30.40	0.54	0.00	0.00	1.12
mean	15.08	7.95	1.11	486.92	44.94	70.81	1.47	0.01	0.00	7.58
median	14.90	7.96	0.85	466.00	45.65	63.30	1.39	0.008	0.003	7.58
max	26.00	8.69	3.94	712.00	67.40	143.60	2.98	0.04	0.02	14.40
st.dev	7.41	0.26	0.97	85.23	10.73	23.70	0.48	0.01	0.00	3.52
skewness	0.03	0.11	1.50	0.97	−0.03	0.76	0.98	1.56	2.24	0.18
kurtosis	−1.48	0.64	1.83	0.07	−0.46	0.66	1.64	2.60	6.19	−0.73
Sampling point S3										
min	3.20	7.43	0.00	371.00	19.80	36.20	0.02	0.00	0.00	3.32
mean	16.61	7.70	0.56	719.46	71.23	92.93	0.71	0.01	0.00	9.37
median	17.15	7.64	0.24	868.50	89.78	96.44	0.45	0.004	0.002	9.52
max	25.60	8.58	4.00	897.00	96.07	189.80	3.14	0.02	0.02	15.30
st.dev	4.87	0.22	0.83	187.54	24.53	26.27	0.68	0.01	0.00	2.96
skewness	−0.68	1.93	2.68	−0.61	−0.53	1.05	1.34	0.99	2.23	−0.05
kurtosis	0.61	4.93	7.64	−1.29	−1.40	3.97	2.05	−0.28	5.62	−0.77
Sampling point S4										
min	3.20	7.22	0.00	486.00	44.22	50.60	0.08	0.00	0.00	3.22
mean	16.48	7.65	0.47	718.10	72.30	92.21	0.52	0.00	0.00	8.07
median	17.00	7.68	0.32	670.50	76.43	98.35	0.43	0.003	0.001	7.83
max	24.40	8.12	2.19	915.00	93.50	150.30	1.42	0.02	0.01	17.60
st.dev	4.87	0.22	0.59	151.70	18.52	25.07	0.31	0.00	0.00	2.60
skewness	−0.62	0.21	1.42	0.07	−0.16	0.05	0.77	1.42	0.63	0.94
kurtosis	0.37	−0.67	1.69	−1.82	−1.85	−0.96	0.00	2.38	−0.42	2.62

Most values are inside the admissible limits. Exceptions are some turbidity, ammonium, and nitrite values, whose maxima are in bold in Table 1. The median was also computed, since the range (difference between maximum and minimum) of some series values or their standard deviations are high, indicating a significant dispersion of the values around the mean. Significant differences between mean and median were found for the highly skewed series (TUR at S1 and S3, and NO_2^- at S3, for example).

The free residual chlorine had values of 0 throughout the study for the influent—site 1—because it did not undergo pre-treatment before entering the treatment plant. For the groundwater, which is pre-treated, the chlorine values ranged from 0.43 to 1.28 mg/L, with an average value of 0.85 mg/L. For the treated surface water, the chlorine values ranged from 0.25 to 0.90 mg/L, with an average of 0.50 mg/L, while the concentration of chlorine in the drinking water in the effluent obtained values between 0.40 and 0.64 mg/L, with an average of 0.56 mg/L. Although the allowed values for potable water are between 0.1 and 0.5 mg/L, they must be achieved throughout the entire distribution network. Therefore, even if the values obtained at the treatment plant exceed the MAV, they are accepted to ensure proper water disinfection in the storage tanks and a minimum required chlorine level in the supply pipes.

The histograms and boxplots of some water parameters are shown in Figures 3 and 4. The histograms of the free residual chlorine series recorded at the first two sampling points are symmetric, whereas those for the last sampling points are slightly skewed. A positive skewness is noticed for the concentration series of nitrate at the last sampling point. Various skewness values are determined, indicating that most series are not symmetrically distributed. Kurtosis shows platykurtic distributions for different series, as, for example, pH, EC, Cl^-, SO_4^{2-}, and NO_2^- series at S4. The boxplots of pH, turbidity, EC, and ammonia indicate the outliers' presence (represented by stars).

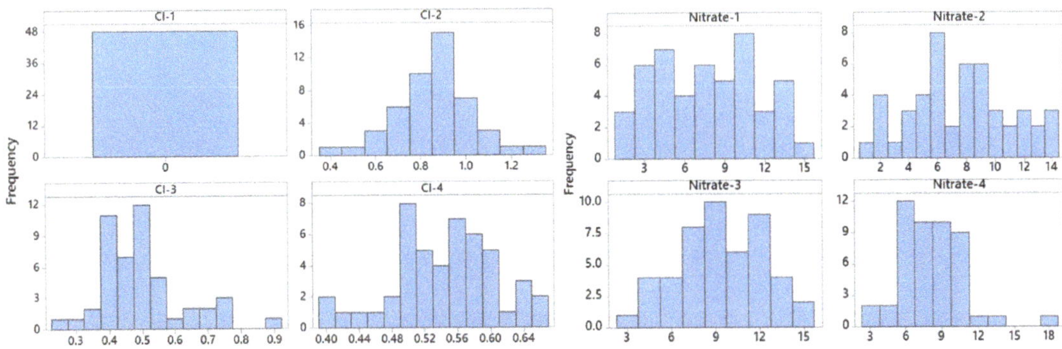

Figure 3. Histograms of free residual chlorine and nitrate at the four sampling points.

By comparison, the other series are more homogenous, with only a few outliers. Figure 4c,d,f point out that there are significant variations in the values of the series recorded at different sampling points, especially for Cl^-, PMI, and EC. The existence of high outliers, especially for the TUR series, will significantly decrease the computed performance indices.

3.2. *Results on the Efficiency Indices*

3.2.1. Results on the Efficiency Indices at a Given Moment *t*

The efficiency indices computation was based on the same output—S4—with respect to the input from S1, S2, and S3, respectively denoted by the corresponding indicator followed by S1, S2, and S3. For example, the MCE_t_S1 means that the input series is from S1. The water temperature and free residual chlorine are not considered in this study because the free residual chlorine is absent in the surface water and groundwater (being added during the purification process), and the temperature does not impact the drinking water quality. Models of free residual concentration series are presented in [48].

The variations in the individual efficiency computed by (1) are represented in Figure 5. Their minimum and maximum values are listed in Table 2.

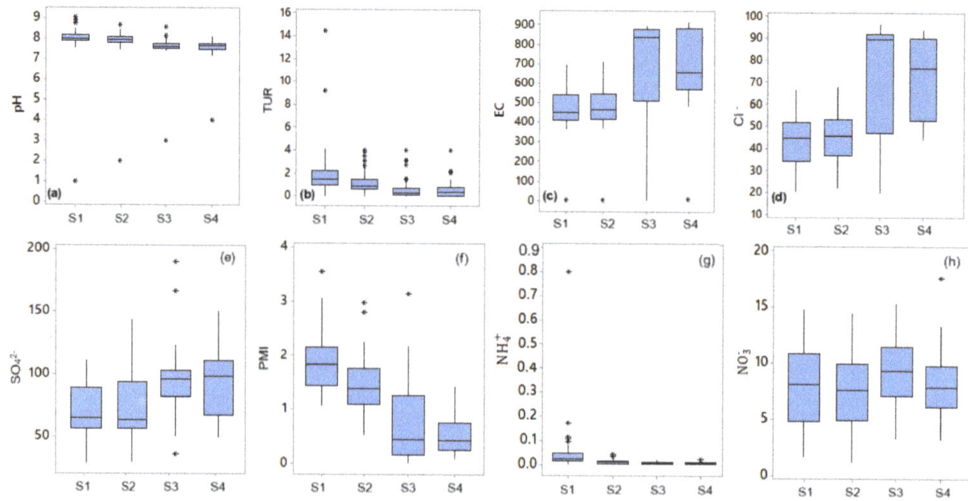

Figure 4. Boxplots of pH, TUR, EC, Cl^-, SO_4^{2-}, PMI, NH_4^+ and NO_3^- at the four sampling points (**a–h**). Stars represent the outliers.

Figure 5. (**a**) ef_{kt}_S1, (**b**) ef_{kt}_S2, and (**c**) ef_{kt}_S3. The extreme values (in bold in Table 2) are not represented in the chart.

Table 2. The minimum and maximum individual efficiencies.

		pH	TUR	EC	Cl^-	SO_4^{2-}	PMI	NH_4^+	NO_2^-	NO_3^-
ef_{kt}_S1	min	−0.51	−73.91	−143.17	−335.61	−202.25	5.41	**−450.00**	0.00	−173.53
	max	17.42	100.00	0.00	3.90	22.45	51.43	100.00	100.00	62.91
ef_{kt}_S2	min	0.00	−250.00	−139.89	−305.91	−189.23	−82.35	**−1600.00**	**−200.00**	−429.46
	max	12.77	100.00	0.72	6.77	49.86	95.86	100.00	100.00	51.80
ef_{kt}_S3	min	−6.73	**−2871.43**	−139.89	−351.01	−201.10	**−2118.75**	**−1100.00**	**−500.00**	−106.93
	max	11.66	100.00	39.59	50.59	56.76	95.94	100.00	100.00	72.24

Some aspects related to individual efficiencies ef_{kt}_S1, are presented below:

- The values of ef_{kt}_S1 varied in very large intervals, from negative values for all but PMI and NO_2^- to the maximum (100%).
- TUR's efficiency values are all positive, half being 100, except for two negative values (−73.91 and −61.67 in May and June 2017).
- The maximum individual efficiency of EC is zero, and more than 80% of chloride efficiencies are negative, meaning that the values recorded in the effluent are lower than those in the influent.
- PMI is the only index whose efficiency values are positive. This means there is good performance in removing the humic materials and organics that could result from the birds and fish exhausts or decomposition.
- The value of −450 for ammonia is due to a jump from 0.01 (mg/L) in the input to 0.120 (mg/L) in the effluent in August 2019. Another negative value (−200 in November 2019) is noticed in the ammonia efficiency ef_{kt_S1}, due to a concentration change from 0.01 to 0.03 mg/L.
- Negative efficiencies were recorded in June 2017 (−173.53), July, August, October, and November 2021 for NO_3^- and June–October 2019 (less than −137.2) and May–August 2017 (less than −81.79) for SO_4^{2-}.

The analysis of ef_{kt}_S2 and ef_{kt}_S3 shows that, generally, the maximum efficiencies increased and the minimums decreased. Specific remarks for ef_{kt}_S2 are as follows:

- All values computed based on the pH are positive.
- The only negative value of efficiency in the TUR series is −250, recorded in March 2016, with half of the values being 100 (excellent efficiency).
- The lowest negative individual values of efficiencies for EC, Cl^-, and SO_4^{2-} (under −79.90, −128.46, and −60.32, respectively) are computed for May–December 2017. Moreover, Cl^- efficiency is mainly negative.
- Only nine values of the individual efficiency for ammonia are noticed, the lowest being −900, −1200, and −600 (recorded in May–September 2019, May and August 2018).
- All NO_2^- efficiency indices are positive except six, recorded, for example, in November 2018, and January and December 2019 (with values of −200 and −100).
- In total, 22 values of NO_3^-'s efficiencies are negative, most of the positive ones being under 30.

The lowest individual efficiencies are ef_{kt}_S3, as explained in the following:

- All values corresponding to pH are in the interval [−6.72, 11.66], with most being negative, so an increase in the water's pH appears when the influent is considered the series at S3 and the effluent is the series at S4.
- The efficiency of TUR recorded unexpected low values (marked with bold letters) in Table 2, as −2871.43 followed by −1918.20, in June and September 2019, respectively. Most values around −500 were also recorded in February, April, June–November 2018. Some explanation of these values are presented in the next section.

- The lowest chloride efficiencies were in the range [−155.84, −113.64] and between (−352.01) and (−117.27) in July, August, October, and November 2016, and in June–August and October–December 2017.
- The lowest values (negative) of the sulfates' efficiencies were recorded during the same period as those of Cl^-.
- PMI recorded extremely low efficiencies in the same months as TUR. The value marked in bold was registered in February 2018.

Comparing the period where some values of the individual efficiencies were extremely low, we think that this situation is the consequence of the malfunctioning of the PCTP during May–December 2017.

The series of mean (and weighted) cumulated efficiencies with respect to all water parameters MCE_t (WCE_t) are represented in Figure 6.

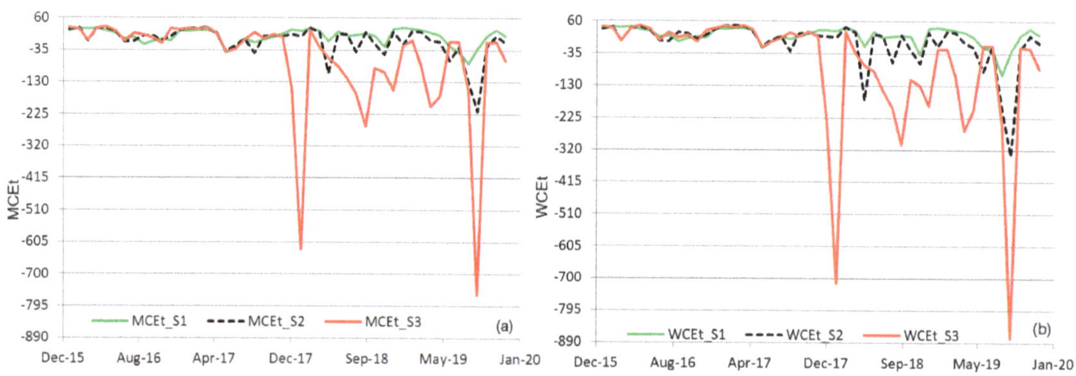

Figure 6. (**a**) MCE_t and (**b**) WCE_t computed with the influent from S1, S2, or S3, and the effluent from S4.

Given that the values −2871.43 and −2118.75 are outliers, we removed them from the computation. Due to the weights assigned to the water parameters as a function of their contribution to the water quality, when most values of the individual indices were positive, $WCE_t > MCE_t$. In the case of the negative values, the inequality is the opposite. Comparisons of the extreme values value of both indices are given in Table 3.

Table 3. Extreme values of MCE_t and WCE_t.

	MCE_t_S1	MCE_t_S2	MCE_t_S3	WCE_t_S1	WCE_t_S2	WCE_t_S3
min	−74.27	−215.32	−762.81	−97.33	−336.86	−875.15
max	30.95	31.24	31.97	43.38	47.86	47.82

The charts from Figure 6 indicate mainly negative cumulated efficiencies computed when the input data series were S3, compared to the case when the input was from S2 or S1, respectively. The best cumulated efficiencies were recorded in the last case—columns 2 and 5 in Table 3. Still, even when using the weighted indices, the cumulated efficiency remains under 50.

3.2.2. Efficiency Indices of the Treatment Process during the Study Period

The individual average efficiencies permit determining the water parameters whose efficiency should be improved considering the recorded values during the entire study period. The efficiencies of the PCTP with respect to each water parameter—(4) and (5)—are presented in Table 4.

Table 4. Values of (a) AE_k and (b) JAE_k.

	pH	TUR	EC	Cl^-	SO_4^{2-}	PMI	NH_4^+	NO_2^-	NO_3^-
AE_k_S1	5.54	66.19	−51.90	−81.21	−42.27	23.56	62.32	91.05	−27.27
AE_k_S2	3.62	55.43	−49.48	−72.51	−41.73	58.63	−37.86	41.09	−33.08
AE_k_S3	0.40	**−739.52**	−12.24	−29.92	−9.21	**−381.64**	−42.53	17.01	5.97
JAE_k_S1	5.63	77.69	−49.83	−66.69	−32.59	23.03	91.51	95.73	−3.97
JAE_k_S2	3.69	57.69	−47.48	−60.87	−30.23	64.86	53.26	62.19	−6.42
JAE_k_S3	0.64	16.81	0.19	−1.50	0.78	27.66	20.91	54.49	13.86

Based on AE_k and JAE_k, the best performances are those of nitrites, turbidity, and ammonia removal with respect to S1. The same parameters remain positive for PMI and TUR (with respect to S1 and S2).

The values of JAE_k are generally higher than those of AE_k because the average of input and output series are computed, diminishing the difference between the computed values. Therefore, JAE_k_S3 has all values but that for chloride greater than zero. Removing the abovementioned, $AE_{TUR\_S3} = -739.52$ and $AE_{PMI\_S3} = -381.64$ will become 131.62 and −182.46, respectively.

The cumulated average efficiency with respect to the considered water parameters are as follows:

- $\overline{CAE}$_**S**1 = 5.11, $\overline{CAE}$_**S**2 = −8.43, and $\overline{CAE}$_**S**3 = −130.1. When eliminating the highest outlier, $\overline{CAE}$_S3 = −39.84.
- $\overline{JAE}$_**S**1 = 15.61, $\overline{JAE}$_**S**2 = 10.74, and $\overline{JAE}$_**S**3 = 15.61.

The weighted cumulated efficiency with respect to all parameters are as follows:

- $\overline{WCE}$_S1 = 17.61, $\overline{WCE}$_S2 = −4.69, and $\overline{WCE_S}$3 = −46.24. When eliminating the highest outlier, $\overline{WCE_S}$3= −73.10.
- ***WJAE*_S**1 = 30.21, ***WJAE*_S**2 = 21.96, and ***WJAE*_S**3 = 17.55.

Considering the cumulated indices that reflect the global efficiency in time and considering all water parameters, we remark a very low performance of the PCTP in time with respect to each input source. The highest one is with respect to the influent from S1.

4. Discussion

This article proposes different categories of indices for evaluating the efficiency of water purification of a drinking water treatment plant. These provide a synthetic modality for achieving the goal, given that when working with hundreds of values over a long period it is difficult to look at the charts or the individual values for each day, and it is time consuming. Moreover, determining a model that can be used for forecasting is also difficult in the presence of extreme values or outliers. Computing the indices' values can be easily accomplished (in an Excel file, for example), and the obtained values (positive or negative, close to 100, for example) will provide a quick answer related to the efficiency of the cleaning process.

The necessity of obtaining individual indices as high as possible for all water parameters but pH comes from the importance of each water parameter, which will be discussed shortly here.

Maintaining the pH for the drinking water between 6.5 and 8.5 is essential given that increased alkalinity of acidity can lead to pipe damage (favoring the detachment of tiny particles from the pipes' materials) and the impurities' circulation in the distribution system. Therefore, the water becomes unhealthy for the organism [64,65]. Therefore, negative efficiencies with respect to pH during a long period raise an alarm signal that the pH would be above 8.5, whereas an increasing trend of the individual efficiency with respect to pH will indicate possible pH's decay under 6.5.

Turbidity indicates that the water is clean from the viewpoint of its aspect (transparent, without suspensions). It is known that water characteristics can undergo significant changes in a short period; for example, turbidity can be strongly affected by heavy rainfall. Increased water turbidity can result from runoff or soil erosion, especially following heavy rains. Therefore, additional operations should be applied at the water treatment plant to prevent hazards and manage associated risks: (a) rainwater storage and management; (b) construction of retention basins to minimize the effects of heavy rains on water quality; (c) advanced water filtration through the use of activated carbon or membranes; (d) addition of coagulants and flocculants; and (e) adjustment of operational parameters [66].

The values of ef_{TURt}_S1 indicate that the water was clean from the aspect viewpoint. As for the recorded increased values for water turbidity in May and June 2017, or February, April, June, and November 2018, it is considered that they were mainly determined by abundant precipitation leading to massive water runoff that could accumulate high quantities of soil particles, sand, and other impurities at the treatment plant through the alluvial deposits from flood periods in the area where the treatment plant is located. Other secondary contributions could include construction works, intensive agriculture, or soil erosion.

The individual efficiency indices with respect to EC have mostly negative values, showing an increase in the conductivity values in the output with respect to those in the input. Even if high values of EC do not directly impact health, the large amount of dissolved ionizable solids leads to water hardness and consumer dissatisfaction [67].

PMI is the only water quality indicator with respect to which the PCTC's individual efficiency values are positive, indicating the correct removal of dead organic material form water.

The very low ef_{kt}_S3 (k represents the chloride) be explained by the overlap of the station modernization's works and the seasonal variations in the summer months when higher temperatures and exposure to solar radiation can lead to changes in water composition, including more intense biological activity of organisms (e.g., algae). Additionally, drought episodes leading to decreased water levels through evaporation, altered water composition, or the intensified use of fertilizers in agriculture during the late autumn campaign could contribute to these variations. Another possible explanation at the drinking water treatment plant is the potential infiltration from other nearby sources during maintenance works at the station.

The idea of introducing cumulated efficiency indices was issued from the authors' previous studies in the water quality indicators field. These indices reflect the water treatment plant's efficiency with respect to all the considered water parameters. The outliers' existence in any data series impacts the individual efficiencies and the cumulated ones to a certain extent. A high weight assigned to a parameter with a low (high) individual efficiency will lead to a decrease (increase) in the weighted cumulated efficiency with respect to the average cumulated efficiency. Still, the weighted indices better reflect the impact of each water parameter on the water quality and, consequently, on people's health.

Given that there are slight variations in the weights assigned by different authors to the same water parameters, the introduced indicators may incorporate some percentage of subjectivity that might be eliminated by averaging the values of the weights found in the literature. Future studies should be performed in this direction.

5. Conclusions

This article introduces some indicators used in a case study for assessing the efficiency of a water treatment plant. Whereas the individual indicators show the efficiency with respect to a specific water parameter, emphasizing the issues that may appear on a particular period or with respect to a parameter, the cumulated ones evaluate the overall efficiency over time considering all parameters.

It was shown that the individual efficiencies are sensitive to jumps in values in the effluent with respect to those in the influent (even if they are within the MAV limits). Therefore, the cumulated indices will be drastically affected when very small values participate in their computation. Weighted cumulated indices always differ from the average ones. However, given the importance of each water parameter and the necessity of maintaining good water quality, they must also be observed.

The data analysis indicates that there were periods of malfunctioning of the PCTP, leading to very low negative individual efficiencies with respect to some input sampling points (especially S3) and, consequently, a significant decrement in the cumulated efficiencies concerning each water parameter and the influent. It has been expected that the efficiencies with respect to S3 will be higher given that the water passed through the sedimentation and separation processes. The question that arose was if the maintenance of the water storage tank was correctly performed. To answer this question, sampling should also be performed after exiting the storage tank of 6000 m^3. A similar sampling should be performed after the storage in the tank of 10,000 m^3. Unfortunately, at this moment, we do not have such information.

The present study opens the direction for aligning the drinking water quality evaluations with the sustainability objectives (based on objective criteria). Future work will also evaluate the possibility of improving the presented indices and creating a system that will permit the implementation of the necessary corrective measures shortly after they are observed. Moreover, a working methodology must also be determined for the case of outlier existence, given that such values introduce significant biases in the indices' computation.

Author Contributions: Conceptualization, A.B. and L.B.; methodology, A.B. and L.B.; software, A.B.; validation, A.B. and L.B.; formal analysis, A.B. and L.B.; investigation, A.B. and L.B.; resources, A.B. and L.B.; data curation, L.B.; writing—original draft preparation, A.B. and L.B.; writing—review and editing, A.B.; visualization, A.B. and L.B.; supervision, A.B.; project administration, A.B.; funding acquisition, A.B. and L.B. All authors have read and agreed to the published version of the manuscript.

Funding: This research received no external funding.

Institutional Review Board Statement: Not applicable.

Informed Consent Statement: Not applicable.

Data Availability Statement: Data will be available on request from the authors.

Conflicts of Interest: The authors declare no conflict of interest.

References

1. Javier, M.; Zadeh, S.; Turral, H. *Water Pollution from Agriculture: A Global Review*; The Food and Agriculture Organization of the United Nations Rome and the International Water Management Institute on Behalf of the Water Land and Ecosystems Research Program: Colombo, Sri Lanka, 2017; pp. 1–35. Available online: https://www.susana.org/en/knowledge-hub/resources-and-publications/library/details/3508 (accessed on 20 July 2023).
2. Kumar, M.; Gikas, P.; Kuroda, K.; Vithanage, M. Tackling water security: A global need of cross-cutting approaches. *J. Environ. Manag.* **2022**, *306*, 114447. [CrossRef] [PubMed]
3. Al-Mayah, W.T. Chemical and microbial health risk assessment of drinking water treatment plants in Kut City, Iraq. *Mater. Today Proc.* **2021**, *42*, 3062–3067. [CrossRef]
4. Lakshmi, S.; Sankari, S.; Prasanna, S.; Madhurambal, G. Evaluation of Water Quality Suitability for Drinking using Drinking Water Quality Index in Nagapattinam district, Tamil Nadu in Southern India. *Groundw. Sustain. Dev.* **2018**, *6*, 43–49.
5. Bărbulescu, A.; Barbeş, L.; Dumitriu, C.Ş. Assessing the Water Pollution of the Brahmaputra River Using Water Quality Indexes. *Toxics* **2021**, *9*, 297. [CrossRef] [PubMed]
6. Bărbulescu, A.; Barbeş, L.; Dumitriu, C.Ş. Statistical Assessment of the Water Quality Using Water Quality Indicators. A case study from India. In *Water Safety, Security and Sustainability, Advanced Sciences and Technologies for Security Applications*; Vaseashta, A., Maftei, C., Eds.; Springer International Publishing: Cham, Switzerland, 2021; Chapter 26, pp. 599–613.
7. Olukanni, D.O.; Ebuetse, M.A.; Wu, A. Drinking water quality and sanitation issues: A survey of a semi-urban setting in Nigeria. *Int. J. Res. Eng. Sci.* **2014**, *2*, 58–65.
8. Onyango, L.A.; Quinn, C.; Tng, K.H.; Wood, J.G.; Leslie, G. A study of failure events in drinking water systems as a basis for comparison and evaluation of the efficacy of potable reuse schemes. *Environ. Health Insights* **2016**, *9*, 11–18. [CrossRef] [PubMed]
9. Hamid, A.; Bhat, S.U.; Jehangir, A. Local determinants influencing stream water quality. *Appl. Water Sci.* **2020**, *10*, 24. [CrossRef]
10. Bărbulescu, A.; Barbeş, L. Assessing the water quality of the Danube River (at Chiciu, Romania) by statistical methods. *Environ. Earth. Sci.* **2020**, *79*, 122. [CrossRef]

11. Bărbulescu, A.; Barbeş, L.; Dani, A. Statistical analysis of the quality indicators of the Danube river water (in Romania). In *Frontiers in Water-Energy-Nexus—Nature-Based Solutions, Advanced Technologies and Best Practices for Environmental Sustainability*; Naddeo, V., Balakrishnan, M., Choo, K.-H., Eds.; Springer: Cham, Switzerland, 2019; pp. 177–179.
12. Popa, P.; Murariu, G.; Timofti, M.; Georgescu, L.P. Multivariate statistical analyses of water quality of Danube River at Galati, Romania. *Environ. Eng. Manag. J.* **2018**, *17*, 491–509.
13. Aminiyan, M.M.; Aminiyan, F.M.; Heydariyan, A. Study on hydrochemical characterization and annual changes of surface water quality for agricultural and drinking purposes in semi-arid area. *Sustain. Water Resour. Manag.* **2016**, *2*, 473–487. [CrossRef]
14. Bărbulescu, A.; Dani, A. Statistical analysis and classification of the water parameters of Beas river (India). *Rom. Rep. Phys.* **2019**, *71*, 716.
15. Sutadian, A.D.; Muttil, N.; Yilmaz, A.G.; Perera, B.J.C. Development of a water quality index for rivers in West Java Province, Indonesia. *Ecol. Indic.* **2018**, *85*, 966–982. [CrossRef]
16. Uddin, M.G.; Nash, S.; Olbert, A.I. A review of water quality index models and their use for assessing surface water quality. *Ecol. Indic.* **2021**, *122*, 107218. [CrossRef]
17. Iticescu, C.; Georgescu, L.P.; Murariu, G.; Topa, C.; Timofti, M.; Pintilie, V.; Arseni, M. Lower Danube Water Quality Quantified through WQI and Multivariate Analysis. *Water* **2019**, *11*, 1305. [CrossRef]
18. Patil, D.; Kar, S.; Gupta, R. Classification and Prediction of Developed Water Quality Indexes Using Soft Computing Tools. *Water Conserv. Sci. Eng.* **2023**, *8*, 16. [CrossRef]
19. Heddam, S.; Kisi, O.; Sebbar, A.; Houichi, L.; Djemili, L. Predicting Water Quality Indicators from Conventional and Nonconventional Water Resources in Algeria Country: Adaptive Neuro-Fuzzy Inference Systems Versus Artificial Neural Networks. In *Water Resources in Algeria Part II. The Handbook of Environmental Chemistry*; Negm, A.M., Bouderbala, A., Chenchouni, H., Barceló, D., Eds.; Springer: Cham, Switzerland, 2019; Volume 98, pp. 13–34.
20. Icaga, Y. Fuzzy evaluation of water quality classification. *Ecol. Indic.* **2007**, *7*, 710–718. [CrossRef]
21. Soares, S.; Vasco, J.; Scalize, P. Water Quality Simulation in the Bois River, Goiás, Central Brazil. *Sustainability* **2023**, *15*, 3828. [CrossRef]
22. Kim, J.; Yu, J.; Kang, C.; Ryang, G.; Wei, Y.; Wang, X. A novel hybrid water quality forecast model based on real-time data decomposition and error correction. *Process Saf. Environ.* **2022**, *162*, 553–565. [CrossRef]
23. Yu, J.W.; Kim, J.S.; Li, X.; Jong, Y.C.; Kim, K.H.; Ryang, G.I. Water quality forecasting based on data decomposition, fuzzy clustering and deep learning neural network. *Environ. Pollut.* **2022**, *303*, 119136. [CrossRef]
24. Wan, D.; Zeng, H. Water environment mathematical model, mathematical algorithm. *IOP Conf. Ser. Earth Environ. Sci.* **2018**, *170*, 032133. [CrossRef]
25. Water Framework Directive. Directive 2000/60/EC of the European Parliament and of the Council. Available online: https://eur-lex.europa.eu/resource.html?uri=cellar:5c835afb-2ec6-4577-bdf8-756d3d694eeb.0004.02/DOC_1&format=PDF (accessed on 21 July 2023).
26. Commission Directive (EU) 2184/2020 of the European Parliament and of the Council on the Quality of Water Intended for Human Consumption. Available online: https://eur-lex.europa.eu/eli/dir/2020/2184/oj (accessed on 21 July 2023).
27. Bucurica, I.A.; Dulama, I.D.; Radulescu, C.; Banica, A.L. Surface water quality assessment using electro-analytical methods and inductively coupled plasma mass spectrometry (ICP-MS). *Rom. J. Phys.* **2022**, *67*, 802.
28. Voinea, S.; Nichita, C.; Burchiu, E.; Diac, C.; Armeanu, I. Study case of potable water from wells in the metropolitan Bucharest area. Influences on human health–interdisciplinary lab. *Rom. Rep. Phys.* **2022**, *74*, 902.
29. Voinea, S.; Nichita, C.; Armeanu, I.; Solomonea, B. Experimental study of biodegradable materials in environmental physics classes. *Rom. Rep. Phys.* **2021**, *73*, 903.
30. Bărbulescu, A.; Barbeş, L. Statistical methods for assessing water quality after treatment on a sequencing batch reactor. *Sci. Total Environ.* **2021**, *752*, 141991. [CrossRef] [PubMed]
31. Chirilă, E.; Bari, T.; Barbeş, L. Drinking water quality assessment in Constanţa town. *Ovidius Univ. Ann. Chem.* **2010**, *21*, 87–90.
32. Chilian, A.; Tanase, N.-M.; Popescu, V.; Radulescu, C.; Bancuta, O.-R.; Bancuta, I. Long-Term Monitoring of the Heavy Metals Content (Cu, Ni, Zn, Cd, Pb) in Wastewater Before and after the Treatment Process by Spectrometric Methods of Atomic Absorption (FAAS and ETAAS). *Rom. J. Phys.* **2022**, *67*, 804.
33. Sterpu, A.E.; Bărbulescu, A.; Barbeş, L.; Koncsag, C.I. Modeling the Mixing Process of Industrial and Domestic Wastewater Sludge. *Environ. Eng. Manag. J.* **2015**, *14*, 1241–1246.
34. Bărbulescu, A.; Sterpu, A.E.; Barbeş, L.; Koncsag, C.I. New Correlation for the Mixing of Wastewater Sludge. *Rom. J. Phys.* **2017**, *62*, 801.
35. Thomas, O.; Burgess, C. *UV-Visible Spectrophotometry of Waters and Soils*, 3rd ed.; Elsevier: Amsterdam, The Netherlands, 2022.
36. Brar, S.K.; Kumar, P.; Cuprys, A. *Modular Treatment Approach for Drinking Water and Wastewater*; Elsevier: Amsterdam, The Netherlands, 2022.
37. Vara Prasad, M.N. *Disinfection Byproducts in Drinking Water: Detection and Treatment*; Butterworth-Heinemann: Oxford, UK, 2020.
38. Caratar, J.F.; Cano, R.E.; Garcia, J.I. Model of a drinking water treatment process and the variables involved using Coloured Petri Nets. *Ingeniare. Rev. Chil. Ing.* **2020**, *28*, 424–433. [CrossRef]
39. Brusseau, M.L.; Pepper, I.A.; Gerba, C.P. *Environmental and Pollution Science*, 3rd ed.; Elsevier: Amsterdam, The Netherlands, 2019.

40. Farhaoui, M.; Derraz, M. Review on Optimization of Drinking Water Treatment Process. *J. Water Res. Prot.* **2016**, *8*, 777–786. [CrossRef]
41. Mihăilescu, M.; Negrea, A.; Ciopec, M.; Negrea, P.; Duțeanu, N.; Grozav, I.; Svera, P.; Vancea, C.; Bărbulescu, A.; Dumitriu, C.Ș. Full Factorial Design for Gold Recovery from Industrial Solutions. *Toxics* **2021**, *9*, 111. [CrossRef] [PubMed]
42. Fighir, D.; Teodosiu, C.; Fiore, S. Environmental and Energy Assessment of Municipal Wastewater Treatment Plants in Italy and Romania: A Comparative Study. *Water* **2019**, *11*, 1611. [CrossRef]
43. Aonofriesei, F.; Bărbulescu, A.; Dumitriu, C.-S. Statistical analysis of morphological parameters of microbial aggregates in the activated sludge from a wastewater treatment plant for improving its performances. *Rom. J. Phys.* **2021**, *66*, 809.
44. Negrea, A.; Gabor, A.; Davidescu, C.-M.; Ciopec, M.; Negrea, P.; Duteanu, N.; Barbulescu, A. Rare Earth Elements Removal from Water Using Natural Polymer. *Sci. Rep.* **2018**, *8*, 316. [CrossRef] [PubMed]
45. Teodosiu, C.; Barjoveanu, G.; Sluser, B.M.; Ene Popa, S.A.; Trofin, O. Environmental assessment of municipal wastewater discharges: A comparative study of evaluation methods. *Int. J. Life Cycle Assess.* **2016**, *21*, 395–411. [CrossRef]
46. Paun, I.; Chiriac, F.L.; Iancu, V.I.; Pirvu, F.; Niculescu, M.; Vasilache, N. Disinfection by-products in drinking water distribution system of Bucharest City. *Rom. J. Ecol. Environ. Chem.* **2021**, *3*, 13–18. [CrossRef]
47. Vîrlan, C.-M.; Toma, D.; Stătescu, F.; Marcoie, N.; Prăjanu, C.-C. Modeling the chlorine-conveying process within a drinking water distribution network. *Environ. Eng. Manag. J.* **2021**, *20*, 487–494.
48. Bărbulescu, A.; Barbeș, L. Modeling the Chlorine Series from the Treatment Plant of Drinking Water in Constanta, Romania. *Toxics* **2023**, *11*, 699. [CrossRef]
49. Iordache, A.; Woinaroschy, A. Analysis of the efficiency of water treatment process with chlorine. *Environ. Eng. Manag. J.* **2020**, *19*, 1309–1313.
50. *SR ISO 10523:2012*; Water Quality. Determination of pH. Asociația de Standardizare din România (Romanian Standardization Association): București, Romania, 2023. Available online: https://magazin.asro.ro/ro/standard/200485 (accessed on 4 November 2023). (In Romanian)
51. *SR EN 27888:1997*; Water Quality. Determination of Electrical Conductivity. Asociația de Standardizare din România (Romanian Standardization Association): București, Romania, 2023. Available online: https://magazin.asro.ro/ro/standard/22665 (accessed on 4 November 2023). (In Romanian)
52. *SR EN ISO 7027-1:2016*; Water Quality. Determination of Turbidity-Part 1: Quantitative Methods. ISO: Geneva, Switzerland, 2023. Available online: https://www.iso.org/standard/62801.html (accessed on 4 November 2023).
53. *SR ISO 6059:2008*; Water Quality. Determination of the Sum of Calcium and Magnesium-EDTA Titrimetric Method. Asociația de Standardizare din România (Romanian Standardization Association): București, Romania, 2023. Available online: https://magazin.asro.ro/ro/standard/168610 (accessed on 4 November 2023). (In Romanian)
54. *SR EN ISO 8467:2001*; Water Quality. Determination of Permanganate Index. Asociația de Standardizare din România (Romanian Standardization Association): București, Romania, 2023. Available online: https://magazin.asro.ro/ro/standard/26286 (accessed on 4 November 2023). (In Romanian)
55. *SR EN ISO 7393-2:2018*; Water Quality. Determination of Free Chlorine and Total Chlorine-Part 2: Colorimetric Method Using N, N-Dialkyl-1,4-Phenylenediamine, for Routine Control Purposes. Asociația de Standardizare din România (Romanian Standardization Association): București, Romania, 2023. Available online: https://magazin.asro.ro/ro/standard/261837 (accessed on 4 November 2023). (In Romanian)
56. *SR ISO 9297:2001*; Water Quality. Determination of Chloride. Silver Nitrate Titration with Chromate Indicator (Mohr's Method). Asociația de Standardizare din România (Romanian Standardization Association): București, Romania, 2023. Available online: https://magazin.asro.ro/ro/standard/26186 (accessed on 4 November 2023). (In Romanian)
57. *STAS 3069-87*; Drinking Water. Sulphates Content Determination. Asociația de Standardizare din România (Romanian Standardization Association): București, Romania, 2023. Available online: https://magazin.asro.ro/ro/standard/15243 (accessed on 4 November 2023). (In Romanian)
58. *SR ISO 7150-1:2001*; Water Quality. Determination of Ammonium. Part 1: Manual Spectrometric Method. Asociația de Standardizare din România (Romanian Standardization Association): București, Romania, 2023. Available online: https://magazin.asro.ro/ro/standard/26536 (accessed on 4 November 2023). (In Romanian)
59. *SR EN 26777:2006*; Water Quality. Determination of Nitrite. The Method by Molecular Absorption Spectrometry. ISO: Geneva, Switzerland, 2023. Available online: https://www.iso.org/standard/13273.html (accessed on 4 November 2023).
60. *SR ISO 7890-3:2000*; Water Quality. Determination of Nitrate. Part 3: Spectrometric Method Using Sulfosalicylic Acid. ISO: Geneva, Switzerland, 2023. Available online: https://www.iso.org/standard/14842.html (accessed on 4 November 2023).
61. Romanian Law 458/2002 Regarding the Quality of Drinking Water. Available online: https://www.aspms.ro/documente/legislatie/Legea%2520458-republicata.pdf (accessed on 20 July 2023). (In Romanian).
62. Jucherski, A.; Walczowski, A.; Bugajski, P.; Jóźwiakowski, K. Technological reliability of domestic wastewater purification in a small Sequencing Batch Biofilm Reactor (SBBR). *Sep. Purif. Technol.* **2019**, *224*, 340–347. [CrossRef]
63. Pesce, S.F.; Wunderlin, D.A. Use of water quality indices to verify the impact of Cordoba city (Argentina) on Suquia River. *Water Resour.* **2000**, *34*, 2915–2926. [CrossRef]

64. WHO/SDE/WSH/07.01/1. pH in Drinking-Water. Revised Background Document for Development of WHO Guidelines for Drinking-Water Quality. Available online: https://cdn.who.int/media/docs/default-source/wash-documents/wash-chemicals/ph.pdf?sfvrsn=16b10656_4 (accessed on 25 November 2023).
65. Adams, K. Does the pH Level of Your Drinking Water Really Matter. Available online: https://intermountainhealthcare.org/blogs/does-the-ph-level-of-your-drinking-water-really-matter (accessed on 25 November 2023).
66. Interreg Danube Transnational JOINTISZA, Program Manual for Knowledge Development Tools and Knowledge Transfer in Urban Hydrology WP4–Activity 4.4 Deliverable 4.4.1. 2019. Available online: https://www.gwp.org/globalassets/global/gwp-cee_files/projects/jointisza/jointisza-manual-knowledge-development-tools-and-knowledge-transfer-in-urban-hydrology.pdf (accessed on 25 November 2023).
67. Jones, S. Conductivity. 2020. Available online: https://www.h2olabcheck.com/blog/view/conductivity (accessed on 25 November 2023).

toxics

Article

Machine Learning-Based Early Warning Level Prediction for Cyanobacterial Blooms Using Environmental Variable Selection and Data Resampling

Jin Hwi Kim [1], Hankyu Lee [1], Seohyun Byeon [1], Jae-Ki Shin [2], Dong Hoon Lee [3], Jiyi Jang [4], Kangmin Chon [5,6] and Yongeun Park [1,*]

1 School of Civil and Environmental Engineering, Konkuk University, Gwangjin-gu, Seoul 05029, Republic of Korea; jinhwi25@naver.com (J.H.K.); haeckel@konkuk.ac.kr (H.L.); shbyeon1@gmail.com (S.B.)
2 Busan Region Branch Office of the Nakdong River, Korea Water Resources Corporation (K-Water), Saha-Gu, Busan 49300, Republic of Korea; shinjaeki@gmail.com
3 Department of Civil and Environmental Engineering, Dongguk University-Seoul, 30, Pildong-ro 1-gil, Jung-gu, Seoul 04620, Republic of Korea; leedonghoon@dongguk.edu
4 Division of Atmospheric Sciences, Korea Polar Research Institute, 26, Songdomirae-ro, Yeonsu-gu, Incheon 21990, Republic of Korea; j.jiyi19@kopri.re.kr
5 Department of Environmental Engineering, Kangwon National University, Gangwon-do, Chuncheon 24341, Republic of Korea; kmchon@kangwon.ac.kr
6 Department of Integrated Energy and Infra System, Kangwon National University, Gangwon-do, Chuncheon 24341, Republic of Korea
* Correspondence: yepark@konkuk.ac.kr; Tel.: +82-2-2049-6106

Abstract: Many countries have attempted to mitigate and manage issues related to harmful algal blooms (HABs) by monitoring and predicting their occurrence. The infrequency and duration of HABs occurrence pose the challenge of data imbalance when constructing machine learning models for their prediction. Furthermore, the appropriate selection of input variables is a significant issue because of the complexities between the input and output variables. Therefore, the objective of this study was to improve the predictive performance of HABs using feature selection and data resampling. Data resampling was used to address the imbalance in the minority class data. Two machine learning models were constructed to predict algal alert levels using 10 years of meteorological, hydrodynamic, and water quality data. The improvement in model accuracy due to changes in resampling methods was more noticeable than the improvement in model accuracy due to changes in feature selection methods. Models constructed using combinations of original and synthetic data across all resampling methods demonstrated higher prediction performance for the caution level (L-1) and warning level (L-2) than models constructed using the original data. In particular, the optimal artificial neural network and random forest models constructed using combinations of original and synthetic data showed significantly improved prediction accuracy for L-1 and L-2, representing the transition from normal to bloom formation states in the training and testing steps. The test results of the optimal RF model using the original data indicated prediction accuracies of 98.8% for L0, 50.0% for L1, and 50.0% for L2. In contrast, the optimal random forest model using the Synthetic Minority Oversampling Technique–Edited Nearest Neighbor (ENN) sampling method achieved accuracies of 85.0% for L0, 85.7% for L1, and 100% for L2. Therefore, applying synthetic data can address the imbalance in the observed data and improve the detection performance of machine learning models. Reliable predictions using improved models can support the design of management practices to mitigate HABs in reservoirs and ultimately ensure safe and clean water resources.

Keywords: harmful algal blooms; alert level; feature selection; data resampling; machine learning; early warning

check for updates

Citation: Kim, J.H.; Lee, H.; Byeon, S.; Shin, J.-K.; Lee, D.H.; Jang, J.; Chon, K.; Park, Y. Machine Learning-Based Early Warning Level Prediction for Cyanobacterial Blooms Using Environmental Variable Selection and Data Resampling. *Toxics* **2023**, *11*, 955. https://doi.org/10.3390/toxics11120955

Academic Editors: Alina Barbulescu and Lucica Barbes

Received: 21 October 2023
Revised: 14 November 2023
Accepted: 20 November 2023
Published: 23 November 2023

1. Introduction

Toxic harmful algal blooms (HABs) cause various environmental problems in aquatic ecosystems, including public health threats, massive fish deaths, drinking water safety problems, increased wildlife mortality, and the destruction of aquatic habitats [1,2]. The recent rise in water temperature owing to climate change and the increase in nutrient discharge caused by human activity have promoted the growth of HABs in aquatic ecosystems [3–5]. In 2007, excessive algal blooms in Lake Taihu, China, affected the supply of drinking water for approximately two million people in nearby cities [6]. Furthermore, European countries, such as France, the United Kingdom, and Italy, suffer from social, economic, and environmental problems caused by HABs in coastal and inland areas [7–9]. These events suggest that toxic HABs can threaten public health and regional economies by contaminating drinking water, fish, and shellfish.

Therefore, the excessive growth of HABs across all regions is a significant global concern related to water quality management [10]. Therefore, many countries around the world, including South Korea, have conducted research and introduced policies and activities to solve the algal bloom problem to protect aquatic ecosystems, reduce public health threats, and secure safer water resources. As part of this, algal alert warning systems have been introduced and used in many countries to respond quickly to high-level algal blooms with HABs [11–13]. Recently, the Food and Agriculture Organization of the United Nations (FAO), in collaboration with the Intergovernmental Oceanographic Commission (IOC) and the International Atomic Energy Agency (IAEA), developed technical guidelines for implementing early warning systems for HABs that affect food safety or security [14]. Furthermore, algal alert warning systems serve as an important indicator for monitoring and managing algal blooms in terms of water quality management. They provide monitoring and management sequences to government officials, drinking water treatment plant operators, and water quality managers to help them make decisions [15].

In South Korea, a large-scale national project was implemented to dredge rivers and install eco-friendly weirs to increase the water storage capacity and restore the ecosystems of the country's four major rivers. However, since 2012, the flow velocity of rivers between weirs has decreased, leading to an increase in the frequency and intensity of algal blooms with HABs and risk of drinking water pollution [16,17]. In South Korea, an algal alert warning system is currently in place at 29 stations along four major rivers and reservoirs. The algal alert level of the system is determined based on the concentration of harmful algal cells. Therefore, the system focuses on the postblooming response rather than predicting the algal alert level. If the algal alert level can be predicted, it would be possible to respond before the occurrence of HABs with proactive water quality management.

In recent years, various studies have been conducted on data-driven models, which are easier to construct than numerical models [18]. However, the frequency of water quality monitoring for HABs is typically weekly or monthly [19,20], which makes it challenging to acquire sufficient data to train machine learning models. In addition, the occurrence of algal blooms typically has a seasonal pattern; algal blooms rarely occur in cold winters when the temperature is low and usually occur from spring to autumn when the temperature rises [21]. The magnitude of algal blooms has an uneven distribution and is characterized by sporadic occurrences [11]. For this reason, the distribution of data is imbalanced when classified based on the concentration or alert level of harmful algae. Shin et al. [22] collected and analyzed the distribution of algal alert levels at 13 monitoring stations in a reservoir and reported that the distribution was imbalanced at nine of the stations. Training machine learning models using imbalanced data can lead to accurate predictions at the majority alert level and inaccurate predictions at the minority level. However, accurate predictions for algal blooms with high concentrations that occur infrequently can be utilized as more important information than predictions for low concentration blooms in terms of water quality [23]. Furthermore, the results of supervised machine learning models are dominated by the quality and quantity of the data used in the training step. Therefore, class imbalance data in classification models reduce the ability to predict minority classes, and basic machine learning algorithms designed to improve overall prediction performance more accurately predict instances of the majority class than the minority class [24,25].

Recently, several studies have been conducted to solve the problem of data imbalance in statistical models. Choi et al. [26] solved the data imbalance problem using the synthetic minority oversampling technique (SMOTE), an oversampling method, and predicted the chlorophyll-a concentration in the Daechung reservoir in South Korea using a convolutional neural network model. Jeong et al. [27] considered SMOTE to solve data imbalance and predicted cyanobacterial cell density in eight water supply reservoirs in South Korea using machine learning models such as random forest (RF) and extreme Gradient Boosting. Bourel et al. [28] considered three under- and oversampling methods, including SMOTE, and predicted fecal coliforms on 21 beaches in Uruguay using various machine learning models. Despite the existence of such studies, studies specifically addressing imbalanced data related to harmful algal blooms, chlorophyll-a, nutrients, and specific environmental problems in the field of aquatic ecosystems are limited. In addition, few comprehensive studies have simultaneously addressed imbalanced data related to harmful algal blooms and feature selection for input variables.

Therefore, the impact of feature selection and data imbalance on machine learning models must be evaluated. The specific objectives of this study were to (1) acquire environmental variables, including water quality, hydrologic, and meteorological data, as input variables and apply feature selection methods to identify appropriate environmental variables, (2) solve data imbalance by generating synthetic data for minority classes using various resampling methods based on measurement data, (3) develop the algal alert warning system that can predict the algal alert levels in advance using artificial neural network (ANN) and RF models, and (4) evaluate differences in feature selection and resampling methods for improving prediction accuracy in minority classes.

2. Materials and Methods

2.1. Site Description

The Geum River is a major river in South Korea with agricultural and industrial functions. The shape and flow system of the river has changed since the construction of a multifunctional weir in 2012 [11], increasing the retention time of the flow rate [29]. As a result, blooms have expanded to the middle and upper reaches of rivers [30]. In addition, algal blooms, including HABs, have been continuously reported in the BJR [31]. The study area was the Baekje reservoir (BJR), located at the mid-stream of Geum River between 126°56′20″ E and 127°05′55″ E longitude and 36°19′07″ N and 36°27′45″ N latitude. The river width between the BJR and Gongju Reservoir (GJR) is 290–570 m, which is relatively large in South Korea (Figure 1). The BJR weir, which is a prediction point for algal alert levels, is located downstream of the study area and the GJR, which collects the cell density of cyanobacteria as an input variable, is located upstream of the study area. The main land-use type in the environs of the BJR is agricultural.

2.2. Data Acquisition

Seven water quality variables such as cyanobacteria cell density, total dissolved nitrogen concentration (TDN), nitrate concentration (NO_3-N), ammonium concentration (NH_4-N), total dissolved phosphorus concentration (TDP), phosphate concentration (PO_4-P), and conductivity (Cond) in the BJR were collected by the Korea Ministry of Environment from a monitoring station which was located 500 m upstream of the weir (Table 1). The average monitoring interval was 8 days and ranged from 4 to 62 days owing to irregular sampling caused by weather conditions, sampling management officers, and reservoir conditions. The algal alert levels of the BJR as an output variable were classified into three levels according to the classification criteria of the algal alert warning system implemented in South Korea based on cyanobacteria cell density [32]: normal level (<1000 cells/mL, L-0), caution level (≥1000 cells/mL and <10,000 cells/mL, L-1), warning level (≥10,000 cells/mL and <1,000,000 cells/mL, L-2), and blooming level (≥1,000,000 cells/mL, L-3). Four hydrological and three meteorological variables were monitored by the Korea Water Resource Corporation and the Korea Meteorological Administration. Daily hydrological and meteorological data, including air temperature, wind speed, water level, total inflow, total discharge, and total hydropower plant discharge, were used as average values between water quality monitoring events, and precipitation was used as the cumulative precipitation. In addition, cyanobacterial cell density measured at the GJR water quality monitoring station upstream of the BJR was used as an input variable to consider the connectivity between the two reservoirs for predicting algal alert levels. In this study, a total of 429 datasets were collected over 9 years from 2013 to 2021 (Figure 2A), but 345 datasets were chosen to develop a machine learning model. Datasets were excluded from the winter season (January, February, and December) in South Korea because of the impossibility of monitoring frozen rivers and the lack of cyanobacterial growth at low temperatures. During this period, the algal alert levels corresponding to L-1 and L-2 were zero. Therefore, we developed a machine learning model using 345 monitoring data points from March to November. All data were collected over 9 years (2013–2021) (Figure 2A). Table 1 lists the 14 environmental variables considered as input variables.

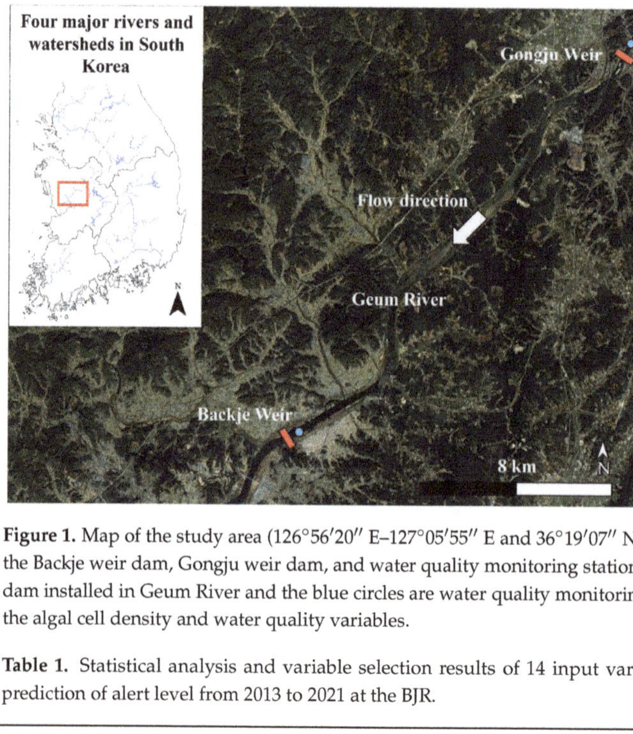

Figure 1. Map of the study area (126°56′20″ E–127°05′55″ E and 36°19′07″ N–36°27′45″ N) showing the Backje weir dam, Gongju weir dam, and water quality monitoring stations. The red box is a weir dam installed in Geum River and the blue circles are water quality monitoring stations that measure the algal cell density and water quality variables.

Table 1. Statistical analysis and variable selection results of 14 input variables collected for the prediction of alert level from 2013 to 2021 at the BJR.

Variables		Description	Unit	Descriptive Analysis		Variable Selection Method	
				Range	Mean	Dependence Test (*p*-Value)	MI Score
Water quality	TDN	Total dissolved nitrogen concentration	mg/L	1.17 to 6.92	2.79	<0.001	0.128
	NO_3-N	Nitrate concentration	mg/L	0.72 to 3.91	2.13	<0.001	0.118
	NH_4-N	Ammonium concentration	mg/L	0.01 to 2.24	0.19	0.005	0.015
	TDP	Total dissolved phosphorus concentration	mg/L	0.01 to 0.16	0.04	0.001	0.085
	PO_4-P	Phosphate concentration	mg/L	0 to 0.15	0.02	<0.001	0
	Cond	Conductivity	µmhos/cm	125 to 639	348.23	0.001	0.012
	GJ-cell	Cyanobacteria cell density in GJR	cells/mL	0 to 50970	1077	<0.001	0.161

Table 1. *Cont.*

Variables		Description	Unit	Descriptive Analysis		Variable Selection Method	
				Range	Mean	Dependence Test (*p*-Value)	MI Score
Hydro-dynamic	Wlevel	Average water level of the BJR	m	1.21 to 5.01	3.71	0.396	0.026
	Inflow	Average inflow rate of the BJR	m^3/s	20.25 to 2536.23	145.59	0.011	0.044
	Discharge	Average total discharge rate of the BJR	m^3/s	20.10 to 2555.66	145.70	0.011	0.049
	Dhydro	Average discharge rate by the hydropower plant of the BJR	m^3/s	0 to 124.60	43.85	0.050	0.043
Meteorological	Atemp	Average air temperature	°C	−1.30 to 30.46	16.85	<0.001	0.207
	Precip	Accumulated precipitation	mm	0 to 352.80	28.38	0.934	0.016
	Wspeed	Average wind speed	m/s	0.56 to 2.39	1.30	0.587	0

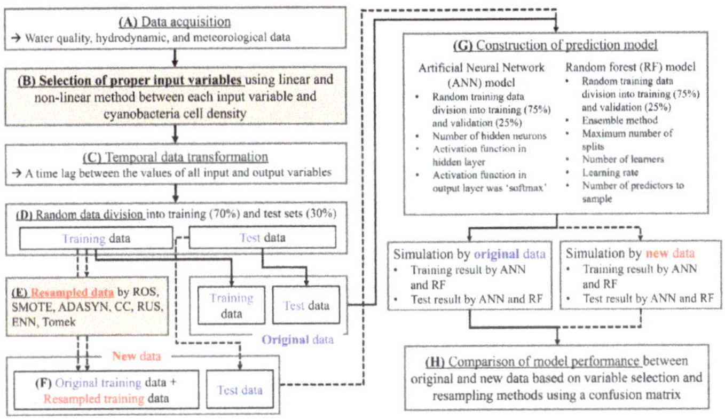

Figure 2. Flow chart for the construction of artificial neural network (ANN) and random forest (RF) models to predict algal alert levels using original and synthetic data according to variable selection and resampling methods.

2.3. Feature Selection for Algal Alert Levels

We statistically identified the relationship between cyanobacteria cell density, which determines algal alert levels, and input variables, including water quality, hydrodynamics, and meteorological variables, using linear and non-linear variable selection methods (Figure 2B). For the linear variable selection method, a simple linear regression analysis was used to analyze the input and output variables one-to-one. Simple linear regression was used to statistically test the dependence between variables [33] and the dependent input variables were selected based on statistical significance ($p < 0.05$). For the non-linear variable selection method, mutual information (MI) was used to measure the degree of relatedness between the output and input variables. MI is interpreted as the amount of information shared between variables, regardless of the average value and variance, and is based on information theory on a methodologically established basis [34]. The larger the MI, the higher the dependence on the probability distribution between variables. At an MI of 0, the relationship between variables is independent. Table 1 shows the statistical significance and MI scores for each of the 14 input variables.

2.4. Resampling Methods for Imbalanced Datasets

In data-driven models, including machine learning, deep learning, and linear statistical models, the imbalanced distribution of the output variable to be predicted results in the biased learning of the model because the accuracy is dominated by the amount and quality of the original dataset [35]. A total of 345 cyanobacteria cell density data collected from the BJR with algal alert level criteria were classified into L-0 (269; 78.0%), L-1 (47; 13.6%), and L-2 (29; 8.4%), respectively. The distribution of algal alert levels was sufficiently unbalanced to affect the model training. In a previous study [36], we used adaptive synthetic sampling (ADASYN) to generate synthetic data for algal alert levels corresponding to L-1 and L-2 based on observational data, addressing the imbalance of the data and improving the accuracy of the machine learning model. In the previous study, the amount of data was increased using synthetic data to resolve the data imbalance. In the present study, the amount of data increased and a method of reducing the majority class to the minority class was considered.

Oversampling involves creating copies of existing samples or adding more samples with values similar to those of a minority class [37]. However, oversampling can increase the size of the training dataset, resulting in additional computation time and potential overfitting of the model [38]. Undersampling involves the removal of samples from the majority class until a balance is achieved between the minority and majority classes. Therefore, during the training step, the reduced amount of data can improve the computation time for weight calculation and address storage-related issues, making the overall model implementation more efficient, which may improve the predictive accuracy of the model [39]. However, using undersampling, it may be challenging to improve the imbalance in predicting algal alert levels for relatively small datasets, such as that used in this study. To address these issues, hybrid sampling methods, such as ENN, which combine oversampling and undersampling have been proposed [40]. The resampling methods used were as follows: (1) random oversampling (ROS), SMOTE, and ADASYN as oversampling methods; (2) cluster centroid undersampling (CC) and random undersampling (RUS) as undersampling methods; and (3) synthetic minority oversampling technique–edited nearest neighbor (ENN) and synthetic minority oversampling technique–Tomek link (Tomek) as hybrid sampling methods. The detailed resampling methods are described in Appendix A of the Supplementary Materials.

2.5. Construction of Machine Learning Models and Evaluation of Model Accuracy

Figure 2 shows a flowchart of the study process in the order of data preparation, synthetic data generation and application, two machine learning model constructions, and model comparison. During the data acquisition and preprocessing stages, data on algal alert levels were collected as output variables, and water quality and hydrodynamic and meteorological data were collected as potential environmental variables affecting algal blooming (Figure 2A). We determined the input variables using the linear and non-linear variable selection method between each input variable and output (Figure 2B). We modified the selected input and output variables to focus on predicting future algal alert levels (Figure 2C). In other words, the measured value of the output variable for a specific algal alert level was matched with the values of previously measured input variables in the monitoring conducted at an average interval of eight days. For example, the algal alert level measured on 23 April 2013, was considered the output variable of the input variables measured on 15 April 2013. These variables comprised a single dataset. In this preprocessing, the prediction of future algal alert levels using the current input variables was reflected in the training steps of the two machine learning models.

In the dataset reconstruction stage, all datasets were randomly extracted into training (70%) and test (30%) datasets (Figure 2D). For each resampling method, synthetic data generated based on the training dataset were added to the dataset used in the training step (Figure 2E). The dataset at the test step for all of the prediction models was used as the original dataset without adding synthetic data. Therefore, a total of 24 cases, each possessing different sets of data, were generated considering variable selection methods and resampling methods for training and testing of the models: (1) eight cases consisted of the original dataset without variable selection and seven datasets with generated synthetic data based on the original data for each resampling method, (2) eight consisted of the original dataset with the linear variable selection method and seven datasets with seven resampling methods, and (3) eight consisted of the original dataset with the non-linear variable selection method and seven datasets with seven resampling methods (Figure 2F).

In the construction and evaluation stages of the algal alert level prediction model, the datasets, excluding the respective test datasets from the 24 generated cases, were randomly extracted into training (75%) and validation (25%) datasets. The training datasets were used to train the model and optimize the hyperparameters (Figure 2G). Test datasets were used to evaluate the performance of each model constructed using the resampling methods. In this study, two prominent machine learning models, ANN and RF, were utilized to predict the alert levels for harmful algal blooms. Machine learning models, known for their powerful computational techniques, are useful for predicting specific phenomena and interpreting complex relationships in the environment [41,42]. In addition, ANN and RF models are representative machine learning models which assess the impact of imbalanced data on predictive performance, making it more convenient for other researchers to utilize the approach presented in this study. ANN and RF were optimized based on the hyperparameters of each model (Figure 2G). For ANN hyperparameters, such as the number of hidden neurons and the activation function in the hidden layer, the number of hidden neurons was optimized using a pattern search algorithm and the activation functions in the hidden layer were experimentally optimized. The activation function in the output layer was 'softmax.' For RF hyperparameters, such as the ensemble aggregation method, the number of ensemble learning cycles, learning rate for shrinkage, minimum leaf size, the maximum number of decision splits, and the number of predictors to select at random for each split, a random search optimization algorithm was used to optimize these hyperparameters. The ANN model structure is described in the Supplementary Materials in our previous study [36], whereas the structure of the RF model is described in Appendix B of the Supplementary Materials.

Finally, the classification performances of the two models on each dataset were compared using a confusion matrix (Figure 2H). The confusion matrix is described in the Supplementary Materials in our previous study [36]. We selected the optimized model from 100 repeated executions for each model using variable selection and resampling methods. We calculated the average accuracy of the models for each method to evaluate the overall classification performance of the two machine learning models. All processes, including statistical analysis, machine learning model configuration, and model optimization, were performed in a MATLAB (MathWorks Inc., Natick, MA, USA) environment.

3. Results and Discussion

3.1. Descriptive Analysis of Cyanobacteria and Nutrients in the BJR

Table 2 shows the results of monthly descriptive analysis for weekly cyanobacteria cell density, Chl-a concentration, and nutrient concentration in the BJR from March to November. Out of 345 events issued by the early warning system, caution (43 events) and warning (29) levels were mostly announced between July and October. The formation of algal blooms in reservoirs in East Asia, including South Korea, with monsoon climate characteristics, occurs most actively in summer [11], and these climate characteristics were reflected in the BJR. As a result of calculating the N:P ratio to identify nutrients that affect the algal growth in the BJR, the range and average value for the entire period were 5.26–240.79 and 42.5, respectively (Table 2). The N:P ratios in about 85% of samples were higher than 17, which, according to Forsberg and Ryding [43], means that primary productivity in the BJR is limited by phosphorus. Nitrogen and phosphorus are essential and influential in regulating the structure, function, and processes of ecosystems [44]. However, imbalances in the N:P ratio resulting from excessive nutrient inputs can exacerbate eutrophication in reservoirs, altering ecological structure and function and deteriorating aquatic ecosystems [45]. Therefore, the management and control of phosphorus loadings into the BJR can help suppress the occurrence of harmful algal blooms. Chl-a and phosphate concentrations from July to October, which were predominantly associated with algal bloom events corresponding to the caution and warning levels, ranged from 5.3–177.7 (an average of 50.5 μg/L) and 1–153 (an average of 31.9 μg/L), respectively. Based on Carlson [46], the nutritional status of the BJR from July to October was classified as eutrophic (Table S1).

A detailed description of the N:P ratio, Chl-a, and phosphate concentrations is given in Appendix C of the Supplementary Materials.

Table 2. Results of descriptive analysis for monthly cyanobacteria cell density and nutrient concentration measured in the BJR.

Month	Algal Alert Level (Number of Events)			Cyanobacteria Cell (Cells/mL)		N:P Ratio		Chl-a (μg/L)		Phosphate (μg/L)	
	L-0	L-1	L-2	Range	Average	Range	Average	Range	Average	Range	Average
March	40	0	0	0 to 140	4	29.1 to 240.8	74.8	7.6 to 105.3	42.3	1 to 29	6.7
April	39	0	0	0 to 625	34	16.4 to 139.2	55.3	16.3 to 162.8	56.6	0 to 43	7.5
May	36	2	0	0 to 1950	117	8.7 to 123.4	39.9	20.2 to 176.1	64.5	1 to 113	10.9
June	38	1	0	0 to 2920	131	12.2 to 50.6	30.0	11.9 to 185.1	70.8	0 to 33	9.0
July	25	5	7	0 to 95,500	7684	6.5 to 46.8	23.8	7.4 to 165.1	46.2	2 to 140	32.2
August	12	10	17	0 to 398,820	27,391	5.3 to 42.9	19.6	5.3 to 144.3	51.1	2 to 135	40.4
September	16	15	5	0 to 95,355	7206	6.8 to 84.5	27.9	6.4 to 177.7	55.1	2 to 153	34.1
October	25	13	0	0 to 6565	1071	7.4 to 84.1	43.6	9.3 to 123.0	49.8	1 to 141	20.8
November	38	1	0	0 to 1160	51	14.9 to 135.8	65.2	5.1 to 128.4	35.7	1 to 97	15.5
Total	269	47	29	0 to 398,820	4828	5.26 to 240.8	42.5	5.1 to 185.1	52.4	0 to 153	19.5

3.2. Selection of Input Variables and Generation of Synthetic Data

Table 1 shows the *p*-values and MI results for the 14 input variables according to the variable selection method. In the case of the dependence test, 11 variables, excluding average water level of the BJR (Wlevel), accumulated precipitation (Precip), and average wind speed (Wspeed), had a statistically significant linear dependence ($p < 0.05$) on cyanobacteria cell density; total dissolved nitrogen concentration (TDN), nitrate concentration (NO_3-N), ammonium concentration (NH_4-N), and conductivity (Cond) were negatively correlated, and total dissolved phosphorus concentration (TDP), phosphate concentration (PO_4-P), average inflow rate of the BJR (Inflow), average total discharge rate of the BJR (Discharge), average discharge rate by the hydropower plant of the BJR (Dhydro), average air temperature (Atemp), and cyanobacteria cell density in the GJR (GJ-cell) were positively correlated (Figure S1). The dependence test results for phosphorus as a limiting factor for eutrophication in the BJR showed that phosphorus-related variables were positively correlated with cyanobacterial cell density, whereas nitrogen-related variables were negatively correlated. This implies that nitrogen is more abundant in the BJR than phosphorus and that a favorable N:P ratio for harmful algal blooms is formed by the inflow of phosphorus or a decrease in nitrogen in the water body. For the MI score, 12 variables were selected as input variables with statistical correlation considering nonlinearity for cyanobacteria cell density; TDN, NO_3-N, NH_4-N, TDP, Cond, GJ-cell, Wlevel, Inflow, Discharge, Dhydro, Atemp, and Precip had MI scores above 0 and PO_4-P and Wspeed had scores of 0. In both variable selection methods, Wspeed, without a statistical correlation, was excluded from the input variables for predicting algal alert levels. Wong et al. [47,48] reported that wind speed affects the growth, transport, and diffusion of algal blooms. However, these studies were conducted in oceans over a wider area than the present study. Zhang et al. [49] reported that the annual average wind speed has a statistically significant correlation with the occurrence of algal blooms via regression analysis using 25 years of long-term observational data from Lake Taihu (2338 km^2) in China. However, the yearly average wind speed was higher than that of this study area and the regression coefficient was low (-0.023~-0.027).

Considering these results, it is necessary to evaluate whether wind speed should be included as an input variable when constructing statistical models for small-scale reservoirs with characteristics similar to those in the study area. Various variable selection methods based on linear and non-linear methods can determine appropriate input variables, and the selected input variables can assist in constructing statistical models with high prediction accuracy [50]. Finally, from a total of 14 water quality, meteorological, and hydrological variables, 11 variables for the linear method and 12 variables for the non-linear method were selected as input variables to predict algal alert levels using machine learning models.

3.3. Improvement in Data Imbalance Using Synthetic Data

The distribution of the monitored algal alert levels used in the training step of the model from the original data was 189 (77.8%) for L-0, 33 (13.6%) for L-1, and 21 (8.6%) for L-2 (Figure 3). The overall distribution of algal alert levels was imbalanced and skewed toward the L-0. Traditionally, classification algorithms in machine learning have been used to increase the overall accuracy of the classifiers. While maximizing the overall accuracy, the model tended to focus on the majority class because of its higher weight in the distribution of the entire class [51]. For this reason, classification models can achieve high accuracy for the majority class or entire dataset, whereas they can predict poorly for minority classes. Therefore, when a dataset is imbalanced, maximizing the overall accuracy without considering the accuracy of the minority classes may not be optimal. We applied seven resampling methods of different types to improve the data imbalance: oversampling methods—ROS, SMOTE, and ADASYN; undersampling methods—CC and RUS; and hybrid sampling methods—ENN and Tomek.

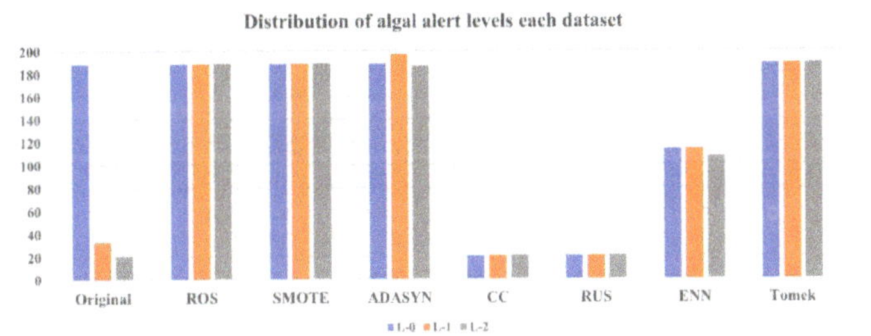

Figure 3. Comparison of algal alert level distribution between the original data and the new dataset with resampled data via each sampling method.

Figure 3 shows the distribution of the datasets obtained using each resampling method. The datasets newly constructed using ROS, SMOTE, ADASYN, and Tomek achieved a balance between the majority class (L-0) and its data, whereas the datasets constructed using CC and RUS balanced the minority class (L-2) with the fewest samples. For ENN, oversampling was performed for L-1 and L-2 to match the data with the majority class and undersampling was performed for all classes, resulting in balanced data with 114 for L-0, 114 for L-1, and 107 for L-2. Therefore, all of the new datasets, excluding the original, were generated using a balanced distribution of algal alert levels. In the case of rare occurrence problems, identifying the minority class is often more significant than identifying the majority class and an imbalance in the dataset can lead to the generation of misleading information regarding the minority class in classification algorithms [52]. Problems such as harmful algal blooms, droughts, floods, and chemical accidents in the environment typically have a low occurrence frequency but a significant socioeconomic impact. Therefore, when analyzing these problems, it is necessary to adequately consider minority classes.

3.4. Comparison of Model Performance According to the Feature Selection and Resampling Methods

To assess the impact of the feature selection method on the prediction of algal alert levels, the predictive performances of the original data, original data with a linear approach, and original data with a non-linear approach were compared. Table 3 presents the performances of the ANN and RF models obtained via 100 iterations using three different datasets: the original dataset considering 14 input variables, the original dataset considering 11 input variables extracted from dependency tests, and the original dataset considering 12 input variables derived from MI scores. The key results showed that there was no clear distinction in predictive performance among the models, regardless of whether feature selection methods were applied. A detailed comparison of their performance values is provided in Appendix B of the Supplementary Materials.

Table 3. Comparison of overall and optimal performance for accuracy, recall, and precision for algal alert levels according to the feature selection methods in ANN and RF.

Machine Learning Model		Feature Selection	Training (Including Validation)							Test						
			Performance Index							Performance Index						
			Accuracy	Recall			Precision			Accuracy	Recall			Precision		
				L-0	L-1	L-2	L-0	L-1	L-2		L-0	L-1	L-2	L-0	L-1	L-2
Overall model performance	ANN	No feature selection	92.6 (±6)	98.3 (±2)	66.6 (±30)	82.8 (±17)	94.1 (±5)	82.5 (±17)	87.6 (±12)	82.4 (±3)	93.3 (±4)	35.6 (±12)	55.3 (±17)	88.9 (±2)	55.6 (±16)	56.3 (±16)
		Linear method	91.7 (±6)	98.1 (±1)	62.7 (±27)	79.3 (±16)	93.4 (±5)	80.8 (±16)	85.0 (±12)	83.0 (±3)	93.7 (±4)	38.2 (±13)	54.9 (±16)	89.2 (±3)	58.0 (±17)	60.0 (±16)
		Non-linear method	91.2 (±6)	98.2 (±1)	59.8 (±29)	78.5 (±17)	93.1 (±5)	78.8 (±16)	84.2 (±11)	82.2 (±4)	93.4 (±5)	34.9 (±13)	53.3 (±19)	89.0 (±2)	54.0 (±18)	56.7 (±16)
	RF	No feature selection	92.5 (±4)	98.1 (±1)	67.9 (±18)	80.6 (±15)	94.2 (±3)	83.1 (±13)	88.5 (±10)	84.4 (±2)	95.8 (±2)	36.4 (±11)	54.4 (±16)	89.7 (±2)	53.8 (±15)	67.5 (±17)
		Linear method	92.4 (±4)	98.1 (±1)	67.5 (±17)	80.5 (±14)	94.2 (±3)	82.8 (±12)	87.2 (±11)	85.2 (±2)	96.0 (±2)	40.1 (±11)	56.9 (±16)	90.2 (±2)	60.2 (±15)	66.8 (±15)
		Non-linear method	93.5 (±4)	98.3 (±2)	72.1 (±18)	83.1 (±14)	95.0 (±3)	85.5 (±13)	89.5 (±11)	84.5 (±3)	95.3 (±3)	38.3 (±13)	56.8 (±17)	89.9 (±2)	55.6 (±14)	65.4 (±15)
Optimal model performance	ANN	No feature selection	90.5	96.3	63.6	81.0	93.3	75.0	85.0	90.2	96.3	64.3	75.0	92.8	75.0	85.7
		Linear method	84.0	96.8	27.3	57.1	88.0	50.0	70.6	88.2	98.8	42.9	62.5	89.8	85.7	71.4
		Non-linear method	84.8	96.3	24.2	76.2	87.5	53.3	80.0	88.2	98.8	50.0	50.0	88.8	87.5	80.0
	RF	No feature selection	88.1	96.8	48.5	71.4	92.4	69.6	68.2	88.2	98.8	50.0	50.0	89.8	87.5	66.7
		Linear method	94.7	98.9	75.8	85.7	94.9	89.3	100	89.2	98.8	50.0	62.5	89.8	100	71.4
		Non-linear method	94.7	97.9	81.8	85.7	95.9	87.1	94.7	92.2	100	42.9	100	92.0	100	88.9

The ANN and RF models, using data generated by different resampling methods, were compared to evaluate the impact of data imbalance. All data used in the comparison were subjected to resampling methods without applying feature selection methods. Figure 4 shows the overall performance of each model, which was performed 100 times independently. In the training step, the overall accuracies of ANN and RF on the original dataset were relatively high. However, the overall recall for each algal alert level was unbalanced. From these results, it can be observed that the predictions for each algal alert level were unbalanced, and the accuracy was primarily influenced by L-0, indicating that it had a dominant impact on the overall performance. Furthermore, imbalanced predictions between classes in models that utilize imbalanced data can diminish the statistical reliability of the overall model accuracy [53]. Therefore, to evaluate classifiers for imbalanced data, it is essential to appropriately reflect the predictive ability of minority classes [54]. In the ANN and RF models, the predictive performance for L-1 and L-2 in the models with applied resampling methods in the training and test steps exhibited improvements compared with models utilizing the original dataset. Figure 5 shows the results of the optimal model among the models that were iteratively performed 100 times. For the ANN model, the accuracy for L-1 and L-2 improved in the training step; however, in the test step, the accuracy for L-1 improved significantly, whereas that for L-2 improved but not significantly. In the RF model, there was an enhancement in the accuracy of L-1 and L-2, with a notable improvement in accuracy, particularly for L-2. A detailed comparison of their performance values is provided in Appendix E of the Supplementary Materials.

3.5. Comparison of Model Performance According to Both Feature Selection and Resampling Methods

A total of 28 case datasets were applied to the ANN and RF models to evaluate the combined effects of the feature selection and resampling methods. Tables S2 and S3 show the overall performances of the ANN and RF models, respectively, which were iteratively performed 100 times for each data type.

The overall accuracy of the two machine learning models in the training step was similar to that of the models using the original data, and the accuracy for each algal alert level was improved compared to the models using the original data. The overall recall for the model using original data was, in ANN, 66.6% for L-1 and 82.8% for L-2 and, in RF, 67.9% for L-1 and 80.6% for L-2. The range of variation in recall for L-1 and L-2 based on feature selection methods was, in ANN, 3.9–6.8% for L-1 and 3.5–4.3% for L-2 and, in RF, 0.4–4.2% for L-1 and 0.1–2.5% for L-2. The range of variation in recall based on resampling methods was, in ANN, on average, 23.3–27.4% for L-1 and 14.5–17.4% for L-2 and, in RF, on average, 19.9–24.9% for L-1 and 13.7–16.3% for L-2.

Based on the preceding results, it is evident that, for predicting algal alert levels, the improvement in predictive accuracy via resampling methods surpassed that achieved by feature selection methods. Balanced predictions are made for each class. Despite the results of this study, feature selection methods can efficiently describe the input data while reducing the influence of noise or irrelevant variables, thereby providing better predictive results [55]. Moreover, in classification problems, using variables with a low statistical correlation to classes as pure noise can introduce bias in the prediction of classes and degrade the classification performance [56]. However, feature selection methods can be effective in improving the predictive performance of datasets with numerous features. In the present study, the number of features was 14, which is relatively small compared to the number of features used in previous studies. For example, Bolón-Canedo et al. [57] compared the predictive performance of various feature selection methods for 64 different datasets, with the number of features ranging from 918 to 41,151, and Wei et al. [58] studied 14 different datasets, with the number of features ranging from 72 to 400. Xue et al. [59] demonstrated that applying feature selection methods improved predictive performance with a reduced number of features. However, they also reported that the application of feature selection methods did not significantly enhance the predictive performance of models that already exhibited high accuracy. Therefore, in terms of data with imbalances and fewer features, the application of resampling methods may be more effective than feature selection methods in improving model predictive performance.

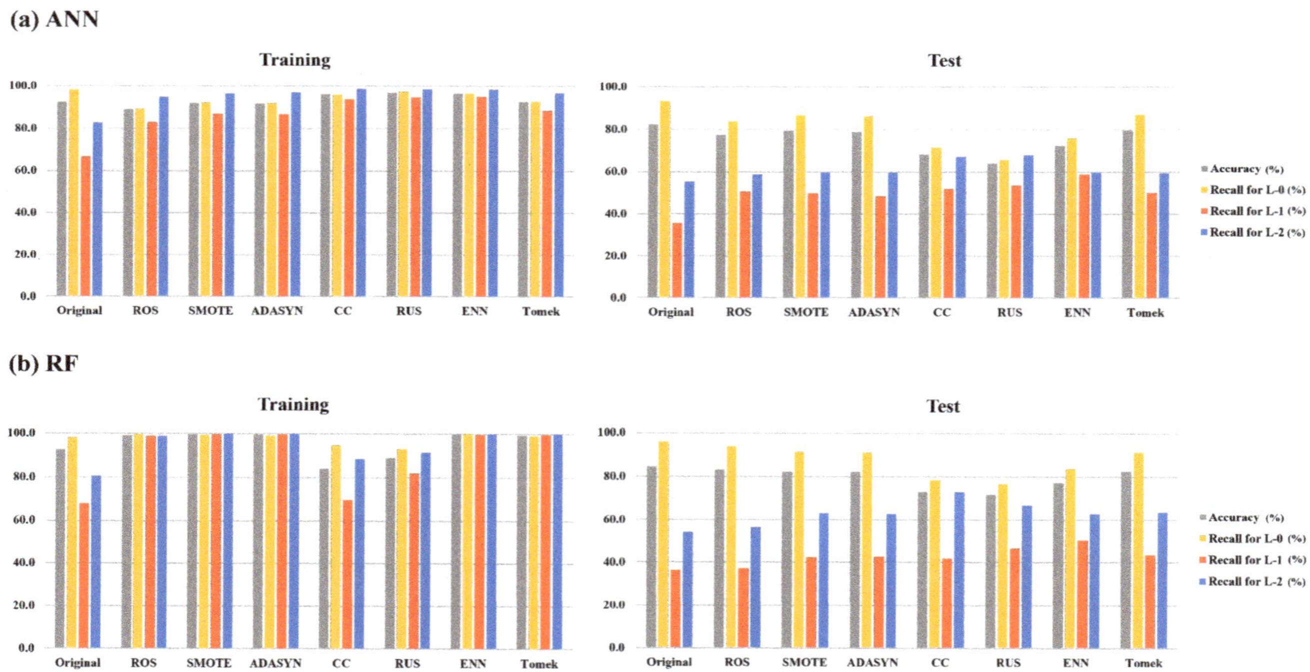

Figure 4. Comparison of overall prediction performance for algal alert levels of applied resampling methods based on original dataset. (**a**) is the overall prediction performance in ANN and (**b**) is the overall prediction performance in RF. Original is a model in which the original data were used, ROS is a model with random oversampling method, SMOTE is a model with synthetic minority oversampling technique, ADASYN is a model with adaptive synthetic sampling, CC is a model with cluster centroid undersampling, RUS is a model with random undersampling, ENN is a model with synthetic minority oversampling technique–edited nearest neighbor, and Tomek is a model with synthetic minority oversampling technique–Tomek link.

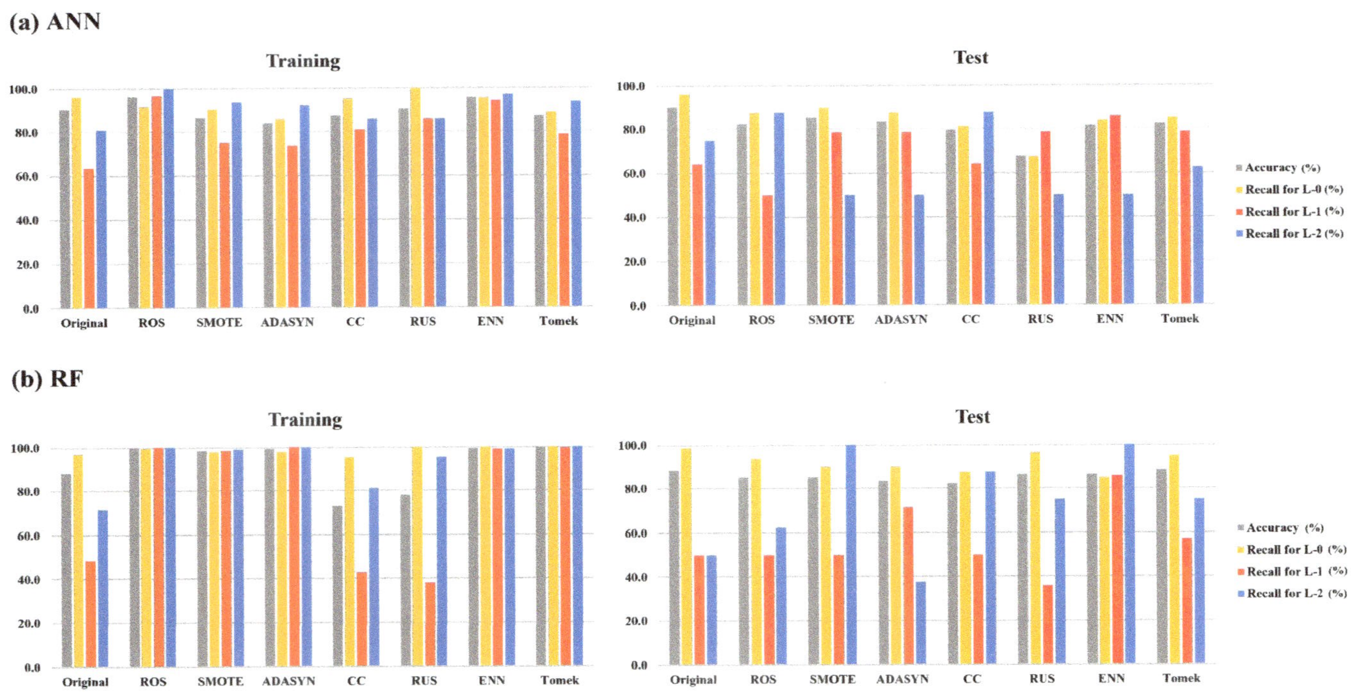

Figure 5. Comparison of prediction performance for algal alert levels of the optimal model by applied resampling methods based on the original dataset. (**a**) is prediction performance in ANN optimal model and (**b**) is prediction performance in RF optimal model. Original is a model in which the original data were used, ROS is a model with random oversampling method, SMOTE is a model with synthetic minority oversampling technique, ADASYN is a model with adaptive synthetic sampling, CC is a model with cluster centroid undersampling, RUS is a model with random undersampling, ENN is a model with synthetic minority oversampling technique–edited nearest neighbor, and Tomek is a model with synthetic minority oversampling technique–Tomek link.

Tables S4 and S5 present the results of the optimal models selected from the models that were iteratively performed for ANN and RF, respectively. The comparison of the performance of the optimal models exhibited results similar to the overall performance comparison. Based on these results, the best model was identified for predicting algal alert levels among the ANN and RF models. Figure 6 compares the confusion matrix between the selected optimal ANN model and the model using the original data. Non-linear feature selection and ENN sampling were applied to data from the selected ANN model. In the training step, accuracy was similar and the recall for L-1 increased from 63.6 to 86.2%, while the recall for L-2 increased from 81.0 to 90.5%. In the test step, despite a decrease in accuracy by 9.8%, the recall for L-1 increased from 64.3 to 71.4%, achieving balanced predictions at each algal alert level. Figure 7 shows the results of the confusion matrix comparison of the optimal RF model selected from the RF models with various data types.

The selected optimal RF model was constructed from data obtained using the ENN sampling method without feature selection. In the training step, the optimal model showed an increase of 11.3% in accuracy compared with the model using original data. During the test step, although the accuracy decreased by 1.9%, the recall for L-1 and L-2 increased by 35.7% and 50.0%, respectively. The recall of each algal alert level was as follows: 85.0% for L-0, 85.7% for L-1, and 100% for L-2. The hyperparameters of the optimal ANN and RF models are listed in Table S6 in Supplementary Materials. In this study, the model with ENN sampling was selected as the optimal model for predicting algal alert levels. A comparison of the optimal models revealed that the models using non-linear feature selection and the CC sampling method exhibited balanced predictions compared with the models using the original data (Figure S2). All performance indices for both the training and test steps were higher for the optimal RF model than for the optimal ANN model (Figures 6 and 7). Therefore, the RF model was deemed more suitable for predicting the algal alert levels.

Optimal model of ANN confusion matrix

ANN — Original data

Training

Predicted \ Observed	L-0	L-1	L-2	Precision
L-0	182	10	3	93.3
L-1	6	21	1	75.0
L-2	1	2	17	85.0
Recall	96.3	63.6	81.0	90.5

Test

Predicted \ Observed	L-0	L-1	L-2	Precision
L-0	77	4	2	92.8
L-1	3	9	0	75.0
L-2	0	1	6	85.7
Recall	96.3	64.3	75.0	90.2

ANN — Non-linear feature selection, ENN sampling method

Training

Predicted \ Observed	L-0	L-1	L-2	Precision
L-0	108	8	0	93.1
L-1	4	94	9	87.9
L-2	1	7	86	91.5
Recall	95.6	86.2	90.5	90.9

Test

Predicted \ Observed	L-0	L-1	L-2	Precision
L-0	66	2	1	95.7
L-1	9	10	1	50.0
L-2	5	2	6	46.2
Recall	82.5	71.4	75.0	80.4

Figure 6. Comparison of confusion matrices between ANN model using original data and selected optimal ANN model.

Optimal model of RF confusion matrix

RF — Original data

Training

Predicted \ Observed	L-0	L-1	L-2	Precision
L-0	183	12	3	92.4
L-1	4	16	3	69.6
L-2	2	5	15	68.2
Recall	96.8	48.5	71.4	88.1

Test

Predicted \ Observed	L-0	L-1	L-2	Precision
L-0	79	6	3	89.8
L-1	0	7	1	87.5
L-2	1	1	4	66.7
Recall	98.8	50.0	50.0	88.2

RF — No feature selection, ENN sampling method

Training

Predicted \ Observed	L-0	L-1	L-2	Precision
L-0	114	1	0	99.1
L-1	0	111	1	99.1
L-2	0	0	99	100
Recall	100	99.1	99.0	99.4

Test

Predicted \ Observed	L-0	L-1	L-2	Precision
L-0	68	1	0	98.6
L-1	10	12	0	54.5
L-2	2	1	8	72.7
Recall	85.0	85.7	100	86.3

Figure 7. Comparison of confusion matrices between RF model using original data and selected optimal RF model.

4. Conclusions

This study presented a series of processes for improving the prediction of algal alert levels in the BJR. Based on the observed data, feature selection and resampling methods were applied and two machine learning models were constructed. The following major conclusions were drawn from this study:

- Applying resampling methods to the imbalanced classes observed in the original data allowed the collection of data with balanced distributions for all classes, thereby preventing biased learning of the model and improving its accuracy.
- Resolving the class imbalance via resampling methods proved to be more effective in improving the accuracy of the model than adjusting the input variables via feature selection methods.
- In the RF model, the accuracy of the model with the resampling method demonstrated the highest performance, whereas in the ANN model, the predictive performance of the model incorporating both feature selection and resampling methods appeared to be superior.
- When considering non-linear models such as machine learning for prediction, it is important to evaluate the availability of feature selection and resampling methods according to the model type.
- The characteristics and quantity of the original data can serve as important factors when selecting the feature selection and resampling methods. In addition, appropriate feature selection and resampling methods can be applied as useful tools for constructing machine learning models.

This study aimed to construct a prediction model for algal alert levels in reservoirs using readily available data from national monitoring stations and to provide a machine learning model that improves accuracy via feature selection and resampling methods. The proposed model is expected to be useful to engineers and decision makers involved in the management of algal blooms in watershed areas, including inland weirs, facilitating the establishment of effective strategies and regulations for their construction and operation.

Supplementary Materials: The following supporting information can be downloaded at: https://www.mdpi.com/article/10.3390/toxics11120955/s1, Appendix A: Description of structure for each resampling method; Appendix B: Description of the RF model structure; Appendix C: Descriptive statistics for cyanobacteria and nutrients in the BJR; Appendix D: Comparison of model performance according to the feature selection; Appendix E: Comparison of model performance according to resampling methods; Figure S1: Results of the dependence test between each input variable and cyanobacteria cell density; Figure S2: Comparison of confusion matrices between the RF model using original data and the optimal RF model with non-linear feature selection and CC sampling method; Table S1: Eutrophication standards for single parameter index in Trophic state index (Carlson, 1997); Table S2(A): Overall performance of the ANN model according to applied feature selection and resampling methods in training step; Table S2(B): Overall performance of the ANN model according to applied feature selection and resampling methods in test step; Table S3(A): Overall performance of the RF model according to applied feature selection and resampling methods in training step; Table S3(B): Overall performance of the RF model according to applied feature selection and resampling methods in test step; Table S4: Optimal performance of the ANN model according to applied feature selection and resampling methods; Table S5: Optimal performance of the RF model according to applied feature selection and resampling methods; Table S6: Descriptive statistics for optimized hyperparameters obtained from the optimal model in ANN and RF using randomly chosen training data.

Author Contributions: Conceptualization, Y.P. and J.H.K.; methodology, Y.P. and J.H.K.; data collection, H.L. and S.B.; formal analysis, J.-K.S. and D.H.L.; investigation, J.-K.S. and D.H.L.; writing—original draft preparation, J.H.K.; writing—review and editing, J.J., K.C. and Y.P. All authors have read and agreed to the published version of the manuscript.

Funding: This work was supported by Konkuk University in 2020.

Institutional Review Board Statement: Not applicable.

Informed Consent Statement: Not applicable.

Data Availability Statement: Water quality data can be downloaded from the Water Environment Information System (https://water.nier.go.kr/web; accessed on 22 November 2023). Meteorological data can be downloaded from the Korea Meteorological Administration (https://www.weather.go.kr/w/index.do; accessed on 22 November 2023). Hydrological data can be downloaded from the Korea Water Resource Corporation (https://www.water.or.kr/kor/realtime/sumun/index.do?mode=mult&menuId=13_91_93_98; accessed on 22 November 2023).

Acknowledgments: We would like to thank the Ministry of Environment, the Korea Meteorological Administration, and the Korea Water Resource Corporation for sharing data.

Conflicts of Interest: The authors declare no conflict of interest.

References

1. Anderson, D.M.; Glibert, P.M.; Burkholder, J.M. Harmful algal blooms and eutrophication: Nutrient sources, composition, and consequences. *Estuaries* **2002**, *25*, 704–726. [CrossRef]
2. Grattan, L.M.; Holobaugh, S.; Morris, J.G. Harmful algal blooms and public health. *Harmful Algae* **2016**, *57*, 2–8. [CrossRef] [PubMed]
3. Gobler, C.J. Climate change and harmful algal blooms: Insights and perspective. *Harmful Algae* **2020**, *91*, 101731. [CrossRef] [PubMed]
4. O'Neil, J.M.; Davis, T.W.; Burford, M.A.; Gobler, C.J. The rise of harmful cyanobacteria blooms: The potential roles of eutrophication and climate change. *Harmful Algae* **2012**, *14*, 313–334. [CrossRef]
5. Chen, C.; Liang, J.; Yang, G.; Sun, W. Spatio-temporal distribution of harmful algal blooms and their correlations with marine hydrological elements in offshore areas, China. *Ocean. Coast. Manag.* **2023**, *238*, 106554. [CrossRef]
6. Qin, B.; Zhu, G.; Gao, G.; Zhang, Y.; Li, W.; Paerl, H.W. A Drinking Water Crisis in Lake Taihu, China: Linkage to Climatic Variability and Lake Management. *Environ. Manag.* **2010**, *45*, 105–112. [CrossRef] [PubMed]
7. Perrot, T.; Rossi, N.; Menesguen, A.; Dumas, F. Modelling green macroalgal blooms on the coasts of Brittany, France to enhance water quality management. *J. Mar. Syst.* **2014**, *132*, 38–53. [CrossRef]
8. Scanlan, C.M.; Foden, J.; Wells, E.; Best, M.A. The monitoring of opportunistic macroalgal blooms for the water framework directive. *Mar. Pollut. Bull.* **2007**, *55*, 162–171. [CrossRef]
9. Viaroli, P.; Bartoli, M.; Azzoni, R.; Giordani, G.; Mucchino, C.; Naldi, M.; Nizzoli, D.; Taje, L. Nutrient and iron limitation to Ulva blooms in a eutrophic coastal lagoon (Sacca di Goro, Italy). *Hydrobiologia* **2005**, *550*, 57–71. [CrossRef]
10. Paerl, H.W.; Otten, T.G. Harmful Cyanobacterial Blooms: Causes, Consequences, and Controls. *Microb. Ecol.* **2013**, *4*, 995–1010. [CrossRef]
11. Cha, Y.; Park, S.S.; Kim, K.; Byeon, M.; Stow, C.A. Probabilistic prediction of cyanobacteria abundance in a Korean reservoir using a Bayesian Poisson model. *Water Resour. Res.* **2014**, *50*, 2518–2532. [CrossRef]
12. Newcombe, G.; House, J.; Ho, L.; Baker, P.; Burch, M. *Management Strategies for Cyanobacteria (Blue-Green Algae): A Guide for Water Utilities*; Research Report No. 74; Water Quality Research Australia: Adelaide, Australia, 2010; ISBN 18766 16245.
13. Zamyadi, A.; Choo, F.; Newcombe, G.; Stuetz, R.; Henderson, R.K. A review of monitoring technologies for real-time management of cyanobacteria: Recent advances and future direction. *Trends Anal. Chem.* **2016**, *85*, 83–96. [CrossRef]
14. Gamarro, E.G.; Englander, K. Joint FAO-IOC-IAEA technical guidance for the implementation of early warning systems for harmful algal blooms. *FAO Fish. Aquac. Tech. Pap.* **2023**, *690*, I-202.
15. Izydorczyk, K.; Carpentier, C.; Mrówczyński, J.; Wagenvoort, A.; Jurczak, T.; Tarczyńska, M. Establishment of an Alert Level Framework for cyanobacteria in drinking water resources by using the Algae Online Analyser for monitoring cyanobacterial chlorophyll a. *Water Res.* **2009**, *43*, 989–996. [CrossRef]
16. Park, Y.; Pyo, J.; Kwon, Y.S.; Cha, Y.; Lee, H.; Kang, T.; Cho, K.H. Evaluating physico-chemical influences on cyanobacterial blooms using hyperspectral images in inland water, Korea. *Water Res.* **2017**, *126*, 319–328. [CrossRef] [PubMed]
17. Shin, J.K.; Park, Y. Spatiotemporal and longitudinal variability of hydro-meteorology, Basic water quality and dominant algal assemblages in the eight weir pools of regulated river (Nakdong). *Korean J. Ecol. Environ.* **2018**, *51*, 268–286. [CrossRef]
18. Shiffrin, R.M. Drawing causal inference from big data. *Proc. Natl. Acad. Sci. USA* **2016**, *113*, 7308–7309. [CrossRef]
19. Coad, P.; Cathers, B.; Ball, J.E.; Kadluczka, R. Proactive management of estuarine algal blooms using an automated monitoring buoy coupled with an artificial neural network. *Environ. Model. Softw.* **2014**, *61*, 393–409. [CrossRef]
20. Park, Y.; Cho, K.H.; Park, J.; Cha, S.M.; Kim, J.H. Development of early-warning protocol for predicting chlorophyll-a concentration using machine learning models in freshwater and estuarine reservoirs, Korea. *Sci. Total Environ.* **2015**, *502*, 31–41. [CrossRef]
21. Fu, B.; Wu, B.; Lü, Y.; Xu, Z.; Cao, J.; Niu, D.; Yang, G.; Zhou, Y. Three gorges project: Efforts and challenges for the environment. *Prog. Phys. Geogr.* **2010**, *34*, 741–754. [CrossRef]
22. Shin, J.; Yoon, S.; Kim, Y.; Kim, T.; Go, B.; Cha, Y. Effects of class imbalance on resampling and ensemble learning for improved prediction of cyanobacteria blooms. *Ecol. Inform.* **2021**, *61*, 101202. [CrossRef]

23. Avila, R.; Horn, B.; Moriarty, E.; Hodson, R.; Moltchanova, E. Evaluating statistical model performance in water quality prediction. *J. Environ. Manag.* **2018**, *206*, 910–919. [CrossRef] [PubMed]
24. Chawla, N.V.; Japkowicz, N.; Kotcz, A. Editorial: Special issue on learning from imbalanced data sets. *Assoc. Comput. Mach.* **2004**, *6*, 1–6. [CrossRef]
25. Sun, Y.; Wong, A.K.C.; Kamel, M.S. Classification of imbalanced data: A review. *Int. J. Pattern Recognit. Artif. Intell.* **2009**, *23*, 687–719. [CrossRef]
26. Choi, J.; Kim, J.; Won, J.; Min, O. Modelling Chlorophyll-a concentration using deep neural networks considering extreme data imbalances and skewness. In Proceedings of the 2019 21st International Conference on Advanced Communication Technology (ICACT), Pyeongchang, Republic of Korea, 17–20 February 2019; pp. 631–634.
27. Jeong, B.; Chapeta, M.R.; Kim, M.; Kim, J.; Shin, J.; Cha, Y. Machine learning-based on prediction of harmful algal blooms in water supply reservoirs. *Water Qual. Res. J.* **2022**, *57*, 304–318. [CrossRef]
28. Bourel, M.; Segura, A.M.; Crisci, C.; López, G.; Sampognaro, L.; Vidal, V.; Kruk, C.; Piccini, C.; Perera, G. Machine learning methods for imbalanced data set for prediction of faecal contamination in beach waters. *Water Res.* **2021**, *202*, 117450. [CrossRef] [PubMed]
29. Cha, Y.J.; Shim, M.P.; Kim, S.K. The Four Major Rivers Restoration Project. In Proceedings of the Water in the Green Economy in Practice: Towards Rio+20, UN-Water International Conference, Zaragoza, Spain, 3–5 October 2011; pp. 1–10.
30. Kang, Y.J.; Lee, K.-L.; Im, T.H.; Lee, I.J.; Kim, S.; Han, K.-Y.; Ahn, J.M. Evaluation of water quality for the Nakdong River watershed using multivariate analysis. *Environ. Technol. Innov.* **2016**, *5*, 67–82.
31. Back, S.; Pyo, J.; Pachepsky, Y.; Park, Y.; Ligaray, M.; Ahn, C.; Kim, Y.; Chun, J.; Cho, K.H. Identification and enumeration of cyanobacteria species using a deep neural network. *Ecol. Indic.* **2020**, *115*, 106395. [CrossRef]
32. NIER (National Institute of Environmental Research). *Annual Report on Algae (Green Algae) Occurrence and Response*; Technical Report (11-1480000-001363-10); NIER: Incheon, Republic of Korea, 2021.
33. Croxton, F.E.; Cowden, D.J. *Applied General Statistics*; Prentice-Hall: Hoboken, NJ, USA, 1939.
34. Ross, B.C. Mutual information between discrete and continuous data sets. *PLoS ONE* **2014**, *9*, e87357. [CrossRef]
35. Xu, T.; Coco, G.; Neale, M. A predictive model of recreational water quality based on adaptive synthetic sampling algorithms and machine learning. *Water Res.* **2020**, *177*, 115788. [CrossRef]
36. Kim, J.H.; Shin, J.; Lee, H.; Lee, D.H.; Kang, J.; Cho, K.; Lee, Y.; Chon, K.; Baek, S.; Park, Y. Improving the performance of machine learning models for early warning of harmful algal blooms using an adaptive synthetic sampling method. *Water Res.* **2021**, *207*, 117821. [CrossRef] [PubMed]
37. Shelke, M.S.; Deshmukh, P.R.; Shandilya, V.K. A review on imbalanced data handling using undersampling and oversampling technique. *Int. J. Recent Trends Eng. Res.* **2017**, *3*, 444–449.
38. Liu, A.; Ghosh, J.; Martin, C. Generative Oversampling for Mining Imbalanced Datasets. *DMIN* **2007**, *7*, 66–72.
39. Tahir, M.A.U.H.; Asghar, S.; Manzoor, A.; Noor, M.A. A classification model for class imbalance dataset using genetic programming. *IEEE Access* **2019**, *7*, 71013–71037. [CrossRef]
40. Colton, D.; Hofmann, M. Sampling techniques to overcome class imbalance in a cyberbullying context. *J. Comput.-Assist. Linguist. Res.* **2019**, *3*, 21–40. [CrossRef]
41. Chatterjee, S.; Sarkar, S.; Dey, N.; Sen, S.; Goto, T.; Debnath, N.C. Water quality prediction: Multi objective genetic algorithm coupled artificial neural network based approach. In Proceedings of the 2017 IEEE 15th International Conference on Industrial Informatics (INDIN), Emden, Germany, 24–26 July 2017; pp. 963–968.
42. Mirzaei, M.; Jafari, A.; Gholamalifard, M.; Azadi, H.; Shooshtari, S.J.; Moghaddam, S.M.; Gebrehiwot, K.; Witlox, F. Mitigating environmental risks: Modeling the interaction of water quality parameters and land use cover. *Land Use Policy* **2020**, *95*, 103766. [CrossRef]
43. Forsberg, C.; Ryding, S.O. Eutrophication parameters and trophic state indices in 30 Swedish water-receiving lakes. *Arch. Hydrobiol.* **1980**, *89*, 189–207.
44. Elser, J.J.; Dobberfuhl, D.R.; MacKay, N.A.; Schampel, J.H. Organism size, life history, and N:P stoichiometry: Toward a unified view of cellular and ecosystem processes. *BioScience* **1996**, *46*, 674–684. [CrossRef]
45. Peñuelas, J.; Poulter, B.; Sardans, J.; Ciais, P.; Van Der Velde, M.; Bopp, L.; Janssens, I.A. Human-induced nitrogen-phosphorus imbalances alter natural and managed ecosystems across the globe. *Nat. Commun.* **2013**, *4*, 2934. [CrossRef]
46. Carlson, R.E. A trophic state index for lakes 1. *Limnol. Oceanogr.* **1977**, *22*, 361–369. [CrossRef]
47. Wong, K.T.M.; Lee, J.H.W.; Hodgkiss, I.J. A simple model for forecast of coastal algal blooms. *Estuar. Coast. Shelf Sci.* **2007**, *74*, 175–196. [CrossRef]
48. Wong, K.T.M.; Lee, J.H.W.; Harrison, P.J. Forecasting of environmental risk maps of coastal algal blooms. *Harmful Algae* **2009**, *8*, 407–420. [CrossRef]
49. Zhang, Y.; Qin, B.; Zhu, G.; Shi, K.; Zhou, Y. Profound changes in the physical environment of Lake Taihu from 25 years of long-term observation: Implications for algal bloom outbreaks and aquatic macrophyte loss. *Water Resour. Res.* **2018**, *54*, 4319–4331. [CrossRef]
50. Jie, C.; Jiawei, L.; Shulin, W.; Sheng, Y. Feature selection in machine learning: A new perspective. *Neurocomputing* **2018**, *300*, 70–79.
51. Thabtah, F.; Hammoud, S.; Kamalov, F.; Gonsalves, A. Data imbalance in classification: Experimental evaluation. *Inf. Sci.* **2020**, *513*, 429–441. [CrossRef]

52. Chawla, N.V. Data mining for imbalanced datasets: An overview. In *Data Mining and Knowledge Discovery Handbook*; Springer: Berlin/Heidelberg, Germany, 2010; pp. 875–886.
53. Fernández-Navarro, F.; Hervás-Martínez, C.; Gutiérrez, P.A. A dynamic over-sampling procedure based on sensitivity for multi-class problems. *Pattern Recognit.* **2011**, *44*, 1821–1833. [CrossRef]
54. Tanha, J.; Abdi, Y.; Samadi, N.; Razzaghi, N.; Asadpour, M. Boosting methods for multi-class imbalanced data classification: An experimental review. *J. Big Data* **2020**, *7*, 70. [CrossRef]
55. Guyon, I.; Elisseeff, A. An introduction to variable and feature selection. *J. Mach. Learn. Res.* **2003**, *3*, 1157–1182.
56. Chandrashekar, G.; Sahin, F. A survey on feature selection methods. *Comput. Electr. Eng.* **2014**, *40*, 16–28. [CrossRef]
57. Bolón-Canedo, V.; Sánchez-Marono, N.; Alonso-Betanzos, A.; Benítez, J.M.; Herrera, F. A review of microarray datasets and applied feature selection methods. *Inf. Sci.* **2014**, *282*, 111–135. [CrossRef]
58. Wei, G.; Zhao, J.; Feng, Y.; He, A.; Yu, J. A novel hybrid feature selection method based on dynamic feature importance. *Appl. Soft Comput.* **2020**, *93*, 106337. [CrossRef]
59. Xue, B.; Zhang, M.; Browne, W.N. Particle swarm optimisation for feature selection in classification: Novel initialisation and updating mechanisms. *Appl. Soft Comput.* **2014**, *18*, 261–276. [CrossRef]

toxics

MDPI

Article

Evaluating the Contamination by Indoor Dust in Dubai

Yousef Nazzal [1], Alina Bărbulescu [2,*], Manish Sharma [1], Fares Howari [3] and Muhammad Naseem [1]

1 College of Natural and Health Sciences, Zayed University, Abu Dhabi P.O. Box 144534, United Arab Emirates; yousef.nazzal@zu.ac.ae (Y.N.); manish.sharma@zu.ac.ae (M.S.); muhammad.naseem@zu.ac.ae (M.N.)
2 Department of Civil Engineering, Transilvania University of Brașov, 5 Turnului Str., 900152 Brasov, Romania
3 College of Arts and Sciences, Fort Valley State University, Fort Valley, GA 31030, USA; Fares.Howari@fvsu.edu
* Correspondence: alina.barbulescu@unitbv.ro

Abstract: Nowadays, people spend most of their time indoors. Despite constantly cleaning these spaces, dust apparition cannot be avoided. Since dust can contain chemical elements that negatively impact people's health, we propose the analysis of the metals from the indoor dust component collected in different locations in Dubai, UAE. Multivariate statistics (correlation matrix, clustering) and quality indicators (QI)—I_{geo}, PI, EF, PLI, Nemerow—were used to assess the contamination level with different metals in the dust. We proposed two new QIs (CPI and AQI) and compared the results with those provided by the most used indices—PLI and Nemerow. It is shown that high concentrations of some elements (Ca in this case) can significantly increase the values of the Nemerow index, CPI, and AQI. In contrast, the existence of low concentrations leads to the decrement of the PLI.

Keywords: contamination; dust; clustering; pollution index

check for updates

Citation: Nazzal, Y.; Bărbulescu, A.; Sharma, M.; Howari, F.; Naseem, M. Evaluating the Contamination by Indoor Dust in Dubai. *Toxics* **2023**, *11*, 933. https://doi.org/10.3390/toxics11110933

Academic Editor: João Fernando Pereira Gomes

Received: 29 October 2023
Revised: 13 November 2023
Accepted: 14 November 2023
Published: 17 November 2023

1. Introduction

Indoor dust is the settled particulate matter (PM) found on carpets, floors, surfaces, and other objects in an indoor space. Among other pollutants from indoor dust, heavy metals require extensive research due to their non-degradable properties, high toxicity, and adverse effects on humans [1,2]. The United States Environmental Protection Agency (USEPA) has raised the alarm about indoor air quality, considering it a significant concern because it tends to be more polluted than outdoor air. This concern has grown because people spend a significant portion of their time indoors, encompassing homes, workplaces, schools, public spaces like shops, restaurants, and vehicles, amounting to up to 90% of their daily activities [3]. Children, who spend most of their day at home, are particularly vulnerable to environmental stressors because their breathing zone is close to the floor, where residential dust tends to collect, exposing them to potential health risks [4–6].

Carbon dioxide, volatile organic compounds, biocontaminants, fungi, bacteria, and particulate matters are among the indoor air pollutants with damaging potential to human health listed by the European Federation of Allergy and Airway Diseases Patient Associations in their document [7]. Dust intake rates for children are estimated to be between 30 and 140 mg/day, whereas adults consume 2–30 mg/day [8,9].

According to [10,11], indoor dust can be described as tiny particles ($\leq$100 μm) that settle in indoor spaces. These particles can come from various sources situated inside and outside the building. Particles with diameters smaller than 10 μm (PM_{10}) can be inhaled, the coarse fractions being retained in the upper airways, and those particles with diameters less than 2.5 μm can reach the pulmonary system or enter the blood [12]. Particles with diameters from 1 μm to 20 μm are responsible for the apparition of asthma [13]. Tsubata et al. [14] indicate that dust particles with diameters less than 11 μm contain up to 90% of allergens.

Research has indicated that indoor dust is a transporter for inorganic and organic contaminants, including heavy metals, pesticides, polychlorobiphenyls, and polycyclic aromatic hydrocarbons [5,6,15–17]. Indoor dust is a heterogeneous combination of particles that includes synthetic and natural fibers, hair, deposited atmospheric PM, biologically derived material (pollen, molds, bacteria, germs, animal fur, and dander), ash, skin particles, soot, and building and consumer product components [18]. Indoor dust typically contains about 35% outdoor soil, but this can vary widely based on factors like pets, shoe-wearing habits, and specific indoor settings. Indoor dust varies in organic content, typically ranging from 5% to 40%. Finer particles contain more organics, which are vital for absorbing pollutants. The fibrous particle content ranges from 9% to 89%, influenced by room type, furniture, and pet presence [19,20].

Pollutants enter the human body by inhalation, ingestion, and dermal contact [6,21–24]. According to [25], when inhaled, these toxic metals in dust can inflame, sensitize, and even scar the lungs and tissues because they are ubiquitous in the environment. Additionally, exposure to these metals may result in gastrointestinal issues, reproductive system problems, and nervous system disorders. Excessive exposure to Pb, Cd, Zn, and Cu is associated with the risk of cancer [26,27]. In this article, we analyze only the toxic metal found in indoor dust, whereas the dust microbiomes and metatranscriptomes have been studied in [28].

Bio-accessibility of heavy metals in indoor dust has been observed by physiologically based extraction tests or simplified bio-accessibility extraction tests based on the rationale that incidental oral ingestion is the main exposure pathway by which humans take in contaminants in indoor dust, especially for children [29–31].

Indoor air pollution poses a significant global health threat, contributing to around 4.5 million annual deaths worldwide. This pollution is responsible for a range of health issues, including pneumonia (12%), strokes (34%), ischemic heart diseases (26%), chronic obstructive pulmonary diseases (22%), and lung cancer (6%) [32,33]. Therefore, research on indoor air quality concluded that correct ventilation and proper cleaning [34,35] are necessary to avoid such health damage.

The International Agency for Research on Cancer (IARC) has classified Al, Co, Fe, Ni, and Zn as non-carcinogenic elements, whereas arsenic As, Cu, Cd, Cr, and Pb are classified as both carcinogenic and non-carcinogenic elements. The U.S. Environmental Protection Agency classified Cu, Cr, Ni, Zn, Cd, Mn, and Pb as environmental priority pollutants [36]. Moreover, it was shown that Cr, Cu, Ni, Zn, and Fe promote the exchange of electrons [34] and help the apparition of reactive oxygen species in the lungs [37].

On one hand, Cu is a micronutrient, a catalyzer of redox reactions, essential for the organism functioning. On the other hand, released in the atmosphere from anthropic (burning fossil fuel, solid waste management) and natural sources, it can attach to particulate matter and is transported long distances from its source [38].

Heavy metals like As, Cd, Cr, and Pb, which are widespread environmental pollutants, can cause health issues, including cancers, respiratory problems, cardiovascular diseases, nerve damage, and slow growth development [39–42].

Different particulate matter can also contain other elements like Ca, Li, and K transported by the wind, issued from the lithology of the place being studied.

This article presents the analysis of indoor settled dust in Dubai, UAE, which holds significant importance since Dubai's rapid urban development and construction activities are closely linked to indoor dust accumulation. Although studies on dust transportation and outdoor pollution (particularly with heavy metals) in different emirates from the UAE have been carried out [43–47], indoor pollution was less analyzed [28,48–50], with the emphasis on gaseous pollutants. Therefore, in this study, the composition of indoor settled dust from 20 important locations across Dubai is investigated using a complex approach involving a multivariate statistical analysis combined with different indices, two newly proposed here. It is shown that a correct conclusion on contamination cannot be drawn from a single index computation but from a combination of such indices, given that some elements present in high concentrations in the samples can have a significant influence on the classification. Moreover, comparisons of the clustering based on the row data and the quality indicators may highlight the differences between the sites where the samples were collected.

2. Materials and Methods

2.1. Data Series

Dubai, located in the United Arab Emirates (UAE) is a remarkable city known for its unique blend of modernity and tradition. Situated on the Southeastern coast of the Arabian Peninsula (Figure 1), Dubai is one of the most prominent global cities, attracting tourists and business professionals from all over the world. It is bordered by the emirate of Sharjah to its north, while Abu Dhabi, the UAE's capital, lies to the south. The climate of the study area is characteristic of the Arabian Peninsula, with hot and arid conditions prevailing throughout the year. Summers are exceedingly hot, with temperatures often exceeding 40 °C (104 °F). The city receives limited rainfall, and as a result, Dubai's terrain is primarily desert, characterized by rolling dunes and sparse vegetation.

Figure 1. Study area location and sampling map. The red points and the numbers represent the sampling points and their IDs.

2.2. Sampling

Indoor-settled dust samples were collected using Dyson filters from twenty different locations in Dubai Emirates (Figure 1) including residential areas (Al Simmak Street, Bijada Blvd Street, Tulip Street), near heavy traffic junctions (Sheikh Zayed Highway), sports facilities (Sports City, Victory Heights), touristic areas (bars, restaurants), near water bodies (Dubai Marina) and commercial areas (markets, beauty lounges, butchers' shops). Additionally, samples were also taken from specific buildings from Al Mustaqbal Street, Sheikh Mohammed Bin Rashid Blvd, a roundabout in Motor City, and near metro stations, offering a diverse range of environmental sources for analysis. The buildings' characteristics differed, varying from the location, building materials, purpose, and maintenance. The dust samples were collected from undisturbed surfaces. Before sampling, the sites' environmental conditions—temperature and humidity—were measured using a Graywolf Indoor Air Quality Meter (GrayWolf Sensing Solutions, LLC, Shelton, CT, USA) [51]. The measurements were performed when the atmospheric conditions were stable. The temperature inside was between 19 and 20 °C, and the relative humidity (RH%) was in the range of 40–45%. The coordinates of the observation sites were recorded using a South S750 Handheld GPS meter (Guangzhou, China) [52].

A Dyson V15 Detect vacuum machine with two heads (Gurugram, India) (separately collecting dust particles from rugs/carpets with a fluffy brush-bar and filter, and hard floors with a built-in laser light to observe the incoming material from the cleaning surface) was utilized. The Dyson vacuum has a HEPA post-motor filter that can trap particles with dimensions at least of 0.1 microns. Moreover, the dust particles are continuously counted and sized by a piezo-sensor [53].

A representative sampling strategy was adopted to collect the samples, which were transferred into re-sealable plastic bags by gently sweeping with fingers wearing powder-free nitrile gloves. They were safely packed and moved to the laboratory, where they were screened to remove any visible hair, soil, and grit. The samples were then air-dried for 48 h to avoid moisture in a well-protected area. All the results were reported based on dry weight.

2.3. Reagents, Standards and Laboratory Ware

In this research, all experiments were conducted using high-quality analytical reagent (AR) grade chemicals. We sourced the reference standard, check standard, and reagents from Sigma Aldrich (St. Louis, MO, USA). To create a 1:1 acid mixture, concentrated nitric acid (69% v/v) and hydrochloric acid (37% v/v) were combined. The water purity was ensured by using ultra-pure water with a chemical resistivity of 18.2 MΩ·cm from the Merck Millipore(Burlington, MA, USA) water purification system. For sample oxidation, 30% hydrogen peroxide was utilized. The equipment quality was maintained by using Class-A grade glassware for all the analyses. To eliminate potential contaminants, all items of glassware and plasticware were cleaned by washing them 5–6 times with ultra-pure water, and rinsing with 10% nitric acid, then drying them with an air drier. Later, sample digestion was carried out using the Mars-6 system from CEM in Matthews, NC, USA. Finally, ICP-OES analysis was conducted using OH, USA's Perkin Elmer Avio 200 system.

The sample digestion process followed the USEPA 3050B procedure (Washington, DC, USA) [54]. Initially, 0.2 g of each sample was weighed and placed into Teflon vessels for microwave-assisted digestion. Subsequently, 10 mL of a 1:1 HCl: HNO_3 solution was added into the digestion vessel, thoroughly mixed with the sample slurry, and subjected to microwave digestion at 95 °C for 5 min. After digestion, the slurry was allowed to cool, and 5 mL of concentrated HNO_3 was added. This mixture was then heated and refluxed at 95 °C for 5 min, followed by cooling and carefully adding 10% H_2O_2 for oxidation. The resulting solutions were transferred into 100 mL volumetric flasks, adjusted to the markup with water, and subsequently filtered using Whatman 41 filters (Maidstone, UK). The filtered solutions were subsequently subjected to analysis for heavy metals using an ICP-OES system, with eight replicate analyses conducted for each sample.

Quality control and assurance protocols were carefully observed throughout the sample preparation and analysis processes, encompassing laboratory blanks, check standards, and standard spiked samples. Laboratory blanks were prepared utilizing the same reagents employed for digestion but excluding the addition of dust samples. For all metals, the laboratory blank values were under the concentrations of metals in the target samples. The method detection limit (MDL) was calculated using the equation:

$$MDL = X + 2.896 \times SD \quad (1)$$

where X is the mean, SD is the standard deviation of blanks, and 2.896 is the value of the Student statistics at the significance level of 99%, and eight degrees of freedom. This equation has been used according to [55,56] because all the method blanks give either positive or negative numerical results. The MDL values ranged between 0.02 μg/kg (Cd) and 25.2 μg/kg (K). The metals recovery percentage (spiked and standard) was between 95% and 105%. The analytical precision for every metal of repeated analysis was determined by using the coefficient of variation, which was less than 3%.

2.4. Statistical Analysis

The first step in the analysis was the computation of the basic statistics—minimum (min), maximum (max), mean, median, standard deviation (std.dev.), coefficient of variation (CV), skewness coefficient, and kurtosis. The correlation matrix was determined to assess the correlation between the chemical elements in the dust.

After normalizing the data series, the set was submitted to clustering to group the 20 series recorded at different sites according to their common properties. For a better classification, the k-means algorithm [57] and hierarchical clustering [58] were used to cross-validate the results. Before performing the algorithms, the elbow [59] and silhouette [60] methods were utilized to choose the optimum number of clusters, k.

Groups of series formed the output of the first technique, while that of the second one was a dendrogram that shows the series hierarchy and can be constructed by employing a certain distance, like the Euclidean one (utilized in this study). The degree of similarity between the elements in each group was estimated using different methods like "complete", "average", "ward.D2", and "median". The better-performing method was selected based on the highest value of the cophenetic correlation coefficient [61]. After clustering, bootstrapping was conducted to compute the average Jaccard measures, to ensure that the algorithm provided a good representation of the groups. A value of the Jaccard coefficient greater than 0.85 indicates a highly stable clustering, whereas one between 0.60 and 0.85 shows a stable grouping [62].

The next stage was to perform the Principal Component Analysis [63]. PCA is a multivariate statistical technique utilized for reducing the number of the observed parameters by replacing them with a smaller number of components, artificially created, called Principal Components (PC). The extracted PCs incorporate the highest part of the variance of raw parameters (usually above 80%) and are obtained as a linear combination of those parameters [64]. They can be considered independent factors that govern the development of a given process [65]. Among the criteria employed for the PC selection—Explained Variance Criterion [64,65], Catell Scree Plot [66], and Kaiser criterion [67]—the first two were utilized in this research.

The R 4.3.1 software (https://cran.r-project.org/, accessed on 15 October 2023) was the tool for performing the analysis.

2.5. Pollution Indices

To assess the pollution level or enrichment with the metals in the dust, the following indices were computed. They are:

For the metal i, I_{geo} is calculated using the formula [68–70]:

$$I_{geo} = log_2(C_i/(1.5CB_i)), \quad (2)$$

where C_i is the concentration of the i-th element in the dust and CB_i is the value of the i-th element in the background.

The pollution index of the j-th element is given by [68]:

$$PI_j = C_j/CB_j. \tag{3}$$

Values of PI in the intervals less than 1, 1–2, 2–3, 3–5, and greater than 5, respectively, indicate the contamination absence, low, moderate, strong, and very strong pollution, respectively.

The enrichment factor with the j-th element, EF_j, is defined by [68–70]:

$$EF_j = [C_j/LV_s]/[CB_j/LV_b] \tag{4}$$

where C_j is the concentration of the element j in the sample, LV_s is the concentration of the reference element (generally Al, Ca, or Fe) in the sample, CB_j is the reference concentration of j-th element in the background, and LV_b is the concentration of the reference element in the background.

The background values utilized here are those from [71]. Same information can be found in [72] for different regions of the world.

Based on the value of the EF factor—less than 2, between 2 and 5, in the interval 5–20, between 20 and 40, or greater than 40—different classes of pollution are defined as deficient to minimal, moderate, significant, very high, and extremely high, respectively.

Aggregated indices can be computed from the individual ones to assess the contamination with multiple elements at a specific location. Two known indices were computed. The first one is PLI, defined by [73]:

$$PLI = \left(\prod_{j=1}^{n} PI_j\right)^{1/n}. \tag{5}$$

PIs of some elements (As, Ba, Co, Pb, in this case) are very low (of order 10^{-2}), so they will artificially decrease the PLI value. Therefore, to have a correct evaluation of the contamination degree, the PIs corresponding to these elements were removed from the computation of the PLI, the resulting index, denoted by PLI_d, being also computed and compared with PLI.

The second one is the Nemerow index, calculated by [74]:

$$PI_{Nem} = \sqrt{\left[\overline{PI}^2 + PI_{max}^2\right]/2} \tag{6}$$

with

$$PI_{max} = \max(PI_1, \ldots, PI_n) \text{ and } \overline{PI} = \left(\sum_{j=1}^{n} PI_j\right)/n. \tag{7}$$

Values less than 0.7, in the intervals 0.7–1, 1–2, 2–3, and higher than 3 are indicative of the absence of pollution, warning level, slight contamination, moderate pollution, and heavy contamination, respectively.

Two new indices are proposed, analogous to those used in water pollution assessment [75,76]. The first one, called in the following Combined Pollution Index (CPI), is defined by the formula:

$$CPI = \frac{1}{n}\sum_{j=1}^{n} (C_j/CB_j). \tag{8}$$

We propose to keep as reference values those for PI_{Nem}.

The arithmetic weighted index is defined by:

$$AQI = \left(\sum_{j=1}^{n} w_j Q_j\right)/\left(\sum_{j=1}^{n} w_j\right), \tag{9}$$

with w_j the weight associated with the quality index Q_j of ith parameter,

$$Q_j = 100 \times C_j/CB_j, \tag{10}$$

$$w_j = \frac{1}{CB_j} \Big/ \left(\sum_{j=1}^{n} \frac{1}{CB_j} \right). \quad (11)$$

The following classes are associated with the ranges (0–25)—unpolluted, (26–50)—warning level, (51–75)—slight pollution, (76–100)—moderate pollution, and (above 100)—heavy pollution.

3. Results and Discussion

Table 1 contains the basic statistics of the chemical elements series from the samples. The highest concentrations are those of Ca, K, Mg, Al, and Fe, and the lowest are those of Co, As, and Pb. Standard deviations (std.dev.) of most series of elements are high, indicating a high variation around the mean, but the variation coefficients are moderate. Only a few series present an accentuated skewness (Cr, Ba, Na, Mg), indicating a large variation range of the corresponding values.

Table 1. Basic statistics of the series of elements from the dust samples [mg/kg].

	Cu	Ni	Pb	Zn	Co	Cr	Ba	Fe	Mn
min	3.04	29.85	0.05	25.81	0.16	19.14	28.96	568.36	66.38
mean	94.30	52.14	4.62	247.08	1.85	56.96	85.17	997.28	126.58
max	309.58	93.50	28.82	397.11	3.62	298.47	309.94	1572.14	186.24
median	53.54	47.10	2.37	255.84	1.91	33.27	74.36	979.05	133.08
Std.dev.	97.74	18.80	6.72	92.52	0.95	63.60	55.28	263.67	43.47
CV	1.04	0.36	1.46	0.37	0.52	1.12	0.65	0.26	0.34
Skewness coef.	1.53	0.72	2.86	−0.57	0.04	3.06	3.54	0.40	−0.04
Kurtosis	0.65	−0.57	8.47	0.10	−0.76	10.31	14.39	−0.03	−1.71
	Mg	**Sr**	**Na**	**Al**	**Ca**	**K**	**As**	**Cd**	
min	834.32	11.44	188.15	349.33	8033.17	3918.61	0.64	6.26	
mean	1876.22	47.50	561.39	1033.78	14,170.02	9159.12	3.89	6.73	
max	4972.55	120.35	1606.82	1883.38	20,421.29	17,984.38	5.61	7.45	
median	1843.27	44.64	493.99	965.54	14,436.28	8661.69	4.26	6.68	
Std.dev.	928.09	23.21	295.98	391.79	3085.02	2873.00	1.41	0.32	
CV	0.49	0.49	0.53	0.38	0.22	0.31	0.36	0.05	
Skewness coef	1.86	1.43	2.43	0.70	−0.36	1.28	−0.85	0.46	
Kurtosis	5.10	3.52	7.29	0.60	0.04	3.52	−0.01	−0.43	

The high concentrations of Cu, Mg, Fe, and Al in the dust might be explained by their existence in the natural rocks and anthropic activity. For example, there are 120 known occurrences of copper mineralization in the United Arab Emirates, situated in the mountainous region between Kalba and Dibba, or Wadi Hamm [77]. UAE is the seventh exporter of Mg in the world [78] and exported USD 53.6 M in iron ore in 2021 [79]. Moreover, it is the fifth aluminum-producing country in the world [80].

Studies indicate that indoor air quality is significantly affected by the outdoor air [81–85]. Kuo and Shen [83] found a similar increase in the concentrations of $PM_{2.5}$ and PM_{10} in both indoor and outdoor air during a dust-storm event and interpreted the cause to be the extraction of outdoor air from their building's ventilation system. The research of Ai and Mak [86] and Meier et al. [87] has shown that natural ventilation contributes to the deterioration of indoor air quality. Fisk [13] has found that the air in mechanically ventilated buildings enters from a small number of intakes so that the indoor air quality is significantly affected by the intakes' neighboring sources situated outdoors. An extended review of the research on the correlation between indoor and outdoor air quality was performed in [88]. Therefore, in the case study, the high concentration of Mg, Fe, and Al from the indoor dust (highly correlated to that from outdoors), originates from the soil dust composition of a desert area, but one cannot ignore the contribution from industrial activities. The above-mentioned mining operations can introduce additional concentrations of minerals like Mg, Fe, and Al into the environment, and dust storms, frequent in the region, can transport these minerals over broader areas. To assess the minerals' origin in the indoor dust, samples should be analyzed in future studies.

Figure 2 presents the correlation matrix. The colors closer to red indicate a higher positive correlation between elements, and those closer to dark blue show a higher negative correlation.

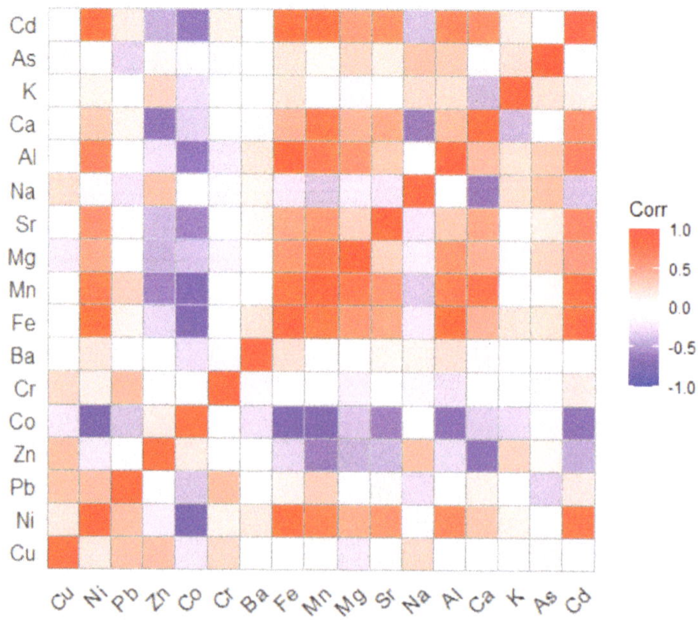

Figure 2. The correlation matrix. The higher the positive correlation, the more intense the nuance of red. The higher the negative correlation, the more intense the nuance of blue is. The nuances of light yellow, light orange, and indigo indicate a low or inexistent correlation.

Table 2 contains the p-values associated with the correlations between the chemical elements in the dust samples. The p-values less than 0.05 indicate a correlation between the elements. The lower the p-value, the higher the correlation is. Significant correlations are between the pairs Co–Ni, Fe–Ni, Mn–Ni, Mn–Mg, Mn–Sr, Mn–Al, Mn–Cd, Zn–Mg, Zn–Ca, Co–Fe, Co–Mn, Co–Sr, Co–Al, Co–Cd, Fe–Mn, Fe–Mg, Fe–Sr, Fe–Al, Fe–Cd, etc. This means that significant correlations are found between the metals in the dust resulting mainly from industrial activities and transported for long distances by the wind.

The optimal number of clusters, *k*, determined by the elbow and silhouette (Figure 3) was two (Figure 3).

After bootstrapping, the calculated average Jaccard values were 0.983 and 0.980, and the corresponding instabilities were 0.005 and 0.014. So, the groups found are highly stable. The first cluster contains the samples collected mainly from Dubai downtown, Burj Khalifa, near crowded zones, and in the vicinity of sandy zones. The second one is formed mainly by locations situated near the seafront, in green zones, and residential areas. The sampling series from the first cluster mainly contains the highest Pb, Zn, and Co concentrations and the lowest concentrations of Ni, Mn, and Mg.

Table 2. p-values related to the correlation coefficients of the elements found in the dust. The shaded cells, containing p-values less than 0.05 (the level of significance), indicate the existence of a significant correlation between the elements from the line and columns that intersect at that cell.

	Cu	Ni	Pb	Zn	Co	Cr	Ba	Fe	Mn	Mg	Sr	Na	Al	Ca	K	As
Ni	0.465															
Pb	0.118	0.104														
Zn	0.105	0.484	0.945													
Co	0.386	0.000	0.153	0.537												
Cr	0.280	0.546	0.101	0.930	0.825											
Ba	0.936	0.446	0.869	0.966	0.300	0.705										
Fe	0.961	0.000	0.605	0.267	0.000	0.955	0.400									
Mn	0.817	0.000	0.226	0.011	0.000	0.862	0.881	0.000								
Mg	0.391	0.035	0.867	0.093	0.146	0.495	0.833	0.012	0.000							
Sr	0.882	0.007	0.763	0.106	0.009	0.924	0.637	0.027	0.011	0.272						
Na	0.285	0.734	0.314	0.109	0.822	0.676	0.590	0.433	0.164	0.469	0.401					
Al	0.937	0.001	0.949	0.302	0.002	0.351	0.387	0.000	0.001	0.011	0.168	0.838				
Ca	0.963	0.159	0.587	0.002	0.202	0.824	0.783	0.068	0.000	0.066	0.030	0.005	0.100			
K	0.880	0.596	0.924	0.257	0.276	0.923	0.957	0.396	0.847	0.717	0.978	0.336	0.372	0.108		
As	0.784	0.816	0.200	0.671	0.705	0.739	0.909	0.416	0.700	0.265	0.516	0.116	0.188	0.804	0.389	
Cd	0.816	0.000	0.461	0.076	0.002	0.493	0.890	0.000	0.000	0.016	0.003	0.125	0.001	0.004	0.493	0.829

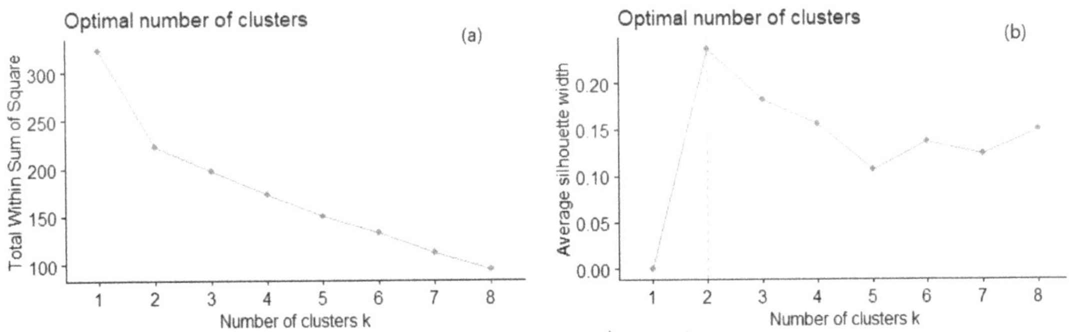

Figure 3. (**a**) Elbow and (**b**) silhouette methods for selecting the number of clusters.

The clusters obtained by the k-means algorithm (k = 2) are presented in Figure 4a. The dissimilarities between the elements in two clusters, in the hierarchical clustering, were assessed by different methods, among which "average" best performed in terms of cophenetic correlation coefficient (which was the highest compared to those of "complete", "average", "ward.D2", and "median" procedures). In this method, all pairwise dissimilarities between the elements in two clusters were computed, and the distance between clusters was calculated by averaging these dissimilarities.

After bootstrapping, the obtained average Jaccard values (instabilities) were 0.828 (0.146) and 0.826 (0.172), showing that the clusters are stable. The dendrogram resulting from the hierarchical clustering is displayed in Figure 4b. Comparing Figure 4a,b, one may observe that both methods provided the same clusters.

PCA found 17 PCs, corresponding to the same number of chemical elements. However, Table 3 provides the computation results of only five PCs, including the proportion of the variance explained by each component, the cumulative proportion, and the standard deviation. The first two (three) PCs explain 80.90% (89.5%) of the variance. So, PC1 explains more than two-thirds of the information provided by the 17 variables, whereas PC2 and PC3 explain, respectively, 11.53% and 8.58% of the total variance.

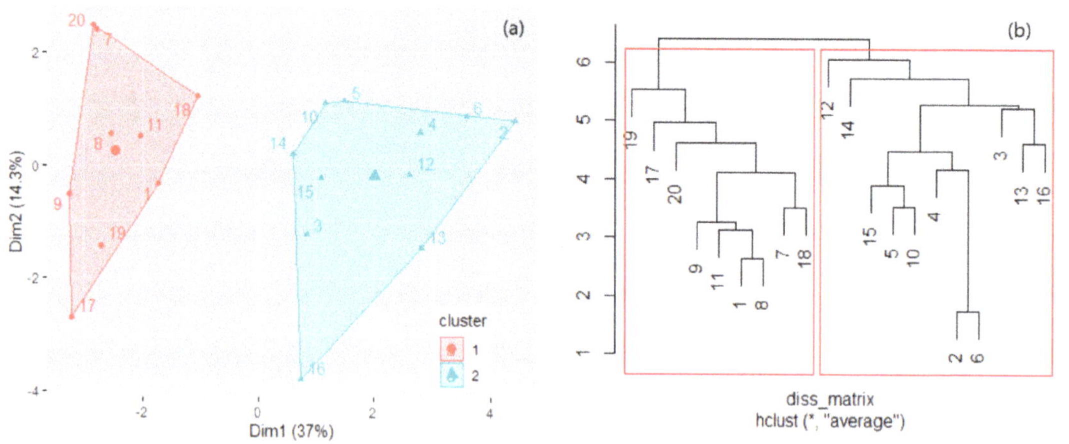

Figure 4. (**a**) The clusters found by k-means with $k = 2$; (**b**) Dendrogram in the hierarchical clustering.

Table 3. PCA.

	PC1	PC2	PC3	PC4	PC5
Standard deviation	1.314	0.536	0.463	0.271	0.249
Proportion of Variance	0.6935	0.1152	0.0858	0.0294	0.0249
Cumulative Proportion	0.6935	0.8087	0.8946	0.9241	0.9489

The cumulative proportion of PC1–PC3 is about 89.46% of the total variance. So, PC1–PC3 (or even only PC1 and PC2) can accurately represent the data set. The screen plot that reflects this information is shown in Figure 5a.

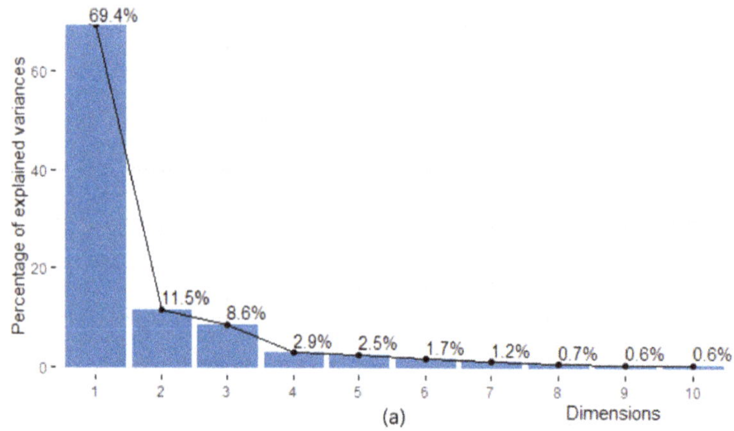

	PC1	PC2	PC3
Cu	0.0599	0.0426	0.4419
Ni	-0.3094	-0.0997	0.2735
Pb	-0.0750	0.2921	0.4816
Zn	0.2542	-0.2197	0.3180
Co	0.2910	0.1926	-0.3095
Cr	0.0338	0.3094	0.3008
Ba	-0.0278	-0.1935	0.0351
Fe	-0.3381	-0.1824	0.0953
Mn	-0.3807	0.0375	-0.0086
Mg	-0.2661	-0.0880	-0.2714
Sr	-0.2667	-0.0003	0.0282
Na	0.1858	-0.4260	0.0850
Al	-0.2902	-0.3033	-0.0215
Ca	-0.3025	0.3189	-0.1832
K	0.0235	-0.3627	0.2110
As	-0.0015	-0.3672	-0.1877

(b)

Figure 5. (**a**) The screen plot and (**b**) loading table.

The PC score (factor loading) of each variable in a PC indicates the processes controlling the variability of the data [89]. The loading table (Figure 5b) shows that the first principal component has high positive values for Co and Na. The values for Mg, Cd, Ca, and Ni are negative. This suggests that sites with a component of Co and Na in the dust are in excess. In PC2, Ca, Cr, and Pb are in excess, while the negative contributions come from Na, As, K, and Al. The highest contributions on PC3 are of Pb, Cu, Zn, Cr, and Ni. Therefore, the main contributions are those of Cr, Cu, Zn, Pb, Ni, and As, resulting mainly from human activities (transportation and industry). The variables' quality representation on the factors map (cos2 representation) is shown in Figure 6. The better the representation, the higher the cos2 is. So, the groups (Mn, Cd, Fe, and Ca), (Pb, Na, K, and Cr), and (Mn, Cd, Fe, and Ni) are, respectively, the best represented on the first three PCs. The variables' contributions in different dimensions are also represented in Figure 7.

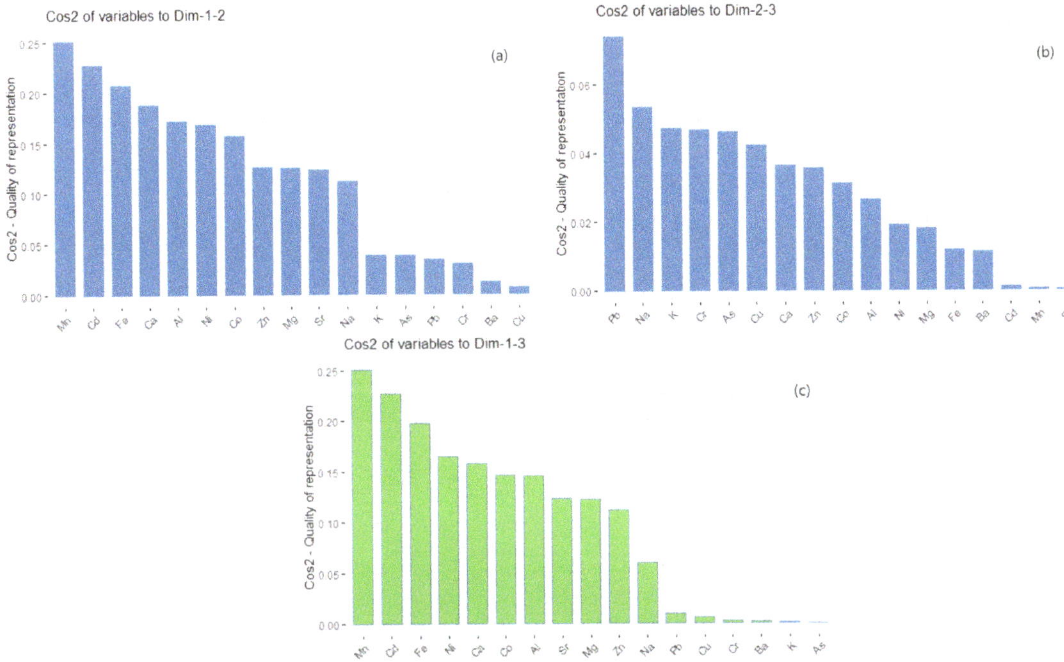

Figure 6. (**a**) Cos2 of variables to Dim 1–2; (**b**) Cos2 of variables to Dim 2–3; (**c**) Cos2 of variables to Dim 1–3.

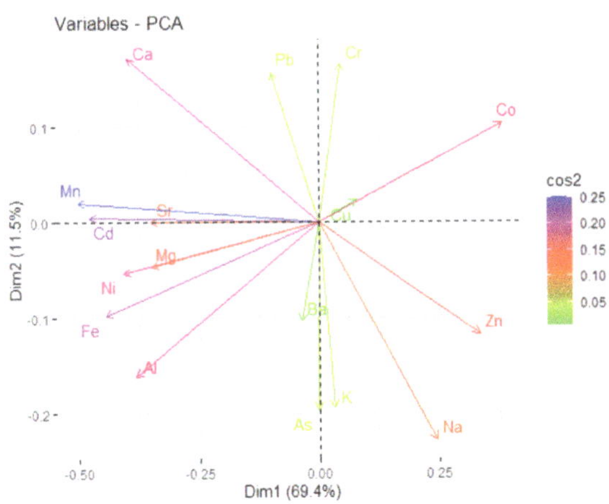

Figure 7. Variables—PCA.

The highest absolute values on PC1 are represented in nuances of blue. They are Mn, Cd, Fe, and Ca. Note that Mn, Cd, and Sr are grouped, indicating their correlation. The same remark stands for Mg, Ni, and Fe. Co is negatively correlated with Fe and Al; the same remark for Na and Pb, etc.

The contamination levels with respect to the I_{geo} values from the literature and the degree of contamination at the studied sites are presented in Table 4. The elements with significant impacts at all sites are Fe, Mg, Ca, and K.

Table 4. I_{geo} values, corresponding contamination levels, and the sites included in each class [90].

Igeo Class	Igeo Value	Contamination Level	Contamination Level at the Study Sites
0	$I_{geo} \leq 0$	Uncontaminated	Cu, Ni, Pb, Co, Cr, Ba, Sr, Mn, Sa Cd—all sitesZn: 1–6, 12–14, 16–19; Na: 1–6, 8–19; Al: 9–19, As, Cd
1	$0 < I_{geo} < 1$	Uncontaminated/Moderately contaminated	Zn: 7–11, 15, 20; Na: 7, 20; Fe: 7, 9, 17; Mg: 17; Al: 1, 3, 7, 8, 10–18, 20
2	$1 \leq I_{geo} < 2$	Moderately contaminated	Fe: 1, 3–5, 8, 10–16, 18–20; Mg: 1, 7–9, 11, 18–20; Al: 2, 4–6; K: 17
3	$2 \leq I_{geo} < 3$	Moderately/Strongly contaminated	Fe: 2, 6; Mg: 2, 4–6, 10, 12–16; K: 1, 3, 4, 8, 9, 11–13, 15, 20
4	$3 \leq I_{geo} < 4$	Strongly contaminated	Mg: 3; K: 2, 5, 7, 9, 15, 18, 19;
5	$4 \leq I_{geo} < 5$	Strongly/Extremely contaminated	Ca: 1, 6–11, 14, 15, 17–20
6	$I_{geo} \geq 5$	Extremely contaminated	Ca: 2–5, 12, 13, 16

With respect to PI, no pollution with Cu, Ni, Pb, Co, Cr, Ba, Mn, Sr, As, or Cd was found. Low pollution with Zn was found at the sites 1, 4, 5, 7–12, 15, 18–20, Al—9, 17, and 20. Moderate pollution was that with Fe (at 7, 9, 17), Mg (at 17), and Al (at 1, 3, 7, 8, 10–16, 18). Strong pollution was noticed with Fe (at 1, 3, 5, 8, 10–16, 18–20), Mg (at 7–9, 11, 19, and 20), and Al (at 2 and 4–6). Very strong pollution was registered with Fe (at 2 and 4) and Mg (1–6, 10, 12–16, and 17). The PI for Na falls between 1 and 2, at sites 1, 5–7, and 18, and between 2 and 3 at 20. PIs for Ca and K are greater than 5 at all sites.

Based on the EF computed with respect to Al, moderate enrichment was seen for Fe (at sites 8, 9, and 19), with Mg (at 1, 2, 4, 6, 8–10, 12–16, and 18–20), and K (at 1–6 and 12–17), whereas significant enrichment was determined only with Mg at site 3, and K (at 7–11, and 18–20).

The EF calculated with respect to Ca shows that all the sites are in the same category of deficient to minimum enrichment. EF computed with respect to Ca indicates a moderate enrichment in K (at all sites but 6, 7, and 18) and Mg (at site 3). Significant enrichment in K was determined at sites 7 and 18 and in Ca at all sites.

PLI values are between 0.26 and 0.58, so less than 1, proving a variation between perfection (indicated by a value of 0) and baseline (shown by a value of 1). Since the PIs corresponding to Co, As, Cd, and Pb are under 0.03, they contribute to the decrease in *PLI* values. Removing these elements from computation, denoted by *PLI_d*, produced values from 0.60 to 1.61 (Figure 8). *PLI_d* is more than two times higher than *PLI*.

PLI_d indicates a variation between perfection and baseline ($0 < PLI < 1$) for sites 1, 8, 9, 16, 17, and 19, and there is a progressive deterioration of the air quality ($1 < PLI < 1.61$) for all sites but those already mentioned. Locations 2, 4, 6, 12, and 20 have the highest *PLI* and *PLI_d*, and so the biggest contamination, as shown in Figure 8. All but site 20 belong to the second cluster in Figure 4a.

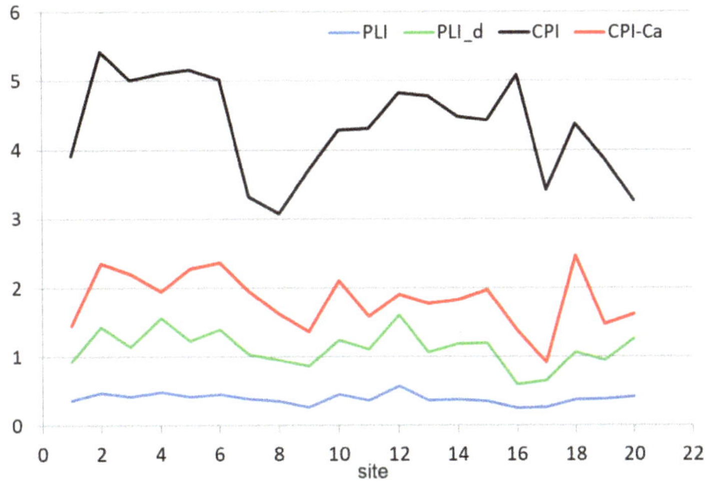

Figure 8. CPI and CPI-Ca.

Taking into account all *PIs*, the Nemerow pollution index obtained values between 18.02 and 45.56, indicating high contamination at all locations. Removing the *PI* for Ca, the values of the index, denoted PI_{Nem-Ca}, varied in the interval 3.67–16.89. Notice the essential influence of the very high PIs on the values of the Nemerow index.

All values of the *CPI* index (Figure 8) were between 3.09 and 5.43, indicating heavy pollution.

Since Ca is an element that has mainly a natural origin, and we did not find essential evidence of another origin in the region, removing it from the *CPI* computation (and denoting the new index by *CPI-Ca*), the variation in *CPI-Ca* was in the interval 0.92–2.46. Therefore, the pollution level from site 17 is graded warning, locations 2, 3, 5, 6, 10, and 18 are moderately polluted, and the rest are slightly contaminated.

The PI_{Nem} and *CPI* are influenced by the highest values of the ratio C_j/CB_j, in contrast with the *PLI*, whose values are more related to the elements' lowest concentrations.

Computation of the *AQI* (Figure 9) taking into account all elements (case a) or without calcium (case b—*AQI-Ca*) resulted in (a) heavy pollution at all sites, respectively, and (b) slight pollution at site 17 and moderate contamination at sites 9, 16, 19. Three of these locations are situated in the same cluster from Figure 4a. The shapes of the *CPI* and *AQI* indices charts are similar. A significant decrement in their values is noticed when Ca is removed from computation.

Figure 9. AQI and AQI-Ca.

As mentioned above, the PIs computed for four elements were under 0.01, and they were removed from the computation of the *PLI*, leading to obtaining *PLI_d*. For consistency, we removed these elements from the initial data set (let us denote it Set1), obtaining set S2 that contained only 13 series of elements. The same analysis as that presented above has been performed for Set2. We only summarize the findings:

- ➢ The k-means algorithm and hierarchical clustering provided the same clusters and dendrogram as in Figure 4, indicating that the removed series does not have a significant importance to lead to a difference between the sites.
- ➢ The Nemerow indices computed with Set2 are the same (up to the third decimal) as those computed using Set1, while the *CPI* and AQI have higher values for Set2.
- ➢ The k-means algorithm performed on the series of indices obtained from Set2 provided the same clusters as in Figure 4a—see Figure 10a.
- ➢ The hierarchical clustering performed on the series of indices obtained from Set2 provided a cluster containing the series 1, 7–9, 11, 17, 19, and 20 which are also in the left-hand-side cluster from Figure 4b.

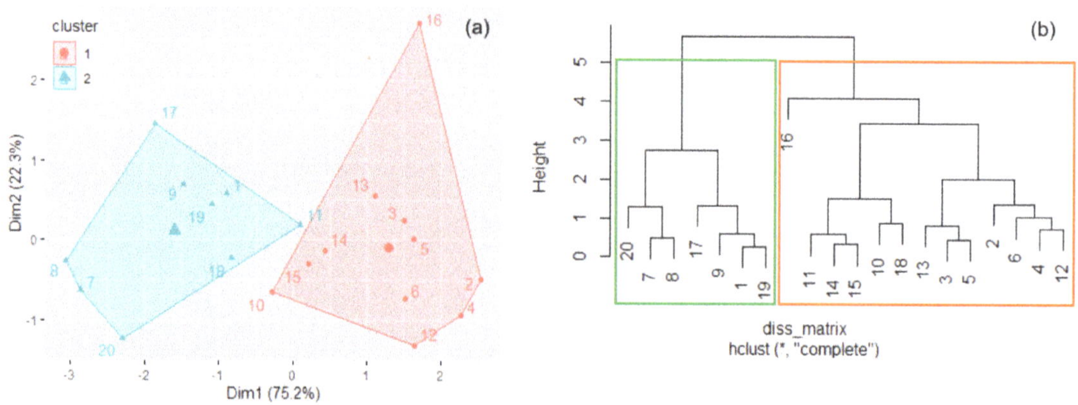

Figure 10. (**a**). The clusters found by k-means with $k = 2$; (**b**) Dendrogram from the hierarchical clustering for Set2.

Figure 11, the biplot obtained performing the PCA for Set2 indicates the positions of the sites in the first cluster from the dendrogram—all grouped at the left-hand side of the biplot, with negative components on PC2. A clear separation line (the red one) can be drawn between the two clusters. The locations from the first cluster are in tourist areas—Dubai Marina, Burj Khalifa, and Dubai Sport City. Site 16 is situated on Dubai Marina Promenade, near the water, in a restricted area for cars, a zone with the lowest recorded concentrations of the study metals. This particular situation is emphasized by its position on the biplot.

We should remember that, generally, the perfect superposition of the clusters determined by both methods is a particular situation given the various mathematical backgrounds on which the algorithms rely. In the case of homogenous sets, it is expected (which is not the case in this study).

Performing the algorithm to determine the clustering by elements, two clusters were obtained, one with only two elements As and Cd (when working with Set1), and Ca and K (when working with Set2). This situation pointed out the elements with the lowest and highest concentrations, respectively, the last ones requiring attention.

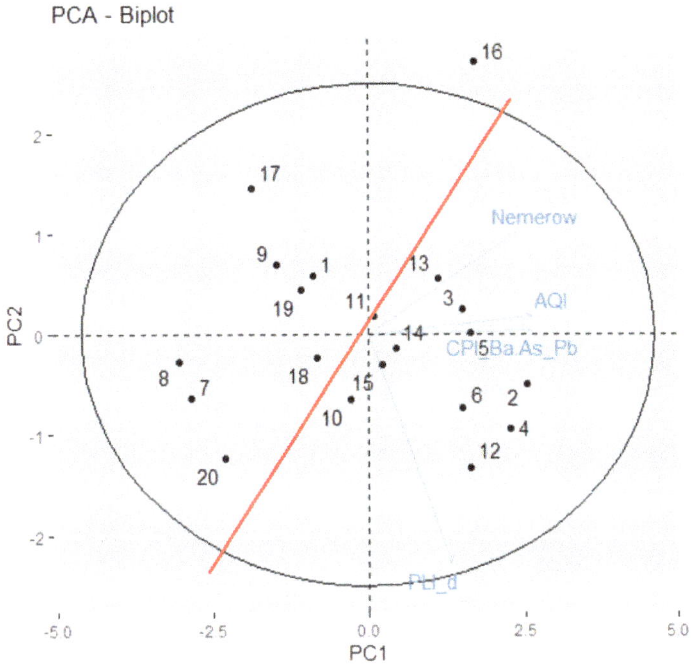

Figure 11. Biplot in PCA for Set2.

4. Conclusions

This article analyzed the degree of enrichment with metals of the dust collected indoors at different locations in Dubai, using multivariate statistics and pollution indices. The study fills a gap in the knowledge concerning indoor pollution due to dust in a region where frequent dust storms appear.

It was shown that the highest enrichment factors (for Ca, Cu, Mg, and Fe) are the consequence of the soil lithology and industrial activities (especially mining), dust being transported for long distances from the emission places during dust storms.

We proposed two new pollution indices—CPI and AWI—and used them for assessing the contamination at the observation places. We classified the sites based on the set formed by the PLI, CPI, AWI, and the Nemerow index and compared it with that built by row data series. It was found that two sites fall into different clusters resulting from these classifications.

Another finding that opens a research direction is using different groups of data sets for classifications in practical applications. It was shown that for the clusters built when eliminating the elements with the lowest concentrations (much under the warning limits) from the data set, the obtained classifications are more realistic.

Employing different clustering algorithms on the raw data series and the pollution indices series, and the use of stability criteria, are important for finding the most similar series in the data set (those that are found all the time together in the same cluster).

In a future study, we intend to present a methodology that will come to cross-validate the clustering findings, using supplementary selection criteria and decision trees.

Author Contributions: Conceptualization, Y.N. and A.B.; methodology, Y.N. and A.B.; software, A.B. and M.S.; validation, A.B. and F.H.; formal analysis, A.B. and M.S.; investigation, A.B. and M.S.; resources, Y.N. and A.B.; data curation, Y.N., A.B. and M.N.; writing—original draft preparation, Y.N., A.B. and M.S.; visualization, A.B. and M.N.; supervision, Y.N. and A.B.; project administration, Y.N. and A.B.; funding acquisition, Y.N. and A.B. All authors have read and agreed to the published version of the manuscript.

Funding: This project was funded by the Research Office, Zayed University, the United Arab Emirates (Project No. R 20115).

Data Availability Statement: Data will be available on request from the author.

Conflicts of Interest: The authors declare no conflict of interest.

References

1. Meza-Figueroa, D.; La O-Villanueva, M.D.; Parra, M.L.D. Heavy metal distribution in dust from elementary schools in Hermosillo, Sonora, Mexico. *Atmos. Environ.* **2007**, *41*, 276–288. [CrossRef]
2. Darus, F.M.; Nasir, R.A.; Sumari, S.M.; Ismail, Z.S.; Omar, N.A. Heavy metals composition of indoor dust in nursery schools building. *Procedia Soc. Behav. Sci.* **2012**, *38*, 169–175. [CrossRef]
3. Höppe, P.; Martinac, I. Indoor climate and air quality. Review of current and future topics in the field of ISB study group 10. *Int. J. Biometeorol.* **1998**, *42*, 1–7. [PubMed]
4. Schweizer, C.; Edwards, R.D.; Bayer-Oglesby, L.; Gauderman, W.J.; Ilacqua, V.; Jantunen, M.J.; Lai, H.K.; Nieuwenhuijsen, M.; Kunzli, N. Indoor time microenvironment-activity patterns in seven regions of Europe. *J. Expo. Sci. Environ. Epid.* **2007**, *17*, 170–181. [CrossRef] [PubMed]
5. Schwarze, P.E.; Ovrevik, J.; Lag, M.; Refsnes, M.; Nafstad, P.; Hetland, R.B.; Dybing, E. Particulate matter properties and health effects: Consistency of epidemiological and toxicological studies. *Human Exper. Toxicol.* **2006**, *25*, 559–579. [CrossRef] [PubMed]
6. Tran, D.T.; Alleman, L.Y.; Coddeville, P.; Gallo, J.C. Elemental characterization and source identification of size resolved atmospheric particles in French classrooms. *Atmos. Environ.* **2012**, *54*, 250–259. [CrossRef]
7. Franchi, M.; Carrer, P.; Kotzias, D.; Rameckers, E.M.A.L.; Seppänen, O.; van Bronswijk, J.E.M.H.; Viegi, G.; Towards Healthy Air in Dwellings in Europe. The THADE Report. Available online: https://ec.europa.eu/health/ph_projects/2001/pollution/fp_pollution_2001_frep_02.pdf (accessed on 15 October 2023).
8. Al-Rajhi, M.A.; Seaward, M.R.D.; Al-Aamar, A.S. Metal levels in indoor and outdoor dust in Riyadh, Saudi Arabia. *Environ. Int.* **1996**, *22*, 315–324. [CrossRef]
9. *Exposure Factors Handbook: 2011 Edition (EPA/600/R-09/052F)*; United States Environmental Protection Agency: Washington, DC, USA, 2011.
10. Wilson, R.; Jones-Otazo, H.A.; Petrovic, S.; Mitchell, I.; Bonvalot, Y.; Williams, D.; Richardson, G.M. Revisiting dust and soil ingestion rates based on hand-to-mouth transfer. *Hum. Ecol. Risk Assess.* **2013**, *19*, 158–188. [CrossRef]
11. Rashed, M.N. Total and extractable heavy metals in indoor, outdoor and street dust from Aswan City, Egypt. *Clean Soil Air Water* **2008**, *36*, 850–857. [CrossRef]
12. Fiordelisi, A.; Piscitelli, P.; Trimarco, B.; Coscioni, E.; Iaccarino, G.; Sorriento, D. The mechanisms of air pollution and particulate matter in cardiovascular diseases. *Heart Fail. Rev.* **2017**, *22*, 337–347. [CrossRef]
13. Fisk, W.J. 10—Impact of ventilation and aircleaning on asthma. In *Clearing the Air: Asthma and Indoor Air Exposure*; National Academies Press: Washington, DC, USA, 2000. Available online: https://www.ncbi.nlm.nih.gov/books/NBK224478/ (accessed on 12 November 2023).
14. Tsubata, R.; Sakaguchi, M.; Yoshizawa, S. Particle size of indoor airborne mite allergens (Der p 1 and Der f 1). *Proc. Indoor Air'96* **1996**, *3*, 155–160.
15. Maertens, R.M.; Bailey, J.; White, P.A. The mutagenic hazards of settled house dust: A review. *Mutat. Res.* **2004**, *567*, 401–425. [CrossRef] [PubMed]
16. Tong, T.Y.; Lam, K.C. Home sweet home? A case study of household dust contamination in Hong Kong. *Sci. Total Environ.* **2000**, *256*, 115–123. [CrossRef] [PubMed]
17. Hassan, S.K.M. Metal concentrations and distribution in the household, stairs and entryway dust of some Egyptian homes. *Atmos. Environ.* **2012**, *54*, 207–215. [CrossRef]
18. Praveena, S.M.; Abdul Mutalib, N.S.; Aris, A.Z. Determination of heavy metals in indoor dust from primary school (Sri Serdang, Malaysia): Estimation of the health risks. *Environ. Forensics* **2015**, *16*, 257–263. [CrossRef]
19. Morawska, L.; Salthammer, T. *Indoor Environment. Airborne Particles and Settled Dust*; Wiley: Hoboken, NJ, USA, 2003.
20. Lioy, P.J.; Freeman, N.C.G.; Millette, J.R. Dust: A metric for use in residential and building exposure assessment and source characterization. Environ. *Health Perspect.* **2002**, *110*, 969–983. [CrossRef]
21. Kang, Y.; Cheung, K.C.; Wong, M.H. Mutagenicity, genotoxicity and carcinogenic risk assessment of indoor dust from three major cities around the Pearl River Delta. *Environ. Int.* **2011**, *37*, 637–643. [CrossRef] [PubMed]

22. Popoola, O.E.; Bamgbose, O.; Okonkwo, O.J.; Arowolo, T.A.; Popoola, A.O.; Awofolu, O.R. Heavy metals content in classroom dust of some public primary schools in metropolitan Lagos, Nigeria. *Res. J. Environ. Earth. Sci.* **2012**, *4*, 460–465.
23. Cao, S.; Duan, X.; Zhao, X.; Wang, B.; Ma, J.; Fan, D.; Sun, C.; He, B.; Wei, F.; Jiang, G. Health risk assessment of various metal(loid)s via multiple exposure pathways on children living near a typical lead-acid battery plant, China. *Environ. Pollut.* **2015**, *200*, 16–23. [CrossRef]
24. Rout, T.K.; Masto, R.E.; Ram, L.C.; George, J.; Padhy, P.K. Assessment of human health risks from heavy metals in outdoor dust samples in a coal mining area. *Environ. Geochem. Health* **2015**, *35*, 347–356. [CrossRef]
25. Gbadebo, A.M.; Bankole, O.D. Analysis of potentially toxic metals in airborne cement dust around Sagamu, Southwestern Nigeria. *J. Appl. Sci.* **2007**, *7*, 35–40. [CrossRef]
26. Wu, G.; Kang, H.; Zhang, X.; Shao, H.; Chu, L.; Ruan, C. A critical review on the bio-removal of hazardous heavy metals from contaminated soils: Issues, progress, eco-environmental concerns and opportunities. *J. Hazard. Mater.* **2010**, *174*, 1–8. [CrossRef] [PubMed]
27. Yousaf, B.; Liu, G.; Wang, R.; Imtiaz, M.; Rizwan, M.S.; Zia-ur-Rehman, M.; Qadir, A.; Si, Y. The importance of evaluating metal exposure and predicting human health risks in urban–periurban environments influenced by emerging industry. *Chemosphere* **2016**, *150*, 79–89. [CrossRef] [PubMed]
28. Nazzal, Y.; Howari, F.M.; Yaslam, A.; Iqbal, J.; Maloukh, L.; Ambika, L.K.; Al-Taani, A.A.; Ali, I.; Othman, E.M.; Jamal, A.; et al. A Methodological Review of Tools That Assess Dust Microbiomes, Metatranscriptomes and the Particulate Chemistry of Indoor Dust. *Atmosphere* **2022**, *13*, 1276. [CrossRef]
29. Chattopadhyay, G.; Lin, K.C.P.; Feitz, A.J. Household dust metal levels in the Sydney metropolitan area. *Environ. Res.* **2003**, *93*, 301–307. [CrossRef] [PubMed]
30. Liu, Y.; Ma, J.; Yan, H.; Ren, Y.; Wang, B.; Lin, C.; Liu, X. Bioaccessibility and health risk assessment of arsenic in soil and indoor dust in rural and urban areas of Hubei province, China. *Ecotox. Environ. Saf.* **2016**, *126*, 14–22. [CrossRef]
31. Rasmussen, P.E.; Beauchemin, S.; Chénier, M.; Levesque, C.; MacLean, L.C.; Marro, L.; Jones-Otazo, H.; Petrovic, S.; McDonald, L.T.; Gardner, H.D. Canadian house dust study: Lead bioaccessibility and speciation. *Environ. Sci. Technol.* **2011**, *45*, 4959–4965. [CrossRef]
32. Household Air Pollution. Available online: https://www.who.int/news-room/fact-sheets/detail/household-air-pollution-and-health (accessed on 27 October 2023).
33. Indoor Air Pollution. Available online: https://www.who.int/news-room/questions-and-answers/item/air-pollution-indoor-air-pollution (accessed on 27 October 2023).
34. Popovici, B.; Postolache, F. MAOS—Oxygen minimum amount calculation software for thermodynamics processes. *Sci. Bull. Naval Acad.* **2018**, *XXI*, 430–433.
35. Popa, I.; Sporiș, A.; Mărășescu, D.; Postolache, F.; Volintiru, O.N. Termodinamically process for atmospheric fresh water production. *Sci. Bull. Naval Acad.* **2022**, *XXV*, 37–44.
36. United States Environmental Protection Agency. Code of Federal Regulations: Priority Pollutants List. 2014. Available online: https://www.gpo.gov/fdsys/pkg/CFR-2014-title40-vol29/xml/CFR-2014-title40-vol29-part423-appA.xml (accessed on 11 November 2023).
37. Sen, S.; Bizimis, M.; Tripathi, S.N.; Paul, D. Lead isotopic finger-printing of aerosols to characterize the sources of atmospheric lead in an industrial city of India. *Atmos. Environ.* **2016**, *129*, 27–33. [CrossRef]
38. Peixoto, M.S.; de Oliveira Galvao, M.F.; Batistuzzo de Medeiros, S.R. Cell death pathways of particulate matter toxicity. *Chemosphere* **2017**, *188*, 32–48. [CrossRef] [PubMed]
39. Morakinyo, O.M.; Mukhola, M.S.; Mokgobu, M.I. Health Risk Analysis of Elemental Components of an Industrially Emitted Respirable Particulate Matter in an Urban Area. *Int. J. Environ. Res. Public Health* **2021**, *18*, 3653. [CrossRef] [PubMed]
40. Sanborn, M.D.; Abelsohn, A.; Campbell, M.; Weir, E. Identifying and managing adverse environmental health effects: 3. Lead exposure. *Can. Med. Assoc. J.* **2002**, *166*, 1287–1292.
41. Faiz, Y.; Tufail, M.; Javed, M.T.; Chaudhry, M.M.; Siddique, N. Road dust pollution of Cd, Cu, Ni, Pb and Zn along Islamabad Expressway, Pakistan. *Microchem. J.* **2009**, *92*, 186–192. [CrossRef]
42. Turner, A.; Hefzi, B. Levels and bioaccessibilities of metals in dusts from an arid environment. *Water Air Soil Poll.* **2010**, *210*, 483–491. [CrossRef]
43. Bărbulescu, A.; Nazzal, Y.; Howari, F. Statistical analysis and estimation of the regional trend of aerosol size over the Arabian Gulf Region during 2002–2016. *Sci. Rep.* **2018**, *8*, 9571. [CrossRef]
44. Nazzal, Y.; Bărbulescu, A.; Howari, F.M.; Yousef, A.; Al-Taani, A.A.; Al Aydaroos, F.; Naseem, M. New insight to dust storm from historical records, UAE. *Arab. J. Geosci.* **2019**, *12*, 396. [CrossRef]
45. Bărbulescu, A.; Nazzal, Y. Statistical analysis of the dust storms in the United Arab Emirates. *Atmos. Res.* **2020**, *231*, 104669. [CrossRef]
46. Nazzal, Y.H.; Bărbulescu, A.; Howari, F.; Al-Taani, A.A.; Iqbal, J.; Xavier, C.M.; Sharma, M.; Dumitriu, C.S. Assessment of metals concentrations in soils of Abu Dhabi Emirate using pollution indices and multivariate statistics. *Toxics* **2021**, *9*, 95. [CrossRef]
47. Nazzal, Y.; Bou Orm, N.; Bărbulescu, A.; Howari, F.; Sharma, M.; Badawi, A.; Al-Taani, A.A.; Iqbal, J.; El Ktaibi, F.; Xavier, C.M.; et al. Study of atmospheric pollution and health risk assessment A case study for the Sharjah and Ajman Emirates (UAE). *Atmosphere* **2021**, *12*, 1442. [CrossRef]

48. Yeatts, K.B.; El-Sadig, M.; Leith, D.; Kalsbeek, W.; Al-Maskari, F.; Couper, D.; Funk, W.E.; Zoubeidi, T.; Chan, R.L.; Trent, C.B.; et al. Indoor air pollutants and health in the United Arab Emirates. *Environ. Health Perspect.* **2012**, *120*, 687–694. [CrossRef] [PubMed]
49. Jung, C.; Alqassimi, N.; El Samanoudy, G. The comparative analysis of the indoor air pollutants in occupied apartments at residential area and industrial area in Dubai, United Arab Emirates. *Front. Built Environ.* **2022**, *8*, 998858. [CrossRef]
50. Mfarrej, B.; Qafisheh, N.A.; Bahloul, M.M. Investigation of Indoor Air Quality inside Houses From UAE. *Air Soil Water Resear.* **2020**, *13*, 1–10. [CrossRef]
51. Greywolf Sensing Solutions. Indoor Air Quality (IAQ) Meters, Monitors for Handheld, Semi-Permanent and Long-Term IAQ Measurement. Available online: https://graywolfsensing.com/iaq/?gad_source=1&gclid=CjwKCAiA6byqBhAWEiwAnGCA4GrTejNDxyRzkNIWFkHzN7VQCwwFmry9m5ksqtSG2hxtb{-}{-}ljiU8txoCU-8QAvD_BwE (accessed on 10 October 2023).
52. High-Precision Handheld GIS Data Collector-SOUTH S750. Available online: https://pdf.directindustry.com/pdf/south-surveying-mapping-instrument-co-ltd/high-precision-handheld-gis-data-collector-south-s750/160571-732546.html (accessed on 11 October 2023).
53. Dyson V15 Detect. Available online: https://www.dyson.com/vacuum-cleaners/cordless/v15/detect/yellow-iron (accessed on 11 October 2023).
54. *Method 3050B: Acid Digestion of Sludges, Sediments, and Soils, Revision 2*; United States Environmental Protection Agency (USEPA): Washington, DC, USA, 1996.
55. Code of Federal Regulations (Annual Edition)—Title 40: Protection of Environment, Part 136: Guidelines Establishing Test Procedures for the Analysis of Pollutants. 2022. Available online: https://www.govinfo.gov/app/search/%7B%22query%22%3A%2240%20CFR%20Part%20136%20Appendix%20B%2C%20Revision%202%22%2C%22offset%22%3A0%7D (accessed on 12 November 2023).
56. Calculating & Using Method Detection Limits. Available online: https://www.wef.org/globalassets/assets-wef/2-resources/online-education/webcasts/presentation-handouts/mdl-webcast-16july20.pdf (accessed on 12 November 2023).
57. Daburra, I. K-means Clustering: Algorithm, Applications, Evaluation Methods, and Drawbacks. Available online: https://towardsdatascience.com/k-means-clustering-algorithm-applications-evaluation-methods-and-drawbacks-aa03e644b48a (accessed on 18 June 2023).
58. Hierarchical Clustering in, R. Available online: https://www.datacamp.com/tutorial/hierarchical-clustering-R (accessed on 18 September 2023).
59. K-Mean: Getting the Optimal Number of Clusters. Available online: https://www.analyticsvidhya.com/blog/2021/05/k-mean-getting-the-optimal-number-of-clusters/ (accessed on 16 June 2023).
60. Rousseeuw, P. Silhouettes: A Graphical Aid to the Interpretation and Validation of Cluster Analysis. *J. Comput. Appl. Math.* **1987**, *20*, 53–65. [CrossRef]
61. Farris, J.S. On the cophenetic correlation coefficient. *Syst. Zool.* **1969**, *18*, 279–285. [CrossRef]
62. Murphy, P. Clustering Data in R. Available online: https://rstudio-pubs-static.s3.amazonaws.com/599072_93cf94954aa64fc7a4b99ca524e5371c.html#Visualize (accessed on 14 October 2023).
63. Jolliffe, I. *Principal Component Analysis*; Wiley: Hoboken, NJ, USA, 2014.
64. Davis, J.C. *Statistics and Data Analysis in Geology*, 2nd ed.; John Wiley and Sons, Inc.: New York, NY, USA, 1982.
65. Kolsi, S.H.; Bouri, S.; Hachicha, W.; Dhia, H.B. Implementation and evaluation of multivariate analysis for groundwater hydrochemistry assessment in arid environments: A case study of Hajeb Elyoun–Jelma, Central Tunisia. *Environ. Earth Sci.* **2013**, *70*, 2215–2224. [CrossRef]
66. Cattell, R.B. The Scree Test for The Number of Factors. *Multivar. Behav. Res.* **1966**, *1*, 245–276. [CrossRef] [PubMed]
67. Kaiser, H.F. The application of electronic computers to factor analysis. *Educ. Psychol. Meas.* **1960**, *20*, 141–151. [CrossRef]
68. Kowalska, J.B.; Mazurek, R.; Gąsiorek, M.; Zaleski, T. Pollution indices as useful tools for the comprehensive evaluation of the degree of soil contamination—A review. *Environ. Geochem. Health* **2018**, *40*, 2395–2420. [CrossRef]
69. Al-Hejuje, M.M.; Al-Saad, H.T.; Hussain, N.A. Application of geo-accumulation index (I-geo) for assessment the sediments contamination with heavy metals at Shatt Al-Arab River-Iraq. *J. Sci. Eng. Res.* **2018**, *5*, 342–351.
70. Sutherland, R.A. Bed sediment-associated trace metals in an urban stream, Oahu, Hawaii. *Environ. Geol.* **2000**, *39*, 611–627. [CrossRef]
71. Lindsay, W.L. *Chemical Equilibrium in Soils*; John Wiley & Sons: New York, NY, USA, 1979.
72. Kabata-Pendias, A. *Trace Elements of Soils and Plants*, 4th ed.; CRC Press: Boca Raton, FL, USA; Taylor & Francis Group: Abingdon, UK, 2011.
73. Selvam, A.P.; Priya, S.L.; Banerjee, K.; Hariharan, G.; Purvaja, R.; Ramesh, R. Heavy Metal Assessment Using Geochemical and Statistical Tools in the Surface Sediments of Vembanad Lake, Southwest Coast of India. *Environ. Monit. Assess.* **2012**, *184*, 5899–5915. [CrossRef] [PubMed]
74. Gong, Q.; Deng, J.; Xiang, Y.; Wang, Q.; Yang, L. Calculating pollution indices by heavy metals in ecological geochemistry assessment and a case study in parks of Beijing. *J. China Univ. Geosci.* **2008**, *19*, 230–241.
75. Pramanik, A.K.; Majumdar, D.; Chatterjee, A. Factors affecting lean, wet-season water quality of Tilaiya reservoir in Koderma District, India during 2013–2017. *Water Sci.* **2020**, *34*, 85–97. [CrossRef]
76. Cude, C.G. Oregon water quality index: A tool for evaluating water quality management effectiveness. *J. Am. Water Resour. Assoc.* **2001**, *37*, 125–137. [CrossRef]

77. Howari, F.M.; Ghrefat, H.; Nazzal, Y.; Galmed, M.A.; Abdelghany, O.; Fowler, A.R.; Sharma, M.; AlAygaroos, F.; Xavier, C.M. Delineation of Copper Mineralization Zones at Wadi Ham, Northern Oman Mountains, United Arab Emirates Using Multispectral Landsat 8 (OLI) Data. *Front. Earth Sci.* **2020**, *8*, 578075. [CrossRef]
78. Manganese Ore in United Arab Emirates. Available online: https://oec.world/en/profile/bilateral-product/manganese-ore/reporter/are (accessed on 14 October 2023).
79. Iron Ore in United Arab Emirates. Available online: https://oec.world/en/profile/bilateral-product/iron-ore/reporter/are (accessed on 14 October 2023).
80. 5 Largest Aluminum Producing Countries in the World. Available online: https://www.insidermonkey.com/blog/5-largest-aluminum-producing-countries-in-the-world-1159098/ (accessed on 14 October 2023).
81. Baek, S.O.; Kim, Y.S.; Perry, R. Indoor air quality in homes, offices, and restaurants in Korean urban areas—Indoor/outdoor relationships. *Atmos. Environ.* **1997**, *31*, 529–544. [CrossRef]
82. Jones, N.C.; Thornton, C.A.; Mark, D.; Harrison, R.M. Indoor/outdoor relationships of particulate matter in domestic. *Atmos. Environ.* **2000**, *34*, 2603–2612. [CrossRef]
83. Kuo, H.W.; Shen, H.Y. Indoor and outdoor PM2.5 and PM10 concentration in the air during a dust storm. *Build. Environ.* **2010**, *45*, 610–614. [CrossRef]
84. Meadow, J.F.; Altrichter, A.E.; Kembel, S.W.; Kline, J.; Mhuireach, G.; Moriyama, M.; Northcutt, D.; O'Connor, T.K.; Womack, A.M.; Brown, G.Z.; et al. Indoor airborne bacterial communities are influenced by ventilation, occupancy, and outdoor air source. *Indoor Air* **2014**, *24*, 41–48. [CrossRef]
85. Fung, C.C.; Yang, P.; Zhu, Y.F. Infiltration of Diesel Exhaust into a Mechanically Ventilated Building. In Proceedings of the Indoor Air 2014—13th International Conference on Indoor Air Quality and Climate, Hong Kong, 7–12 July 2014. Paper#HP0626.
86. Ai, Z.T.; Mak, C.M. From street canyon microclimate to indoor environmental quality in naturally ventilated urban buildings: Issues and possibilities for improvement. *Build. Environ.* **2015**, *94*, 489–503. [CrossRef] [PubMed]
87. Meier, R.; Schindler, C.; Eeftens, M.; Aguilera, I.; Ducret-Stich, R.E.; Ineichen, A.; Davey, M.; Phuleria, H.C.; Probst-Hensch, N.; Tsai, M.Y.; et al. Modeling indoor air pollution of outdoor origin in homes of SAPALDIA subjects in Switzerland. *Environ. Int.* **2015**, *82*, 85–91. [CrossRef] [PubMed]
88. Mohammadi, M.; Calautit, J. Quantifying the Transmission of Outdoor Pollutants into the Indoor Environment and Vice Versa—Review of Influencing Factors, Methods, Challenges and Future Direction. *Sustainability* **2022**, *14*, 10880. [CrossRef]
89. Härdle, W.; Simar, L. *Applied Multivariate Statistical Analysis*, 2nd ed.; Springer: Berlin Heidelberg, Germany, 2007.
90. Müller, G. Heavy Metals in the Sediments of the Rhine: Changes since 1971. A look around. *Sci. Technol.* **1979**, *79*, 778–783.

toxics

MDPI

Article

Modeling the Chlorine Series from the Treatment Plant of Drinking Water in Constanta, Romania

Alina Bărbulescu [1] and Lucica Barbeş [2,3,*]

[1] Department of Civil Engineering, Transilvania University of Braşov, 5 Turnului Str., 500152 Brasov, Romania; alina.barbulescu@unitbv.ro
[2] Department of Chemistry and Chemical Engineering, Ovidius University of Constanța, 124 Mamaia Bd., 900152 Constanta, Romania
[3] Doctoral School of Biotechnical Systems Engineering, Politehnica University of Bucharest, 313, Splaiul Independentei, 060042 Bucharest, Romania
* Correspondence: lucille.barbes2020@gmail.com

Abstract: Ensuring good drinking water quality, which does not damage the population's health, should be a priority of decision factors. Therefore, water treatment must be carried out to remove the contaminants. Chlorination is one of the most used treatment procedures. Modeling the free chlorine residual concentration series in the water distribution network provides the water supply managers with a tool for predicting residual chlorine concentration in the networks. With regard to this idea, this article proposes alternative models for the monthly free chlorine residual concentration series collected at the Palas Constanta Water Treatment Plant, in Romania, from January 2013 to December 2018. The forecasts based on the determined models are provided, and the best results are highlighted.

Keywords: free chlorine residual concentration series; modeling; forecast; water treatment plant

check for updates

Citation: Bărbulescu, A.; Barbeş, L. Modeling the Chlorine Series from the Treatment Plant of Drinking Water in Constanta, Romania. *Toxics* **2023**, *11*, 699. https://doi.org/10.3390/toxics11080699

Academic Editor: Daniel Drage

Received: 1 July 2023
Revised: 10 August 2023
Accepted: 11 August 2023
Published: 13 August 2023

1. Introduction

Drinking water quality is essential, given its impact on the population's health [1]. Therefore, ensuring a sufficient quantity and adequate quality must be a priority of each state/community to improve the health indicators and the population's well-being [2]. The urban population's primary drinking water supply sources are surface water and groundwater, whereas wells are used in rural areas [3]. In an ideal scenario, a water supply system would operate continuously, without changes in flow rate or other special conditions for individual treatment processes, when the raw water quality and quantity are constant. In reality, ideal conditions are not always met [4,5]. Given that various contaminants can affect the drinking water quality, it is crucial to treat the water before its distribution for consumption [6,7].

Due to its effectiveness (in killing viruses, bacteria, etc.), environmental feasibility, and long-lasting effects, chlorine is the primary disinfectant used for drinking water treatment [8,9]. Hypochlorous and hydrochloric acids are produced by adding chlorine or its derivatives to the raw water [10]. The active element in the disinfection process (the hypochlorite ion) results from the dissociation of the hypochlorous acid. During the water treatment, chlorine oxidizes the mineral substances and then produces chloramines by reacting with ammonia. Supplementing the chlorine dose leads to chloramine oxidation, increasing the free chlorine residual level [11,12], which is crucial for effective disinfection. The laboratory analyses performed on water samples taken at the outlet of the water treatment station and the distribution network indicate the disinfection stages and the necessary chlorine doses for ensuring water quality [13,14]. A balance in the chlorine dosing must be kept to protect the population against contamination, on the one hand, and avoid the by-products' formation and pipes' corrosion, on the other hand [14,15]. In these conditions, models that accurately predict the free chlorine residual in the distribution system have been proposed as a first step for optimizing the water treatment plant functioning.

Ghang et al. [16] introduced a chlorine decay model based on potential chlorine decay mechanisms and evaluated its performances on four raw surface and alum-treated waters. The results prove that the proposed model accurately predicts free chlorine residuals (R^2 = 0.98). Gómez-Coronel et al. [17] reported satisfactory results in the chlorine concentration at the input of a water distribution system simulated in EPANET, with a genetic algorithm implemented in MATLAB. The EPANET MSX software was used to model chlorine decay in Algarve's drinking water supply systems [18]. García-Ávila et al. [19] employed the same tool with a built-in first-order equation for modeling chlorine decay for a case study from Ecuador. Nejjari et al. [20] proposed a methodology for efficiently calibrating the free chlorine decay models tested on the Barcelona water transport network. Zhang et al. [21] elaborated a model for integrating water quality and operation for forecasting water production (using a genetic algorithm-enhanced artificial neural network). In contrast, other authors focused on optimizing the chlorine dosing [22,23].

Quantifying chlorine residual, turbidity, standard plate count (SPC), coliforms, etc., was performed using statistical methods in a water distribution system from Pakistan [24]. The correlations between the coliforms' presence in the water and the free chlorine content in the Parisian distribution system were also analyzed based on statistics and econometrics approaches [25]. For Romania, only a few studies provide results on drinking water treatment [13,26,27].

To summarize, most results on the chlorine concentration series in water distribution systems use differential equations and a few other methods, such as artificial intelligence. Despite the last period, econometrics and hybrid methods proved their efficiency for modeling and forecast time series in different research fields, like economics [28–30], signal analysis [31], hydro-meteorology, environmental pollution [32–35], and pharmaceutics [36], they were less utilized in modeling the chlorine series at the outlet of the water treatment plants and in the water distribution systems.

In the above context, this article proposes alternative models (econometrics not based on differential equations) for the free chlorine residual concentrations series collected in the water treatment plant Palas (Constanta, Romania) from January 2013 to December 2018. It also emphasizes the possibility of using them for the forecast. The proposed approaches are univariate, not multivariate, as in most of the above-cited literature. They do not require deep specific knowledge in the modeling field (as in the case of differential equations and artificial intelligence) and are easily understood and utilized. Another advantage is extending the research to an area less explored in Romania, for which only a limited number of studies were performed. The models are compared, and their weaknesses and advantages are highlighted.

2. Materials and Methods

2.1. Data Series and Statistical Analysis

The Palas Constanţa treatment, storage, and pumping complex (PCTC) is located in the industrial area of Constanţa city on the Black Sea Littoral in Romania (Figure 1) and provides water to about 350,000 inhabitants.

The groundwater sources that feed the treatment plant are Cişmea I A, Cişmea I B, Cişmea I C, and Cişmea II. Cişmea I A + B+C are formed of 36 wells with depths from 50 to 120 m, except P35, with a depth of 300 m. They have a total supply capacity of 7657 m^3/h. Cişmea II has 12 wells with depths between 90 and 150 m and a pumping capacity of 1940 m^3/h. The Galeşu surface water source, with 13,050 m^3/h catching capacity, is situated along the banks of Poarta Alba–Midia (on the Channel Danube–Black Sea). It has five intakes equipped with metal sieves for retaining the suspended particles.

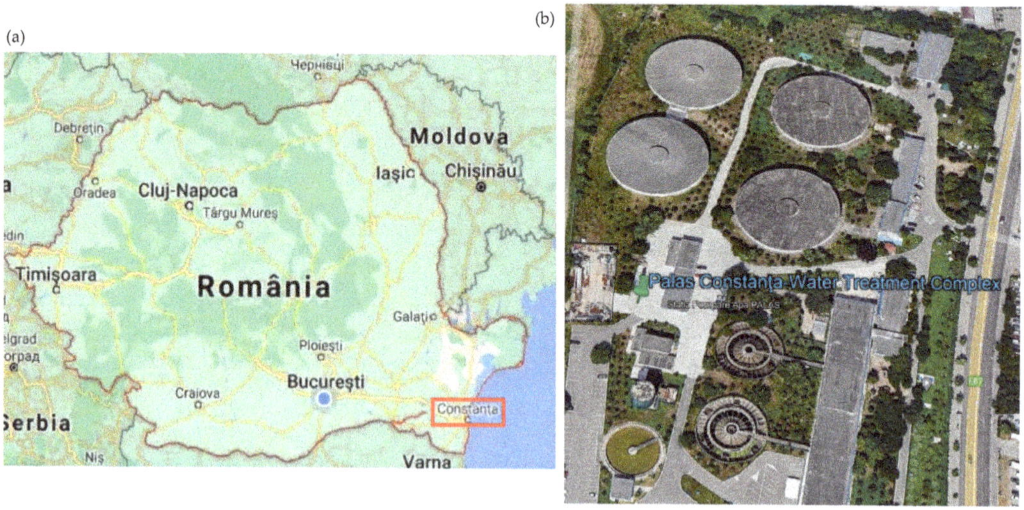

Figure 1. (**a**) Map of Romania; (**b**) the Palas Constanța water treatment complex (PCTC).

This source was created to cope with the high water consumption during the summer and supplement Constanța city's water supply when necessary. The water quality is good even before its treatment, according to [35,37]. After the treatment, the water must satisfy the Directives of the Council of the European Communities [38,39] and the Water Framework Directive [40]. The PCTC stores the water, which is distributed to Constanța and the Littoral water supply system. According to [41], in 2020, the total amount of water supplied to the inhabitants of Constanta was 42,150 m^3 per day.

Generally, for the drinking water distribution networks, there is a risk of insufficient drinking water distributed to consumers caused by phenomena such as the clogging of water sources or the lowering of the surface water level due to drought and lack of precipitation [42,43]. To avoid such situations, there are four water storage stations in Constanța, each of 20,000 m^3, one of 6.000 m^3, and another of 10,000 m^3. The Caragea Dermen groundwater source can also be accessed. It is formed by 18 wells with depths between 35 and 90 m and has a supply capacity of 3.549 m^3/h. The water from different sources undergoes different chlorination processes. Only after chlorination are the streams of water mixed and introduced into the distribution network. The studied data series (Figure 2) is formed of the monthly free chlorine residual concentration collected at the outlet of PCTP during January 2013–December 2018.

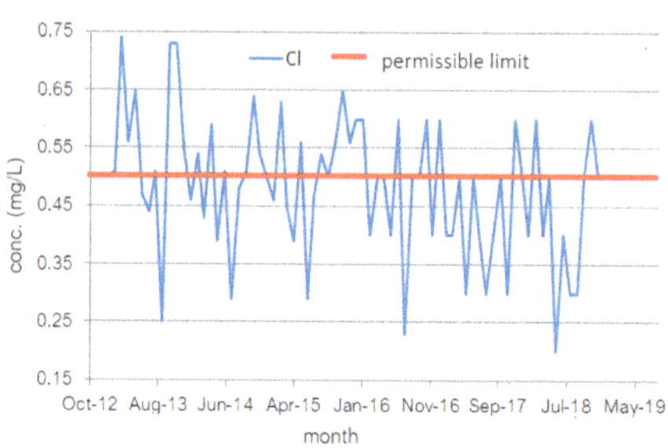

Figure 2. The monthly series of free chlorine residuals from January 2013 to December 2018.

2.2. Statistical Analysis

Basic statistics (mean, median, standard deviation—SD, variation coefficient—CV) were first computed for the monthly series. Then, the following hypotheses were tested: normality against the non-normality (by the Jarque-Bera [44], Shapiro–Wilk [45], and Anderson–Darling [46] tests), homoscedasticity against heteroskedasticity (by the Levene test) [47], the series stationarity vs. its nonstationarity in mean and variance (by the KPSS test) [48]. The null hypothesis that there is no time series trend was tested against the alternative that a monotonic trend exists via the Mann–Kendall and seasonal Mann–Kendall test [49–51]. When the null hypothesis is rejected, Sen's procedure [52] can be used to determine the monotonic trend.

2.3. Mathematical Modeling

Since the preliminary statistical analysis revealed the series seasonality, different approaches have been adopted to model the data series.

In the first approach, the series (y_t) was decomposed using an additive model, of which its components are the trend, the seasonal component, and the random variable. In this case, the steps were the following [53]:

- Determine the trend using the linear trend computed via Sen's method;
- Calculate the detrended series by subtracting the trend from the data series;
- Determine the seasonal component;
- Determine the remainder (random or residual component) as the difference between the detrended series and the seasonal component.

In the multiplicative decomposition, the steps are similar, but the addition is replaced by multiplication and the subtraction by division in the second and fourth steps from the previous method.

In the second approach, the decomposition was conducted following a similar procedure, but the trend was determined using a moving average method of the 12th order.

The third approach was to use the Holt–Winters method, where the series was decomposed using Equations (1)–(4) in the additive model, with a seasonal period $p = 12$ as follows:

$$\hat{y}_{t+h} = a_t + hb_t + s_{t-11+(h-1)mod\ 12}, \tag{1}$$

with

$$a_t = \alpha(y_t - s_{t-12}) + (1-\alpha)(a_{t-1} + b_{t-1}), \tag{2}$$

$$b_t = \beta(a_t - a_{t-1}) + (1-\beta)b_{t-1}, \tag{3}$$

$$s_t = \gamma(y_t - a_t) + (1-\gamma)s_{t-12}, \tag{4}$$

In the multiplicative model, the equations are (5)–(8), which are expressed as follows:

$$\hat{y}_{t+h} = (a_t + hb_t)s_{t-11+(h-1)mod\ 12}, \tag{5}$$

where

$$a_t = \alpha y_t / s_{t-12} + (1-\alpha)(a_{t-1} + b_{t-1}), \tag{6}$$

$$b_t = \beta(a_t - a_{t-1}) + (1-\beta)b_{t-1}, \tag{7}$$

$$s_t = \gamma y_t / a_t + (1-\gamma)s_{t-12}, \tag{8}$$

in the hypothesis that a_t and s_{t-12} are not zero.

In (1)–(8), α, β, γ are smoothing parameters that must be determined for the level, a_t, trend, b_t, and seasonal component, s_t, respectively [54].

The fourth proposed model is a Seasonal Autoregressive Integrated Moving Average model, SARIMA. An ARIMA (*p*,*d*,*q*) process (x_t) with a constant is defined by the following:

$$\phi(L)(1-L)^d y_t = c + \theta(L)\varepsilon_t, \tag{9}$$

where L is the backward operator and

$$\phi(L)y_t = \left(1 - \sum_{i=1}^{p} \phi_i L^i\right) y_t, \tag{10}$$

$$\theta(L)\varepsilon_t = \left(1 + \sum_{i=1}^{q} \theta_i L^i\right) \varepsilon_t, \tag{11}$$

where

$$y_t = x_t - x_{t-d}. \tag{12}$$

p and q are the numbers of autoregressive and moving average terms, respectively, d is the differentiation degree, and (ε_t) is white noise.

A SARIMA (*p*,*d*,*q*) $\times$ $(P,D,Q)_m$ (seasonal ARIMA model) is expressed as the following equation:

$$\phi(L)\Phi(L^m)(1-L)^d(1-L^m)^D y_t = \theta(L)\Theta(L^m)\varepsilon_t, \tag{13}$$

where

$$\Phi(L)y_t = \left(1 - \sum_{i=1}^{P} \Phi_i L^i\right) y_t, \tag{14}$$

$$\Theta(L)\varepsilon_t = \left(1 + \sum_{i=1}^{Q} \Theta_i L^i\right) \varepsilon_t, \tag{15}$$

m, D, P, and Q represent the number of seasonal periods, the seasonal differencing, autoregressive, and seasonal moving average terms, respectively [55].

The residual independence was tested using the Box–Ljung test [56].

In all cases, apart from the residuals' analysis (normality, homoscedasticity, and randomness), the mean absolute deviation (MAD), mean standard deviation (MSD), and mean absolute percentage error (MAPE) were also computed to assess the models' quality. Comparisons of the models, their advantages, and drawbacks are finally discussed.

The MINITAB 17, trial version (https://www.minitab.com/en-us/products/minitab/, accessed on 15 June 2023) and the R software, v.4.3.1 (https://www.r-project.org/, accessed on 15 June 2023) were utilized for testing the statistical hypotheses and mathematical modeling.

3. Results and Discussion

3.1. Results of the Statistical Analysis

The basic statistics of the data series are as follows: minimum = 0.200, maximum = 0.7400, mean = 0.4835, median =0.5000, standard deviation (SD) = 0.1181, coefficient of variance (CV%) = 24.42, skewness = −0.22, and kurtosis = −0.07. Thus, there is a small variation in the series values, and the distribution is left skewed.

Based on the above results, the computed value of the Jarque–Bera statistics was 0.4384, indicating that the normality hypothesis cannot be rejected at a significance level of 0.05. A similar result was obtained by applying the Shapiro–Wilk test. The p-value computed in the Levene test is 0.582 > 0.05, so the homoscedasticity hypothesis cannot be rejected. The statistics of the KPSS test for level (trend) stationarity is 0.59209 (0.03532), and the p-value is 0.02336 (0.1). So, the hypothesis of the level stationary is rejected, and that of the trend stationarity cannot be rejected at the significance level of 0.05. The Mann–Kendall test and its seasonal version rejected the null hypothesis. Therefore, based on Sen's procedure, a linear trend, with the following Equation (16) can be fitted:

$$Y_t = -0.001429(t - 1) + 0.542143, \tag{16}$$

where Y_t is the concentration in the month t.

3.2. Models

When using the first approach, the series decomposition via the additive model (denoted as DECA) is presented in Figure 3a. The recorded (Actual) and the computed (Fitted) values are represented in blue and brown, respectively, and the trend is in green. The violet curve represents the series forecast for the next 48 months. The residuals are normally distributed, according to the Q-Q plot (Figure 3b) and the results of the Shapiro–Wilk test. They are homoscedastic (the p-value of the Levene test is $0.582 > 0.05$) and autocorrelated (the first-order correlation coefficient is -0.3195).

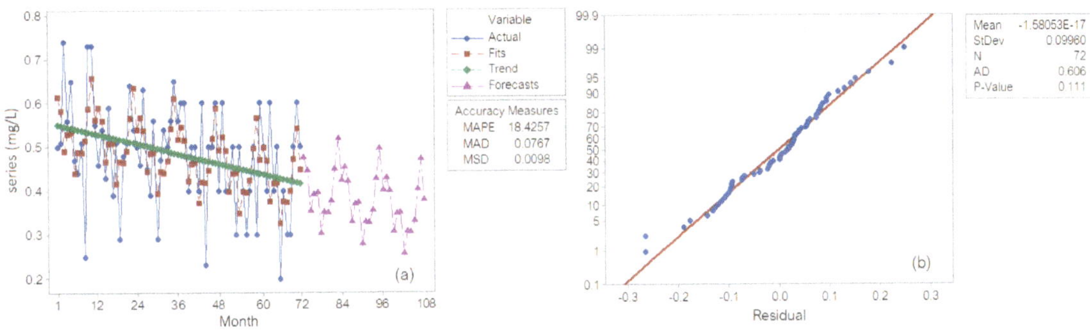

Figure 3. (**a**) Time series decomposition plot for the studied series. DECA; (**b**) the Q-Q plot of the random component. Mean is the average of the residual component's values, StDev is the standard deviation of the residual component's values, N is the number of the values, AD is the value of the Anderson–Darling statistics from the Anderson–Darling applied to the residual component, and P-value is the p-value computed in the Anderson–Darling test on the residual component.

The highest seasonal index corresponds to November, and the lowest to June (Figure 4a). The highest variations of the detrended series (Figure 4b) are those from November and March and the lowest from October.

The highest percentage variations per season (Figure 4c) were in March and November. The highest variation in the residual component (therefore, the worst fitted value) was in March, and the lowest one was in October (Figure 4d).

A similar behavior is noticed in the case of the multiplicative decomposition model (denoted in the following as DECM). Figure 5 shows the original series, the detrended one, the seasonally adjusted series, and the residual one.

Removing the trend from the initial series increases the series range. The seasonally adjusted series presents a lower variance than the original one, indicating that seasonality is a significant component of the series. The multiplicative decomposition model with a linear trend is slightly worse than the additive one since the mean absolute deviation (MAD) of 0.0773, mean standard deviation (MSD) of 0.0114, and mean absolute percentage error (MAPE) of 18.642 are higher than those in the additive model (0.0767, 0.0098, and 18.4257, respectively). Still, the models do not provide significant differences between the seasonal components, percent variation per season, or residuals per season. The hypotheses of the residuals series normality and homoscedasticity could not be rejected, but the randomness could. Therefore, one should look for a model with uncorrelated residuals to avoid the errors' propagation.

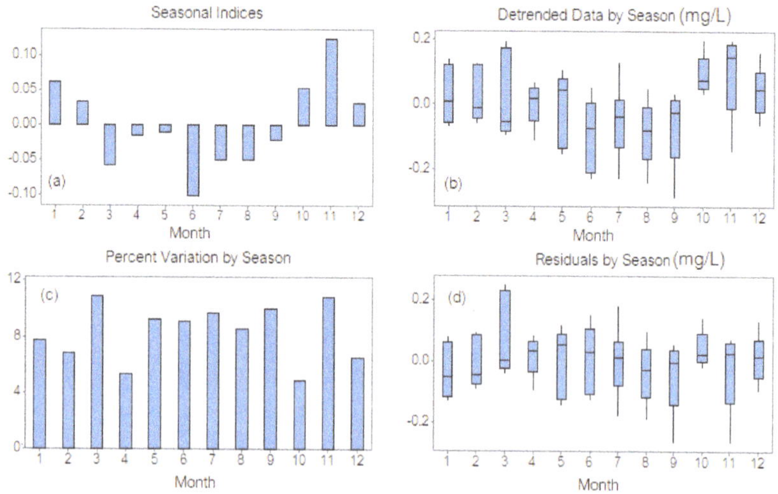

Figure 4. DECA: (**a**) Seasonal indices (1 corresponds to January and 12 to December); (**b**) Detrended data by season; (**c**) Percent variation by season; (**d**) Residuals by season.

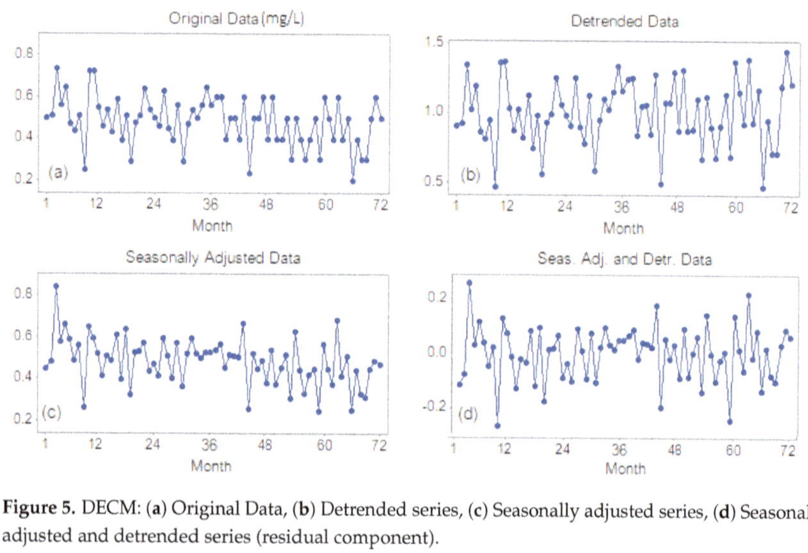

Figure 5. DECM: (**a**) Original Data, (**b**) Detrended series, (**c**) Seasonally adjusted series, (**d**) Seasonally adjusted and detrended series (residual component).

In the second approach (decomposition with a 12th-order moving average trend), the best model was the additive one (denoted as MAA12). Figure 6 shows the initial series (observed), its trend, the seasonal, and the random component (residual). Due to the moving average computation, the trend is not linear or monotonically decreasing.

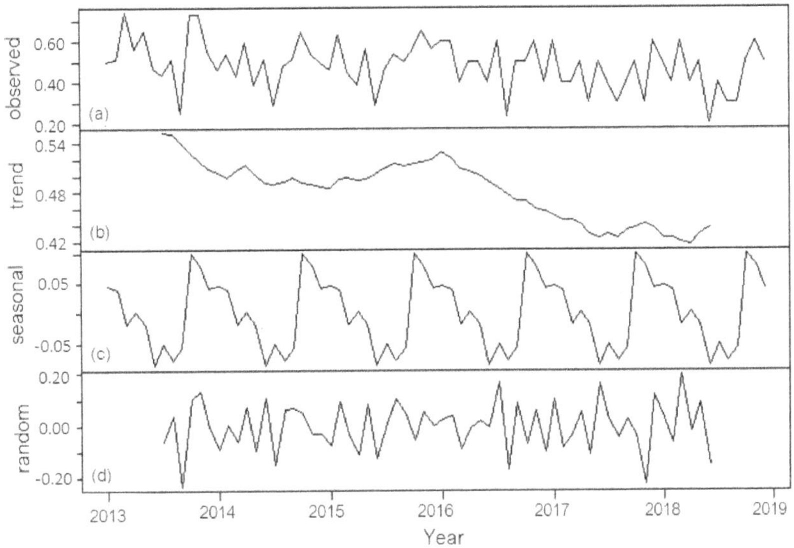

Figure 6. MAA12: (**a**) The initial series. (Observed is the default name given to it by the software); (**b**) Trend; (**c**) Seasonal component; (**d**) Random component.

The seasonal indices are, respectively, Jan = 0.04668, Feb = 0.03876, Mar = −0.01790, Apr = 0.00360, May = −0.01940, June = −0.08790, July = −0.05149, Aug = −0.078569, Sept = −0.05648, Oct = 0.10001, Nov = 0.08059, and Dec = 0.04210. In this case, the highest values of the seasonal component are recorded in October, followed by November, and the lowest in June. The highest seasonal values are correlated to the higher chlorination necessity (in November and December) after the high season and the precipitation absence in summer (to maintain the quality of the drinking water), respectively, to the lowest chlorination necessity in June after the spring season and the high precipitation period.

The random component's analysis provides a p-value of 0.9195 in the Shapiro–Wilk test, so the normality hypothesis cannot be rejected. The correlogram (Figure 7) shows again a first-order autocorrelation of the random component's values.

The hypothesis of the random component's homoscedasticity could not be rejected. For all statistical tests, the significance level was kept at 0.05. In MAA12, which is better than the multiplicative model with a 12th-order moving average trend (denoted as MAM12), MAD = 0.07601, MSD = 0.00870, and MAPE = 18.6546. In terms of MSD and MAD, the MAA12 is the best, while with respect to MAPE, the best is DECA. In both situations, a first-order autocorrelation of the residual series is present, so a third approach, the Holt–Winters method, was proposed to describe the series evolution.

Figure 8a provides the series decomposition using the multiplicative Holt–Winters method (denoted as MHW). The smoothing parameters are $\alpha = 0.04697$, $\beta = 0.07233$, and $\gamma = 0.43818$, and the initial parameters and seasonality indices are $a = 0.38112$, $b = -0.00191$, $s_1 = 0.10025$, $s_2 = 0.03342$, $s_3 = 0.06512$, $s_4 = 0.03010$, $s_5 = 0.01835$, $s_6 = -0.0806$, $s_7 = 0.00810$, $s_8 = -0.09001$, $s_9 = -0.04201$, $s_{10} = 0.11143$, $s_{11} = 0.12542$, and $s_{12} = 0.11005$.

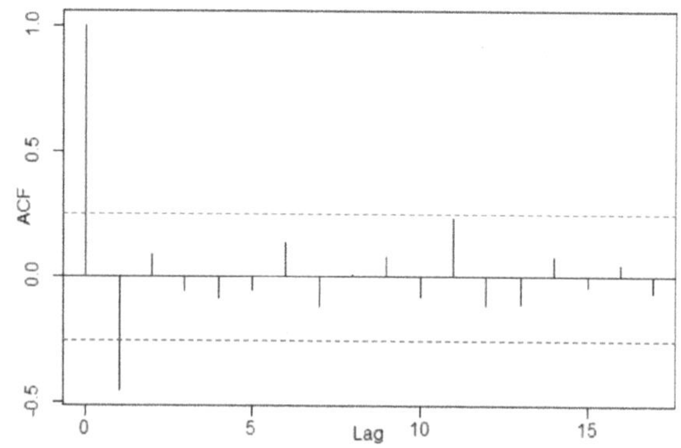

Figure 7. The random component's correlogram in MAA12. The blue dotted line represents the limits of the confidence interval at a 95% confidence level.

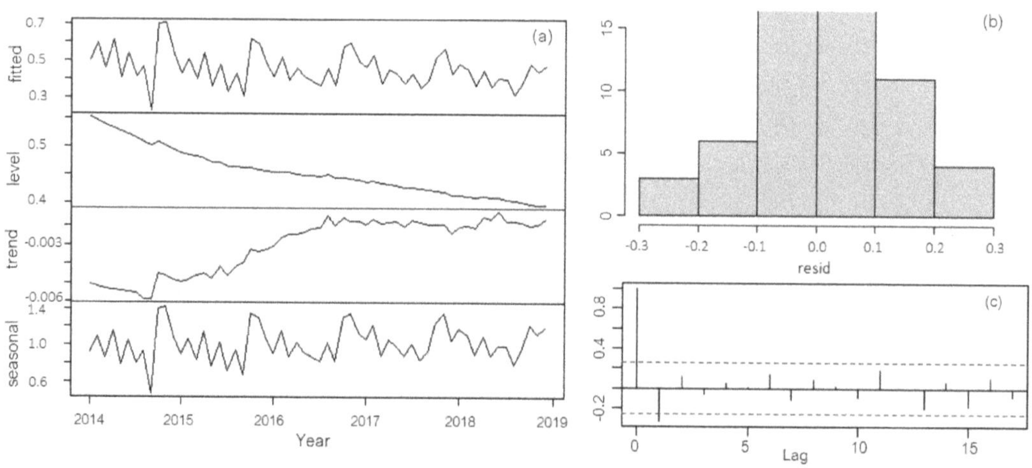

Figure 8. (**a**) Holt–Winters multiplicative model; (**b**) Residuals' histogram; (**c**) Residuals correlogram.

In MHW, the level decreases in time, the trend increases (but not monotonically), and the seasonal component is not constant, according to the regression Equation (4) (or (8) in the multiplicative model).

Adding up the values of the level with the corresponding ones of the trend will result in a decreasing series of values (a decreasing trend in the first approach). Similar results were obtained using the multiplicative Holt–Winters method.

The level compound's shape in MHW is concordant with the time series non-stationarity in level. Among the seasonal components, the highest values are recorded in November and October, followed by December. The seasonal values of the chlorine introduced in water in the treatment station after the high season are higher than in other periods (do not forget that the treatment plant is situated on the Black Sea Littoral in a tourist area, and during summer, the pollution is higher than in the rest of the year) and depends as well on the precipitation record during summer (that can carry the pollutants affecting the source water quality). In the additive Holt–Winters model (denoted as AHW), MAD = 0.0803, MSD = 0.0130, and MAPE = 18.8673, whereas in MHW, the corresponding values are MAD = 0.0772, MSD = 0.0118, and MAPE = 18.2619.

The tests on residuals did not reject their normality (see the histogram in Figure 8b) and homoscedasticity, but the randomness (see the correlogram in Figure 8c).

Figure 9 illustrates the MHW model's forecast for the next 48 months.

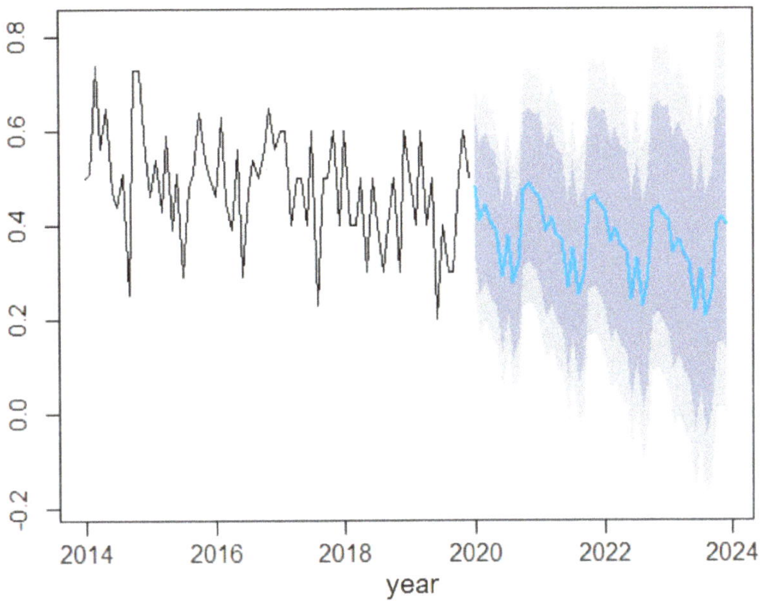

Figure 9. Forecast with the MHW model. The black curve is the series, the blue one is the forecast and the grey backgrounds are the confidence intervals at 95% and 99%, respectively.

The series values are represented in blue, and the confidence intervals at 99% and 95% confidence levels are represented in two nuances of grey. The shape of the forecast curves is similar to that of the data series, confirming the modeling quality.

The advantage of this approach is that the level is considered, and the seasonal indices are updated at each step of the algorithm. The first two models incorporate the level and trend into a single component (trend), which does not reflect the series variation from the base.

The last model is of SARIMA$(0,1,1)(0,1,1)_{12}$ type. For its validation, the residuals' series analysis was performed. The Shapiro–Wilk test indicates that the hypothesis that the series in Gaussian cannot be rejected (p-value > 0.100 > 0.05; Figure 10a), the correlogram (Figure 10b) indicates the correlation absence, and the Levene test (Figure 10c) rejected the heteroskedasticity hypothesis. The p-value associated with the Box–Ljung test is $p = 0.1137$, indicating that the hypothesis of residuals' series independence cannot be rejected. Moreover, MAD = 0.0695, MSD = 0.00868, and MAPE = 16.5426, showing that the SARIMA performs best among all the proposed models.

Figure 11 presents the series forecast based on the built SARIMA model in the blue curve and the confidence intervals at 99% and 95% confidence levels.

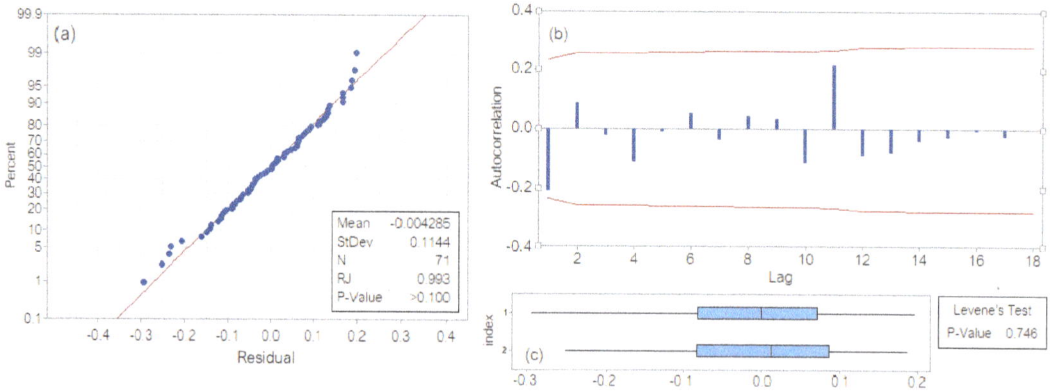

Figure 10. SARIMA model. Residual series analysis (**a**) Results of the Shapiro–Wilk test; (**b**) Correlogram; (**c**) Results of the Levene test.

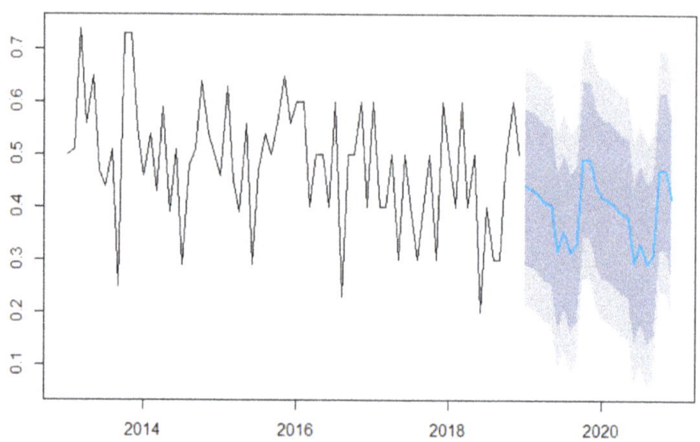

Figure 11. Forecast based on the SARIMA model. The black curve is the series, the blue one is the forecast and the grey backgrounds are the confidence intervals at 95% and 99%, respectively.

To emphasize the performances of the forecast obtained using the MHW and SARIMA, their output was compared with the series values in recorded 2019 (that were not used for modeling). Figure 12 shows that the predicted values obtained using SARIMA are closer to the recorded values via comparison to MHW. The worst forecast was obtained for July and the best one for December.

The goodness of fit indicators for SARIMA (MHW) are MAD = 0.01039 (0.02118), MSD = 0.00016 (0.00068), and MAPE = 2.8181 (5.5738), showing that the SARIMA model is better than MHW.

All the approaches gave good results in modeling the free residual chlorine series, but the best (and more complex one) is the SARIMA(0,1,1)$(0,1,1)_{12}$.

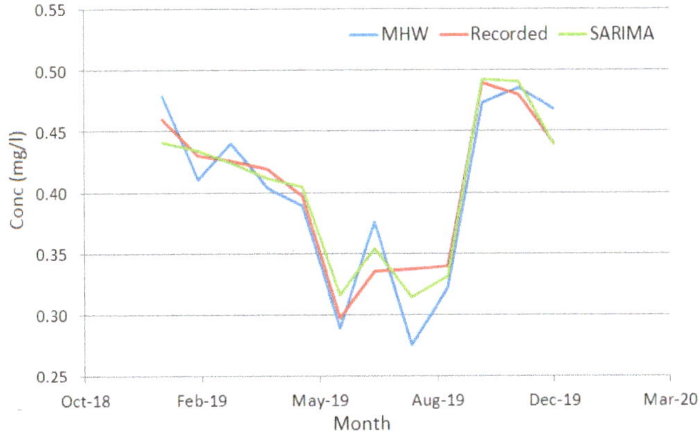

Figure 12. The series recorded in 2019 and the forecast based on MHW and SARIMA models.

As mentioned, the chlorine quantity decreases during the disinfection processes due to the reactions with different substances. Keeping its concentration within optimal limits can be done if this parameter is monitored over time. Traditionally, process-based models to forecast chlorine decay use generally first-order equations [18,23,57]. To build such models, advanced knowledge of the phenomena that appear in the pipes, and accurate and sufficient data on some water parameters in the distribution system are necessary (the last must be experimentally obtained). Often, the coefficients in such models depend on the loading conditions and are not practical for modeling purposes [57]. Therefore, other approaches are required [17,21].

The second approach involves utilizing data-driven statistical models; this means that the forecast of residual chlorine utilizes relationships between the response variable and some regressors. If the experimental data on some variables are difficult to obtain, imprecise, or unavailable, the data-driven models are excellent alternatives to the process-based models [58]. In such models, the knowledge of the processes from the system is less important [14]. Their main advantage is that a deep knowledge of the mathematics and chemistry laws governing chlorine behavior is not necessary [59]. Among the data-driven statistical methods, we mention the linear autoregressive models to predict chlorine concentration and its decay in distribution systems and storage [14,60,61]. This present study falls into this category. Based on our best knowledge, it proposed four models that were first employed for modeling free chlorine monthly series. Therefore, comparisons with the results of similar studies conducted on different series cannot be performed.

4. Conclusions

This article proposed four alternative approaches for modeling monthly free chlorine residual concentration series from PCTP using decomposition, Holt–Winters, and SARIMA models. The novelty of this approach is the use of univariate econometric models in engineering and extending the results of other studies on the water treatment plant in Romania (that previously presented only basic statistical analysis or models of chlorine decay).

In the first approach, the trend was built using a nonparametric Sen's method, which has the advantage that no other restrictions are to be satisfied by the parameters of the linear trend. Another advantage of this method is its simplicity. The second method has the advantage that it can be applied even in a situation when the hypothesis that a monotonic trend exists is rejected. Nevertheless, the twelve values of the series cannot be estimated. In the Holt–Winters method, the seasonality factors and the trend are updated at each step, which gives a more realistic picture of the evolution of each component compared to the classical decomposition. While the first two approaches are simpler, the third one includes a fourth component, the level, as a base from which the series vary. The Holt–Winters model is in concordance with the stationary test results. The SARIMA$(0,1,1)(0,1,1)_{12}$ model is more complex since it involves the first-order differentiation of the series and its seasonal components (to reach its stationarity), and considering the innovation process (by the presence of the moving average, one for both series and seasonality). While the last methodology provides the most accurate results, all the others may be used for modeling and forecast given the easiness and availability of their implementation in MINITAB and R.

In Romania, the studies in the above field are either experimental, present basic statistics of some water parameters series (without correlations to each other) or use the first-order chlorine decay model. Therefore, this article completes the very sparse research in the field. Since the chlorine concentration is regularly monitored, and exceeding the limits imposed by regulation may give birth to protests from the residents that acknowledge the smell and taste of the drinking water, the amount of chlorine must be dosed taking into account the input water quality, resulting from the analyses of chlorine concentrations and the necessity to conform the Romanian regulations.

Despite their performances, the models presented here should be used only for short-time prediction without updating the input given a decreasing trend from the level from which the series' values vary. Updating the input of the models is recommended for improving the forecast. Automating the chlorine concentration monitoring will result in a better dosage and forecast.

Another note is that the models do not include the risk factors and the solution for the situation when the water quality decreases. Therefore, in a future study, these aspects will be considered because there is a need to constantly monitor the water resources and the quality in the water treatment process and to intervene to maintain it.

Author Contributions: Conceptualization, A.B. and L.B.; methodology, A.B. and L.B.; software, A.B.; validation, A.B.; formal analysis, L.B.; investigation, A.B. and L.B.; resources, A.B.; and data curation, A.B.; writing—original draft preparation, A.B. and L.B.; writing—review and editing, A.B. and L.B.; visualization, A.B.; supervision, A.B.; project administration, A.B. All authors have read and agreed to the published version of the manuscript.

Funding: This research received no external funding.

Institutional Review Board Statement: Not applicable.

Informed Consent Statement: Not applicable.

Data Availability Statement: Data will be available upon request from the authors.

Conflicts of Interest: The authors declare no conflict of interest.

References

1. Brandt, M.J.; Johnson, K.M.; Elphinston, A.J.; Ratnayaka, D.D. (Eds.) The Demand for Potable Water. In *Twort's Water Supply*, 7th ed.; Butterworth-Heinemann: Amsterdam, The Netherlands, 2017; Chapter 1; pp. 1–36.
2. Brandt, M.J.; Johnson, K.M.; Elphinston, A.J.; Ratnayaka, D.D. (Eds.) Disinfection of Water. In *Twort's Water Supply*, 7th ed.; Butterworth-Heinemann: Amsterdam, The Netherlands, 2017; Chapter 11; pp. 475–511.
3. Moran, S. (Ed.) Clean water unit operation design: Biological processes. In *An Applied Guide to Water and Effluent Treatment Plant Design*; Butterworth-Heinemann: Amsterdam, The Netherlands, 2018; Chapter 9; pp. 111–116.
4. Mosse, P.; Murray, B. *Good Practice Guide to the Operation of Drinking Water Supply Systems for the Management of Microbial Risk*; Final Report–WaterRA Project 1074; Water Research Australia Limited: Adelaide, Australia, 2015; p. 23. Available online: https://www.wsaa.asn.au/sites/default/files/publication/download/Good%20Practice%20Guide%20April%2015.pdf (accessed on 10 June 2023).
5. Bărbulescu, A.; Barbeș, L. Statistical methods for assessing the water quality after the treatment on a Sequencing Batch Reactor. *Sci. Total Environ.* **2021**, *752*, 141991. [CrossRef] [PubMed]
6. Ecological Risk Models and Tools. U.S. Environmental Protection Agency (EPA). 2021. Available online: https://www.epa.gov/risk/ecological-risk-models-and-tools (accessed on 10 June 2023).
7. Alcayhuamán Guzmán, R.M.; Al-Emam, R.; Alhassan, H.; Ali, A.; Allély-Fermé, D.; Ampomah, B.; Anarna, M.S.S.; Bakir, H.; Bani-Khalaf, R.; Bartram, J.; et al. *Water Safety Plan Manual-Step-by-Step Risk Management for Drinking Water Suppliers*; World Health Organization: Geneva, Switzerland, 2009; Available online: www.who.int (accessed on 10 June 2023).

8. WHO. *Water Safety in Distribution Systems*; World Health Organization Document Production Services: Geneva, Switzerland, 2014; Available online: www.who.int (accessed on 10 June 2023).
9. Basic Information about Chloramines and Drinking Water Disinfection. Available online: https://www.epa.gov/dwreginfo/basic-information-about-chloramines-and-drinking-water-disinfection (accessed on 10 June 2023).
10. Dubey, S.; Gusain, D.; Sharma, Y.C.; Bux, F. The occurrence of various types of disinfectant by-products (trihalomethanes, haloacetic acids, haloacetonitrile) in drinking water. In *Disinfection By-products in Drinking Water*; Priya, T., Mishra, B.K., Prasad, M.N.V., Eds.; Butterworth-Heinemann: Amsterdam, The Netherlands, 2020; Chapter 15; pp. 371–391.
11. Free Chlorine Residual Definition. Available online: https://www.lawinsider.com/dictionary/free-chlorine-residual (accessed on 15 July 2023).
12. Romanian Law no. 458/2002 about the Quality of the Drinking Water. Available online: http://legislatie.just.ro/Public/DetaliiDocument/37723 (accessed on 15 July 2023).
13. Iordache, A.; Woinaroschy, A. Analysis of the efficiency of water treatment process with chlorine. *Environ. Eng. Manag. J.* **2020**, *19*, 1309–1313.
14. Gibbs, M.S.; Morgan, N.; Maier, H.R.; Dandy, G.C.; Nixon, J.B.; Holmes, M. Investigation into the relationship between chlorine decay and water distribution parameters using data driven methods. *Math. Comput. Modell.* **2006**, *44*, 485–498. [CrossRef]
15. Priya, T.; Mishra, B.K.; Prasad, M.N.V. (Eds.) Physico-chemical techniques for the removal of disinfection by-products precursors from water. In *Disinfection By-Products in Drinking Water*; Butterworth-Heinemann: Amsterdam, The Netherlands, 2020; Chapter 2; pp. 23–58.
16. Gang, D.C.; Clevenger, T.E.; Banerji, S.K. Modeling Chlorine Decay in Surface Water. *J. Environ. Inform.* **2015**, *1*, 21–27. [CrossRef]
17. Gómez-Coronel, L.; Delgado-Aguiñaga, J.A.; Santos-Ruiz, I.; Navarro-Díaz, A. Estimation of Chlorine Concentration in Water Distribution Systems Based on a Genetic Algorithm. *Processes* **2022**, *11*, 676. [CrossRef]
18. Monteiro, L.; Figueiredo, D.; Dias, S.; Freitas, R.; Covas, D.; Menaia, J.; Coelho, S.T. Modeling of chlorine decay in drinking water supply systems using EPANET MSX. *Procedia Eng.* **2014**, *70*, 1192–1200.
19. García-Ávila, F.; Avilés-Añazco, A.; Ordoñez-Jara, J.; Guanuchi-Quezada, C.; Flores del Pino, L.; Ramos-Fernández, L. Modeling of residual chlorine in a drinking water network in times of pandemic of the SARS-CoV-2 (COVID-19). *Sustain. Environ. Res.* **2021**, *31*, 12. [CrossRef]
20. Nejjari, F.; Puig, V.; Pérez, R.; Quevedo, J.; Cugueró-Escofet, M.À.; Sanz, G.; Mirats, J. Chlorine Decay Model Calibration and Comparison: Application to a Real Water Network. *Procedia Eng.* **2014**, *70*, 1221–1230. [CrossRef]
21. Zhang, Y.; Gao, X.; Smith, K.; Inial, G.; Liu, S.; Conil, L.B.; Pan, B. Integrating water quality and operation into prediction of water production in drinking water treatment plants by genetic algorithm enhanced artificial neural network. *Water Res.* **2019**, *164*, 114888. [CrossRef]
22. Gámiz, J.; Grau, A.; Martínez, H.; Bolea, Y. Automated Chlorine Dosage in a Simulated Drinking Water Treatment Plant: A Real Case Study. *Appl. Sci.* **2020**, *10*, 4035. [CrossRef]
23. Pérez, R.; Martínez Torrents, A.; Martínez, M.; Grau, S.; Vinardell, L.; Tomàs, R.; Martínez Lladó, X.; Jubany, I. Chlorine Concentration Modelling and Supervision in Water Distribution Systems. *Sensors* **2022**, *22*, 5578. [CrossRef] [PubMed]
24. Farooq, S.; Hasmi, I.; Qazi, I.A.; Qaiser, A.; Rasheed, S. Monitoring of Coliforms and chlorine residual in water distribution network of Rawalpindi, Pakistan. *Environ. Monit. Assess.* **2008**, *140*, 339–347. [CrossRef] [PubMed]
25. Cun, C.; Durand, M.; Leguyader, M.; Martin, J.; Vilagines, R. Statistical study of the relationship between free chlorine levels and bacteriological checks B on systems in the Paris area. *Sci. Total Environ.* **2002**, *284*, 49–59. [CrossRef]
26. Paun, I.; Chiriac, F.L.; Iancu, V.I.; Pirvu, F.; Niculescu, M.; Vasilache, N. Disinfection by-products in drinking water distribution system of Bucharest City. *Rom. J. Ecol. Environ. Chem.* **2021**, *3*, 13–18. [CrossRef]
27. Vîrlan, C.-M.; Toma, D.; Stătescu, F.; Marcoie, N.; Prăjanu, C.-C. Modeling the chlorine-conveying process within a drinking water distribution network. *Environ. Eng. Manag. J.* **2021**, *20*, 487–494.
28. Aivaz, K.-A.; Florea Munteanu, I.; Stan, M.-I.; Chiriac, A. A Links Between Transport Noncompliance and Financial Uncertainty in Times of COVID-19 PandMultivariate Analysis on the emics and War. *Sustainability* **2022**, *14*, 10040. [CrossRef]
29. Aivaz, K.-A. Correlations Between Infrastructure, Medical Staff and Financial Indicators of Companies Operating in the Field of Health and Social Care Services. The Case of Constanta County, Romania. In *Under the Pressure of Digitalization: Challenges and Solutions at Organizational and Industrial Levels*, 1st ed.; Edu, T., Schipor, G.-L., Vancea, D.P.C., Zaharia, R.M., Eds.; Filodiritto Publisher, Inforomatica SRL: Bologna, Italy, 2021; pp. 17–25.
30. Vancea, D.P.C.; Aivaz, K.-A.; Duhnea, A. Political Uncertainty and Volatility on the Financial Markets-the Case of Romania. *Transform. Bus. Econ.* **2017**, *16*, 457–477.
31. Bărbulescu, A.; Dumitriu, C.S. ARIMA and Wavelet-ARIMA models for the signal produced by ultrasound in diesel. In Proceedings of the 2021 25th International Conference on Systems, Theory, Control and Computing (ICSTCC), Iasi, Romania, 20–23 October 2021. [CrossRef]
32. Bărbulescu, A.; Nazzal, Y.; Howari, F. Assessing the groundwater quality in the Liwa area, the United Arab Emirates. *Water* **2020**, *12*, 2816. [CrossRef]
33. Bărbulescu, A.; Barbeș, L. Models for pollutants' correlation in the Romanian littoral. *Rom. Rep. Phys.* **2014**, *66*, 1189–1199.
34. Nazzal, Y.H.; Bărbulescu, A.; Howari, F.; Al-Taani, A.A.; Iqbal, J.; Xavier, C.M.; Sharma, M.; Dumitriu, C.S. Assessment of metals concentrations in soils of Abu Dhabi Emirate using pollution indices and multivariate statistics. *Toxics* **2021**, *9*, 95. [CrossRef]

35. Bărbulescu, A.; Barbeș, L. Assessing the Danube River water quality of the Danube River (at Chiciu, Romania) by statistical methods. *Environ. Earth Sci.* **2020**, *79*, 122.
36. Singh, I.; Juneja, P.; Kaur, B.; Kumar, P. Pharmaceutical Applications of Chemometric Techniques. *Int. Scholarly Resear. Not.* **2013**, *13*, 795178. [CrossRef]
37. Frîncu, R.-M. Long-Term Trends in Water Quality Indices in the Lower Danube and Tributaries in Romania (1996–2017). *Int. J. Environ. Res. Public Health* **2021**, *18*, 1665. [CrossRef] [PubMed]
38. Council Directive 80/778/EEC Relating to the Quality of Water Intended for Human Consumption. Available online: https://www.fao.org/faolex/results/details/en/c/LEX-FAOC037618/ (accessed on 15 July 2023).
39. Council Directive 98/83/EC on the Quality of Water Intended for Human Consumption-Repealed. Available online: https://www.fao.org/faolex/results/details/en/c/LEX-FAOC018700 (accessed on 15 July 2023).
40. Water Framework Directive. Directive 2000/60/EC of the European Parliament and of the Council. 2000. Available online: https://eur-lex.europa.eu/resource.html?uri=cellar:5c835afb-2ec6-4577-bdf8-756d3d694eeb.0004.02/DOC_1&format=PDF (accessed on 15 July 2023).
41. County Report. Drinking Water Quality. Constanta. 2020. Available online: https://dspct.ro/wp-content/uploads/2021/08/RAPORT-JUDETEAN-APA-POTABILA-2020.pdf (accessed on 20 July 2023). (In Romanian).
42. Xie, Y.; Zilberman, D. Theoretical implications of institutional, environmental, and technological changes for capacity choices of water projects. *Water Resour. Econ.* **2016**, *13*, 19–29. [CrossRef]
43. Lanz, B.; Provins, A. The demand for tap water quality: Survey evidence on water hardness and aesthetic quality. *Water Resour. Econ.* **2016**, *16*, 52–63. [CrossRef]
44. Gel, Y.R.; Gastwirth, J.L. A robust modification of the Jarque-Bera test of normality. *Econ. Lett.* **2008**, *99*, 30–32. [CrossRef]
45. Shapiro, S.S.; Wilk, M.B. An analysis of variance test for normality (complete samples). *Biometrika* **1965**, *52*, 591–611. [CrossRef]
46. Anderson, T.W.; Darling, D.A. A Test of Goodness-of-Fit. *J. Am. Stat. Assoc.* **1954**, *49*, 765–769. [CrossRef]
47. Levene, H. Robust test for equality of variances. In *Contributions to Probability and Statistics: Essays in Honor of Harold Hotelling*; Olkin, I., Ed.; Stanford University Press: Stanford, CA, USA, 1960; pp. 278–292.
48. Kwiatkowski, D.; Phillips, P.C.B.; Schmidt, P.; Shin, Y. Testing the null hypothesis of stationarity against the alternative of a unit root: How sure are we that economic time series have a unit root? *J. Econ.* **1992**, *54*, 159–178.
49. Kendall, M.G. *Rank Correlation Methods*, 5th ed.; Oxford University Press: London, UK, 1990.
50. Mann, H.B. Non-parametric tests against trend. *Econometrica* **1945**, *13*, 245–259. [CrossRef]
51. Hipel, K.W.; McLeod, A.I. *Time Series Modelling of Water Resources and Environmental Systems*; Elsevier Science: New York, NY, USA, 1994.
52. Sen, P.K. Estimates of the regression coefficient based on Kendall's tau. *J. Am. Stat. Assoc.* **1968**, *63*, 1379–1389. [CrossRef]
53. 3.2 Time Series Components. Available online: https://otexts.com/fpp3/components.html (accessed on 23 April 2023).
54. 8.3 Methods with Seasonality. Available online: https://otexts.com/fpp3/holt-winters.html (accessed on 23 April 2023).
55. SARIMA Models. Available online: https://real-statistics.com/time-series-analysis/seasonal-arima-sarima/sarima-models/ (accessed on 23 April 2023).
56. Ljung, G.M.; Box, G.E.P. On a Measure of a Lack of Fit in Time Series Models. *Biometrika* **1978**, *65*, 297–303. [CrossRef]
57. Jonkergouw, P.M.R.; Khu, S.-T.; Savic, D.A.; Zhong, D.; Hou, X.Q.; Zhao, H.-B. A Variable Rate Coefficient Chlorine Decay Model. *Environ. Sci. Technol.* **2009**, *43*, 408–414. [CrossRef] [PubMed]
58. Rodriguez, M.J.; West, J.R.; Powell, J.; Serodes, J.B. Application of two approaches to model chlorine residuals in Severn Trent Water Ltd. (STW) distribution systems. *Water Sci. Technol.* **1997**, *36*, 317–324. [CrossRef]
59. Serodes, J.B.; Rodriguez, M.J.; Ponton, A. Chlorcast ©: A methodology for developing decision-making tools for chlorine disinfection control. *Environ. Modell. Softw.* **2001**, *16*, 53–62. [CrossRef]
60. Serodes, J.B.; Rodriguez, M.J. Predicting residual chlorine evolution in storage tanks within distribution systems: Application of a neural network approach. *J. Water Supp. Resear. Techn.-Aqua* **1996**, *45*, 57–66.
61. Rodriguez, M.J.; Serodes, J.B. Assessing empirical linear and non-linear modelling of residual chlorine in urban drinking water systems. *Environ. Modell. Softw.* **1999**, *14*, 93–102. [CrossRef]

toxics MDPI

Article

A New Method for Ecological Risk Assessment of Combined Contaminated Soil

Qiaoping Wang [1], Junhuan Wang [1], Jiaqi Cheng [1,2], Yingying Zhu [1,3], Jian Geng [1,4], Xin Wang [1,4], Xianjie Feng [1,5] and Hong Hou [1,*]

1 State Key Laboratory of Environmental Criteria and Risk Assessment, Chinese Research Academy of Environmental Sciences, Beijing 100012, China; wangqiaoping20@mails.ucas.ac.cn (Q.W.)
2 School of Environmental Science and Engineering, Shaanxi University of Science and Technology, Xi'an 710021, China
3 College of Resources and Environment, Shanxi Agricultural University, Taigu 030801, China
4 College of Land and Environment, Shenyang Agricultural University, Shenyang 110866, China
5 School of Environmental Science and Engineering, Nanjing University of Information Science and Technology, Nanjing 210044, China
* Correspondence: houhong@craes.org.cn

Abstract: Ecological risk assessment of combined polluted soil has been conducted mostly on the basis of the risk screening value (*RSV*) of a single pollutant. However, due to its defects, this method is not accurate enough. Not only were the effects of soil properties neglected, but the interactions among different pollutants were also overlooked. In this study, the ecological risks of 22 soils collected from four smelting sites were assessed by toxicity tests using soil invertebrates (*Eisenia fetida*, *Folsomia candida*, *Caenorhabditis elegans*) as subjects. Besides a risk assessment based on RSVs, a new method was developed and applied. A toxicity effect index (*EI*) was introduced to normalize the toxicity effects of different toxicity endpoints, rendering assessments comparable based on different toxicity endpoints. Additionally, an assessment method of ecological risk probability (*RP*), based on the cumulative probability distribution of *EI*, was established. Significant correlation was found between *EI*−based *RP* and the *RSV*−based Nemerow ecological risk index (*NRI*) ($p < 0.05$). In addition, the new method can visually present the probability distribution of different toxicity endpoints, which is conducive to aiding risk managers in establishing more reasonable risk management plans to protect key species. The new method is expected to be combined with a complex dose–effect relationship prediction model constructed by machine learning algorithm, providing a new method and idea for the ecological risk assessment of combined contaminated soil.

Keywords: combined contaminated soil; toxicity effect index (*EI*); ecological risk probability (*RP*); soil invertebrates

Citation: Wang, Q.; Wang, J.; Cheng, J.; Zhu, Y.; Geng, J.; Wang, X.; Feng, X.; Hou, H. A New Method for Ecological Risk Assessment of Combined Contaminated Soil. *Toxics* **2023**, *11*, 411. https://doi.org/10.3390/toxics11050411

Academic Editors: Alina Barbulescu and Lucica Barbes

Received: 20 March 2023
Revised: 11 April 2023
Accepted: 21 April 2023
Published: 26 April 2023

1. Introduction

Studies on the ecological risk assessment of soil are relatively scarce, especially compared with human health risk assessment [1]. The United States Environmental Protection Agency (USEPA) has formulated guidelines for the derivation of ecological risk screening values (*RSV*) and derived the RSVs of some single pollutants. In China, the toxic effects of cadmium, copper, nickel, zinc, lead, arsenic and antimony on crops, vegetables and invertebrates have been investigated and their thresholds in farmland soils have been estimated [2–7]. However, China has yet to establish a system for managing soil ecological risks.

RSVs can roughly screen out potential high−risk pollutants in soil. A single risk index (*RI*) based on RSVs is commonly used to characterize the ecological risk of single pollutants, and the Nemerow risk index (*NRI*), based on *RI*, is used to characterize the overall ecological risk of combined pollutants [8–13]. However, the deduction process of RSVs neglects the influence of soil properties on toxicity effects, while *NRI* overlooks the complex interaction among different pollutants. Therefore, there is a great deal of uncertainty in the ecological risk assessment of combined contaminated soil by *RI* and *NRI* (Scheme 1a). In fact, most actual sites were contaminated by various pollutants, especially metal smelting, coking and other industries [11,14,15]. Under combined pollution states, complex interactions among pollutants could affect their bioavailability and biotoxicity, making them different from those under a single contamination state. Therefore, these differences should be fully considered in the formulation of risk assessment methods. Although some studies have established the relationship between toxic effects and soil physicochemical properties, and the relationship between toxic effects and bioavailable concentration, such research data are very limited [16–20]. The combined effects of different pollutants have been investigated and concentration addition (CA) and independent action (IA) models were applied in the risk assessment process [21–23]. Nevertheless, the actual contaminated soils were of infinite composition, while the joint effect studies cannot be exhaustive.

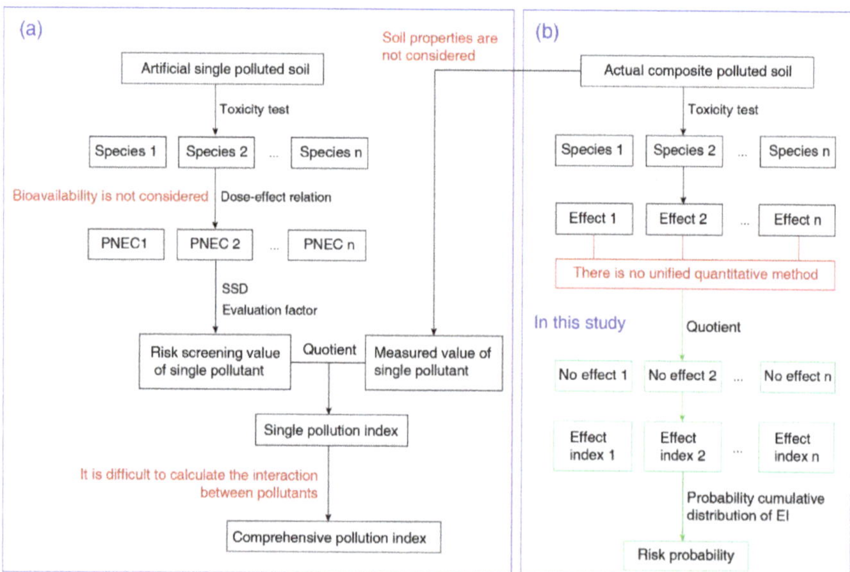

Scheme 1. Two methods to assess ecological risk.

The real toxicity of actual polluted soil can be evaluated by toxicity tests conducted in the actual polluted soil (Scheme 1b) [24,25]. Normally, whether a soil is at risk to a testing organism at a toxicity endpoint can be determined by comparison with background soil. However, it is difficult to compare the risk of different toxicity endpoints due to the lack of a uniform quantification method [26–28].

Given multiple toxicity endpoints, the question of how to comprehensively conduct ecological risk assessment needs further exploration. In this study, toxicity tests using typical invertebrates *Eisenia fetida*, *Folsomia candida* and *Caenorhabditis elegans* as subjects were carried out in 22 actual complex contaminated soils. The purpose of the research was to explore a set of ecological risk assessment methods that can fully consider each of the toxicity endpoints, providing new ideas and schemes for future ecological risk assessment research.

2. Materials and Methods

2.1. Soils

Twenty−two heavy metal polluted topsoil samples (0–20 cm) were collected from four contaminated sites in the metal smelting industry (BY, DQ, QL, ZS). Soil samples were air−dried, screened (2 mm), and thoroughly mixed before use. The detection methods for soil physicochemical properties and pollutant content are as follows.

2.1.1. Soil pH

Soil pH was measured by potentiometric method using an acidometer (FE28, Mettler Toledo Technology (Shanghai, China) Co., Ltd.).

2.1.2. Organic Matter Content

An amount of 0.05–0.5 g soil was placed into a test tube and 10 mL of 0.4 mol/L $K_2Cr_2O_7$ sulfuric acid solution was added. After heating in an oil bath at 170–180 °C for 5 min, the solution was transferred to a conical flask and water was added to 50 mL. The excess $K_2Cr_2O_7$ was titrated with a 0.1 mol/L $FeSO_4$ solution. The organic matter (OM) content of soil was calculated based on the consumption of $K_2Cr_2O_7$.

2.1.3. Clay Content

The determination method of soil clay (d < 0.002 mm) content refers to the pipette method in the Chinese standard HJ 1068−2019 [29].

2.1.4. Cation Exchange Capacity

At (20 ± 2) °C, 3.5 g soil was extracted with a 50.0 mL 1.66 cmol/L $Co(NH_3)_6Cl_3$ solution, and the cations in the soil were exchanged by $Co(NH_3)_6Cl_3$ into the solution. The absorbance of the solution was measured at 475 nm using ultraviolet−visible spectrophotometer (UV756, Shanghai Youke Instrument Co., Ltd. in Shanghai, China). The cation exchange capacity (CEC) of the soil was calculated according to the absorbance difference of the leaching solution before and after leaching.

2.1.5. Mn, Fe, and Al Content

An amount of 0.1–0.2 g soil was placed into a closed digestion tank of polytetrafluoroethylene, amounts of 6 mL hydrochloric acid (1.19 g/mL) and 2 mL nitric acid (1.42 g/mL) were then added before being digested with a microwave at 120, 150 and 185 °C for 2, 5 and 40 min respectively, and the total content of Mn, Fe and Al was determined using an inductively coupled plasma mass spectrometer (7900 ICP−MS, Agilent in Tokyo, Japan).

2.1.6. Hg, As and Sb Content

An amount of 0.1–0.5 g soil was placed into a closed digestion tank, amounts of 6 ML of hydrochloric acid (1.19 g/ML) and 2 ML of nitric acid (1.42 g/ML) were then added before being digested with a microwave at 100, 150, and 180 °C for 2, 3 and 25 min respectively. The total content of Hg, As and Sb was determined using an atomic fluorescence spectrophotometer (AFS−8510, Beijing Haiguang Instrument Co., Ltd. in Beijing, China).

2.1.7. Cr, Pb, Cu, Zn and Cd Content

An amount of 0.25–0.5 g soil was placed into a closed digestion tank. Amounts of 6 mL nitric acid (1.42 g/mL), 3 mL hydrochloric acid (1.19 g/mL) and 2 mL hydrofluoric acid (1.16 g/mL) were added in turn before being digested with a microwave at 120, 160 and 190 °C for 3, 3 and 25 min respectively. The total content of Cr, Pb, Cu and Zn was determined with flame atomic absorption spectroscopy (280FS AA, Agilent in Selangor, Malaysia), and determine the total content of Cd with graphite furnace atomic absorption spectrometer (240Z AA, Agilent in Selangor, Malaysia).

2.2. Test Organisms

Eisenia fetida, bought from Kunlong Farm (Beijing, China), was cultured indoors for at least two weeks before tests and was regularly fed with oats. The temperature was controlled at 20 ± 2 °C, with 16 h of light (light intensity 400–800 lx), 8 h of darkness, and soil water content that was maintained at 50%. Active and similar mature earthworms were selected during the experiment, and their mass was in the range of 300~500 mg. A population of *Folsomia candida*, originally from the Institute of Soil Sciences, Chinese Academy of Sciences, had been cultured in our laboratory for over six years using an artificial climate box (Ningbo Saifu Experimental Instrument Co., Ltd., in Ningbo, China). This population was reared in a moist mixture of gypsum and charcoal at 20 ± 1 °C using a 16−h light/8−h dark light regime and fed small amounts of dry bread yeast twice a week. Distilled water was added weekly to maintain the moisture content of the medium. *Caenorhabditis elegans* (var. Bristol strain N2) and *Escherichia coli* (strain OP50) were purchased from Fujian Shangyuan Biotechnology (Fuzhou, China). *C. elegans* was cultured in Nematode−growth−medium (NGM) agar at 20 ± 1 °C in a constant temperature incubator. *E. coli* OP50 strains were used as the food source of nematode. In order to reduce the influence of individual differences of nematode on the experiment, it was necessary to conduct synchronous culture of nematode before the experiment. See Supplementary Materials for the preparation of NGM agar, *E. coli* culture and synchronous culture of nematode.

2.3. Toxicity Tests

2.3.1. *E. fetida* Toxicity Tests

Before toxicity tests, the earthworms were placed in a beaker with wet filter paper and treated with intestinal cleansing in an incubator at 20 °C for 24 h. An amount of 500 g of air−dried soil was placed in a beaker with 5 g of cow dung and an appropriate amount of deionized water. During the entire exposure period, the soil moisture content was maintained within 10% change by weighing and water replenishing. Three days later, 10 earthworms were added to the soil and cultured in an artificial climate box (Ningbo Saifu Experimental Instrument Co., Ltd., Ningbo, China). Other culture conditions were the same as feeding conditions. Three parallel groups were set up in each soil and observed continuously for 56 days. The number of dead earthworms was recorded every day, and the dead earthworms were promptly removed from the beaker [30,31].

2.3.2. *F. candida* Toxicity Tests

An amount of 30 g of air−dried soil and an appropriate amount of deionized water were placed in a culture tank. During the whole exposure period, water was added by weighing method to keep the soil moisture content within a 10% change. After three days, 10 synchronous cultured *F. candida* and a small amount of yeast were added to the soil. Yeast was supplemented once a week, with 6 parallel groups set for each soil. On the 7th and 28th days, 3 parallel groups were taken from each soil sample, and the whole system in the culture tank was poured into a large 1 L beaker. An appropriate amount of deionized water and a few drops of blue ink were then added. The soil suspension was stirred from bottom to top with a glass rod and left to stand for 1–2 min after stirring. After all the adults and larvae were floating on the water surface, photos were taken for preservation. Image J software was used to count the number of surviving adults and newborn larvae.

2.3.3. *C. elegans* Toxicity Tests

An amount of 0.5 g of air−dried soil was placed in the test hole of culture plate, to which was added 100 µL of *E. coli* culture (15 mg·mL^{-1}). In order to meet the food and water requirements of *C. elegans* during toxic exposure, the soil water content was adjusted to 80% of the maximum field capacity. Ten first−instar nematode larvae were added to the soil samples through capillary tubes, then the culture plates were sealed and exposed at 20 °C for 96 h without light. Three parallel groups were set for each soil sample. After exposure, 500 μL 0.3 g·L^{-1} acid red 94 ($C_{20}H_2C_{14}I_4Na_2O_5$) solution was added to the culture plate to dye the cuticle of nematodes. The culture plate was placed in an electric blast drying oven, and the nematodes were killed at high temperature (80 °C) to terminate the toxicity test. The nematodes in the culture plates were recovered by liquid silica suspension and stored in a centrifuge tube at low temperature (4 °C). The recovered nematodes were poured into the petri dish. Under the microscope, the lengths of the nematodes bodies were checked as well as the fertility (if there was at least one egg in the nematode body, it was considered fertile) and the number of larvae.

2.4. Ecological Risk Assessment Based on RSVs

The ecological risk of a single pollutant in soil can be reflected by the single ecological risk index (*RI*). The calculation formula is as follows:

$$RI = \frac{C}{RSV} \tag{1}$$

where *RI* is the single ecological risk index, C is the measured concentration of single pollutants in soil (mg·kg^{-1}), and *RSV* represents the risk screening values in soil (mg·kg^{-1}). All adopted RSVs are listed in Table 1.

Table 1. Risk screening values of soil invertebrates (mg·kg^{-1}) [32–38].

Zn	Cu	Cr	Pb	Cd	As	Sb	Hg	Mn
120 [1]	80 [1]	150 [2]	1700 [1]	140 [1]	25 [2]	78 [1]	1.3 [2]	450 [1]

[1] Data referenced from the US Environmental Protection Agency. [2] Data referenced from China's "Soil environmental quality Risk control standard for soil contamination of agricultural land".

The Nemerow ecological risk index (*NRI*) can reflect the ecological risk of complex pollutants in soil and highlight the impact of high concentrations of pollutants on ecological risk. The calculation formula is as follows:

$$NRI = \sqrt{\frac{RI_{mean}^2 + RI_{max}^2}{2}} \tag{2}$$

where RI_{mean} is the average *RI* and RI_{max} is the maximum *RI*. *NRI* can be divided into five levels: safe ($NRI \leq 0.7$); ecological warning line ($0.7 < NRI \leq 1.0$); low risk ($1 < NRI \leq 2.0$); medium risk ($2.0 < NRI \leq 3.0$); and high risk ($NRI > 3.0$).

2.5. Ecological Risk Assessment Based on Toxicity Tests

2.5.1. Toxicity Effect Index

Toxicity effect index (*EI*) was defined to represent the toxicity effect of soil on the designated toxicity endpoint of organisms, and the formula is as follows:

$$EI = \frac{E}{NOE} \tag{3}$$

where *E* is the effect of experimental groups and *NOE* is the observed effect of control groups. Specific meanings and values vary by species and toxicity endpoints. In the survival experiment of earthworms/springtails, *E* is the number of living earthworms/springtails after the experiment and *NOE* is the total number of earthworms in the experimental group (*NOE* = 10). In the springtails/nematode propagation test, *E* is the number of larvae in the experimental group and *NOE* is the number of larvae in the control group (springtail *NOE* = 204; nematode *NOE* = 600). In the nematode development experiment, *E* is the number of nematodes with fertility, and *NOE* is the number of nematodes released in the experiment (*NOE* = 10). In the nematode growth test, *E* is the average body length of the experimental group and *NOE* is the average body length of the control group (*NOE* = 1200).

In theory, $0 \leq EI \leq 1$; when $EI = 1$, soil has no toxic effect on organisms at all, when $EI = 0.8$, soil has 20% toxic effect on organisms, and so on.

2.5.2. Cumulative Probability Distribution Curve of *EI* and Risk Probability

The *EI* was ranked from smallest to largest to determine its rank R (the rank of the lowest toxicity value is 1, the rank of the second is 2, and so on. If two or more *EI* are the same, they were arranged into a consecutive rank arbitrarily). The cumulative risk probability (*RP*) of *EI* is calculated by the following formula:

$$RP = \frac{R}{N+1} \tag{4}$$

where *R* is the rank of *EI* and *N* is the total number of toxicity endpoints. Using *EI* as the independent variable *X* and the corresponding *RP* as the dependent variable *Y*, the logistic model is used for fitting. The fitting formula is as follows:

$$Y = \frac{1}{1 + e^{\frac{P_1 - X}{P_2}}} \tag{5}$$

where *e* is a natural constant and P_1 and P_2 are parameters.

To illustrate the model, three groups of rational numbers were randomly generated in the intervals (0, 0.25), (0, 1) and (0.85,1) to simulate the *EI* distribution of the samples of the three soils (Figure 1). Twenty−one numbers in each group represent the *EI* of 21 toxicity endpoints. Samples 1, 2 and 3 simulated the soils of very high, medium and very low risk respectively. The size (range) of *RP* is determined by the position of the curve and the value (range) of *EI*. When determining the risk probability, the risk decision−maker should first set an acceptable safety effect index (*SEI*), and then calculate the *RP* corresponding to the *SEI* according to the curve. If the risk decision maker believes that it is safe for organisms when no toxic effects were observed at all, *SEI* can be set as 1. When the toxicity effect index below 20% is considered to be biosafe, $SEI = 0.8$; If the toxicity effect index less than 50% is considered to be safe for organisms, $SEI = 0.5$, and so on. The soil at the sampling point has a unique and definite risk probability at a definite *SEI* and a given soil sample has a unique and definite risk probability at a definite *SEI*. For example, when $SEI = 0.5$, the *RP* of Sample 2 is a. For soil sample of an extreme high risk (Sample 1) and very low risk (Sample 3), the ecological risk probability can be directly judged without the cumulative probability distribution curve of *EI*: if $EI_{max} < SEI$, $RP = 1$; if $EI_{min} > SEI$, $RP = 0$.

2.6. Data Processing and Statistical Analysis

Two−factor analysis of variance without duplication was performed in Excel 2019 to analyze whether different soils and different toxicity endpoints had significant effects on *EI*. The EEC−SSD software was used to fit the cumulative probability distribution curve, and root mean square error (RMSE) and probability p value of Kolmogorov–Smirnov test (K−S test) were used to evaluate the fitting effect. The closer the RMSE is to 0, the higher the accuracy of model fitting is. If $p > 0.05$, this indicates that the fitting passes the K–S test and the model conforms to the theoretical distribution. The *Hmisc* package of R 4.2.1 software was used to calculate the Spearman correlation coefficient between soil components and the *EI* of different toxicity endpoints, as well as the Pearson and Spearman correlation coefficient between the *EI* of different toxicity endpoints. RStudio software and Origin 2017 software were applied to draw figures and Adobe Illustrator CC 2017 was used to merge images and add annotations.

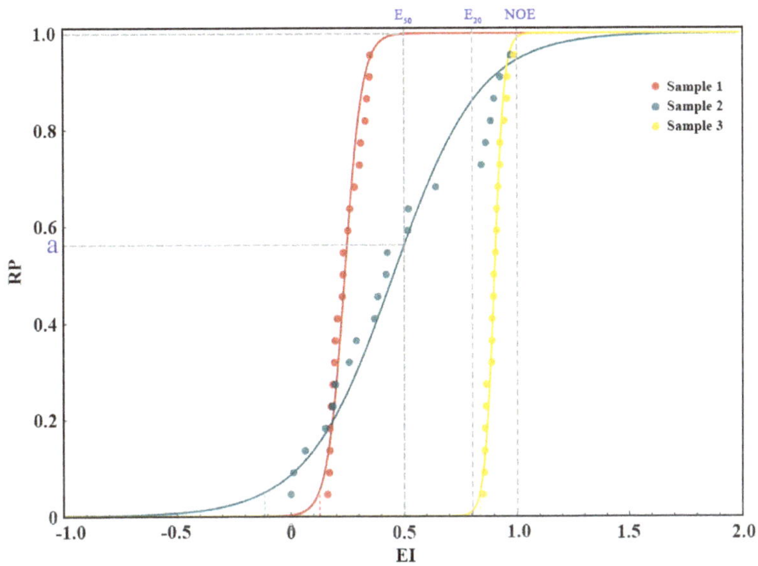

Figure 1. Cumulative probability distribution curve of *EI*—schematic diagram.

3. Results and Discussion

3.1. Ecological Risk Assessment Based on RSVs

Based on the RSVs (Table 1) and the detected concentration of pollutants (Table S1), the *RI* for each of the pollutants in the soil were calculated (Table S2). When $RI \geq 1$, this indicates that the pollutant had potential risk to soil invertebrates. When $RI < 1$, this shows that the pollutant has no risk to soil invertebrates. Figure S1 illustrates the potential risk pollutants in each sampling site for soil invertebrates. The potential risk pollutants in BY were Zn, Cu, Hg, Pb, Cd, As and Mn. DQ has no potential risk pollutants. The potential risk pollutants in QL were As, Sb, Cr, Zn and Cu. The high potential risk pollutants in ZS are Cu, Zn and As.

Based on *RI*, *NRI* was calculated (Table S2, Figure 2). All soil samples collected from site BY were at high risk ($NRI > 3.0$). Except QL−1 (low risk, $1.0 < NRI \leq 2.0$), all soils from site QL were at high risk ($NRI > 3.0$). One soil sample (ZS−4) from site ZS was at high risk ($NRI > 3.0$), one (ZS−1) was at the warning line ($0.7 < NRI \leq 1.0$), while the others were at low risk ($1.0 < NRI \leq 2.0$). All soils samples collected from the site DQ were safe for soil invertebrates.

3.2. Ecological Risk Assessment Based on Toxicity Tests

A total of 11 toxicity endpoints were selected (Table S3). The five toxicity endpoints of earthworm were 7−day survival (ES.7D), 14−day survival (ES.14D), 21−day survival (ES.21D), 28−day survival (ES.28D), and 56−day survival (ES.56D). The three toxicity endpoints of springtails were 7−day survival (TS.7D), 28−day survival (TS.28D), and 28−day reproduction (TR.28D). The endpoints of 4−day pregnancy (NP.4D), 4−day reproduction (NR.4D), 4−day body length (NL.4D) were chosen as toxicity endpoints of nematodes. The unreplicated two−factor analysis of variance indicated that soil samples and toxicity endpoints had significant influence on *EI* ($p < 0.01$).

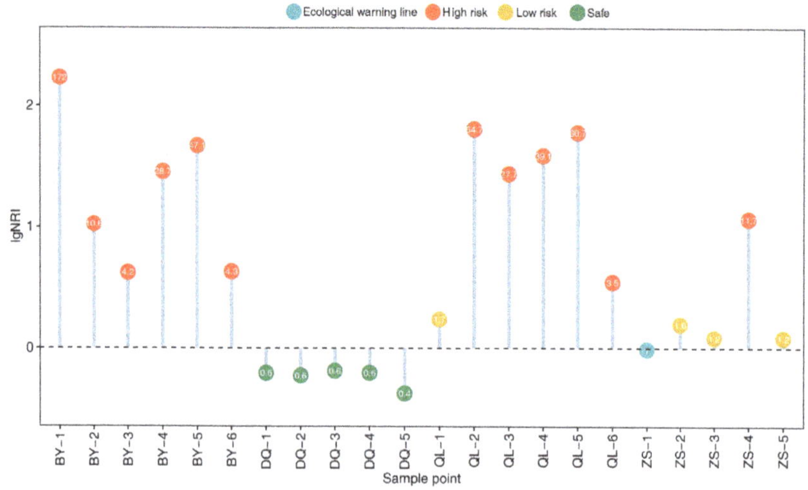

Figure 2. Nemerow risk index of soil samples.

The cumulative probability distribution curves of *EI* for all 21 soil samples were fitted (Table 2). Except QL−3, all fitting curves passed the K–S test ($p > 0.5$), the values of RMSE were small, indicating that the probabilistic cumulative distribution model can fit the *EI* distribution of 11 toxicity endpoints at the tested 21 soils samples. EI_{max} of QL−3 was less than 0.5, meaning that the soil was extremely toxic to soil invertebrates. Additionally, *RP* of QL−3 can be directly determined as 1. *SEI* was set as 0.5 and the *RP* of each soil was calculated (Table 2). Figure 3 illustrates the *EI* cumulative probability distribution curve of some soil samples, and those of the other soils are shown in supporting materials (Figures S2–S5).

Table 2. *EI* cumulative probability distribution curve fitting data of soils.

Sample Point	RMSE	*p* (K−S)	*RP* (*EI* = 0.5)	Sample Point	RMSE	*p* (K−S)	*RP* (*EI* = 0.5)
BY−1	0.098	>0.05	1.000	QL−1	0.061	>0.05	0.759
BY−2	0.060	>0.05	0.877	QL−2	0.154	>0.05	1.000
BY−3	0.075	>0.05	0.697	QL−3	0.230	<0.05	1.000
BY−4	0.079	>0.05	0.755	QL−4	0.061	>0.05	0.550
BY−5	0.051	>0.05	0.971	QL−5	0.053	>0.05	0.777
BY−6	0.057	>0.05	0.711	QL−6	0.058	>0.05	0.759
DQ−1	0.054	>0.05	0.668	ZS−1	0.090	>0.05	0.601
DQ−2	0.068	>0.05	0.696	ZS−2	0.048	>0.05	0.737
DQ−3	0.097	>0.05	0.675	ZS−3	0.086	>0.05	0.469
DQ−4	0.078	>0.05	0.735	ZS−4	0.080	>0.05	0.803
DQ−5	0.068	>0.05	0.940	ZS−5	0.068	>0.05	0.931

The curve can provide risk managers with three aspects of information:

1. What is the overall risk probability of the soil to the subject organism? For example, *RP* = 0.877 for BY−2 in Figure 3;
2. For which toxicity endpoints was the risk of the soil acceptable or unacceptable? For example, the risk of soil QL−6 for ES.7D, NL.4D and TS.7D was acceptable ($EI > 0.5$), while the risk for other toxicity endpoints was unacceptable ($EI < 0.5$);
3. Relative sensitivity distribution among toxicity endpoints in the risk assessment of soil. For different soil, the sensitivity order of the toxicity endpoints varies among species (Figure 3). For instance, the sensitivity order of the toxicity endpoints in soil BY−2 was NR.4D > ES.28D > TR.28D. While in soil ZS−3, the order was TR.28D >

ES.28D > NR.4D. This indicates that if only one toxicity endpoint was used for soil ecological risk assessment, the choice of toxicity endpoint would bring about a great influence on the assessment results.

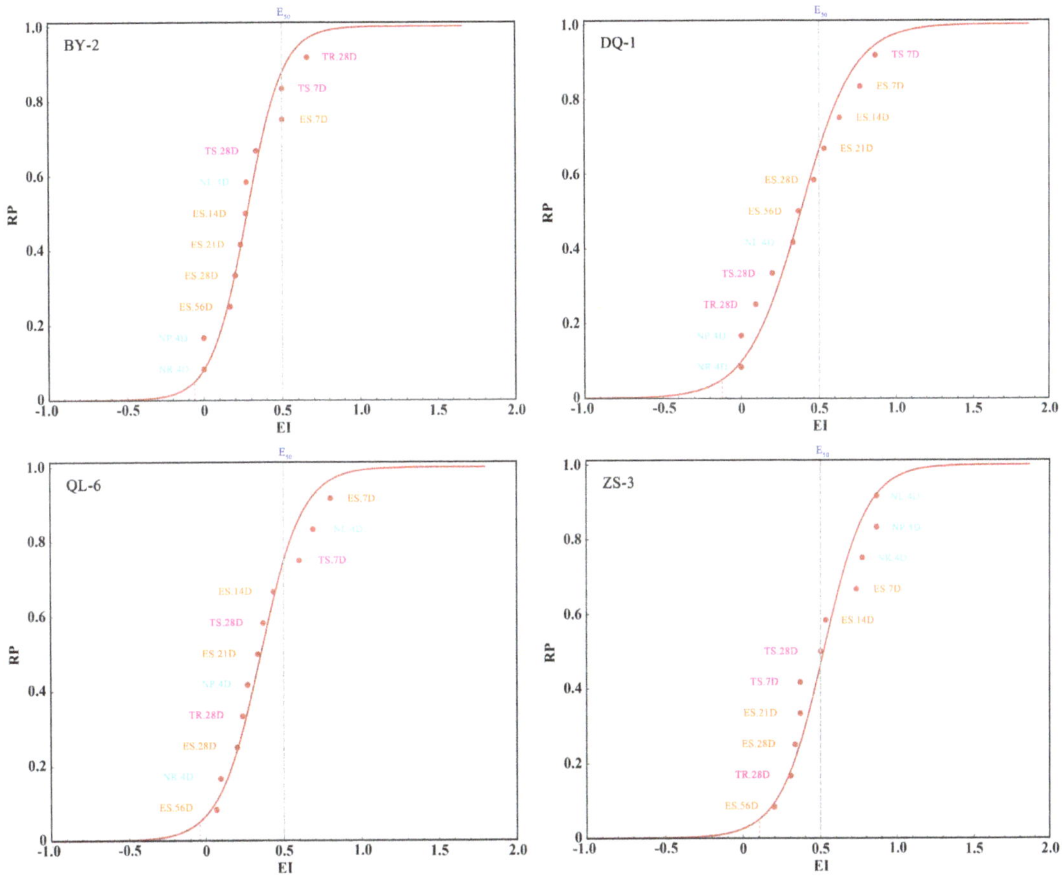

Figure 3. Cumulative probability distribution curve of *EI* for soil BY−2, DQ−1, QL−6 and ZS−3.

The order of sensitivity among toxicity endpoints of the same species was related to the mechanism of toxicity endpoints. It has been widely recognized that the longer earthworms are exposed to contaminated soil, the more significant the toxic effect of contaminated soil on the earthworms will be. Therefore, the sensitivity order of earthworm toxicity endpoints at any sampling point is ES.56D > ES.28D > ES.21D ≥ ES.14D > ES.7D. The sensitivity order of nematodes was also NR.4D ≥ NP.4D ≥ NL.4D for any soil. Because nematodes do not begin to lay eggs until they undergo four molts and reach 1060 μm in length [39], if a nematode cannot reach the required growth stage, fertility inhibition of the nematode will be affected by both reproductive destruction and growth inhibition. Similarly, if nematode fertility is inhibited, reproductive inhibition of nematodes will be affected by both egg damage and fertility inhibition. There was no obvious regularity in the order of sensitivity of the three toxicity endpoints (TS.7D, TS.28D and TR.28D), but the relative sensitivity of TS.28D and TR.28D seemed to be related to *RP* (Figure 4). When *RP* of soil is low, the sensitivity order of the toxicity endpoint is TR.28 > TS.28. Presumably, the springtails would preferentially adapt to the harsh environment and maintain their own survival after experiencing low toxic stress in soil, delaying life activities such as spawning, but the reproductive ability of springtails did not lose at this time. When the *RP* of the soil was moderate, the sensitivity order of the toxicity endpoint was TR.28D < TS.28D. Possibly, higher toxic soil threatened the survival of the springtails, triggering their emergency reproductive strategy and producing more offspring per adult than in a no toxic stress environment. When the *RP* of soil was very high, the sensitivity order of toxicity endpoints was TR.28 > TS.28. Soil with extremely high toxicity could cause reproductive loss of the springtails before death. Pearson correlation and Spearman correlation analyses showed that there was a significant positive correlation between the toxicity endpoints of the same species ($p < 0.05$) (Figure 5), which was consistent with those presented in the sensitivity distribution curve.

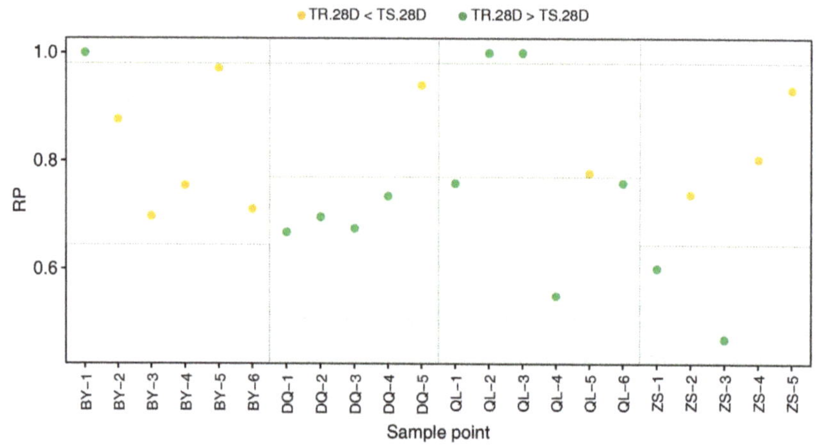

Figure 4. Relationship between *RP* and the sensitivity order of the toxicity endpoints of springtails.

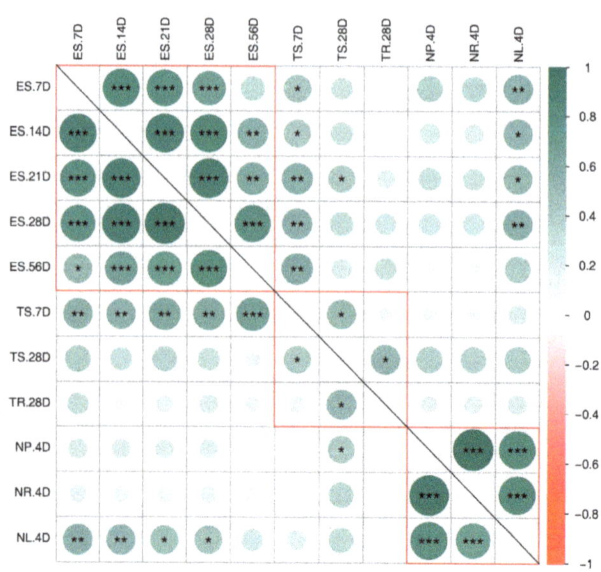

Figure 5. Correlation analysis of toxicity effect index at different toxicity endpoints. The lower−left part is the Pearson correlation, and the upper−right part is the Spearman correlation. (*** $p < 0.001$, ** $0.001 < p < 0.01$, * $0.01 < p < 0.05$).

3.3. Comparative Analysis between NRI and RP

Spearman correlation analyses were conducted between *NRI* and *RP* (Figure 6). *NRI* and *RP* were significantly positive correlated ($p < 0.05$). In general, the assessment results of *NRI* and *RP* were basically consistent, and it is feasible to use the cumulative probability distribution method based on *EI* to assess ecological risks.

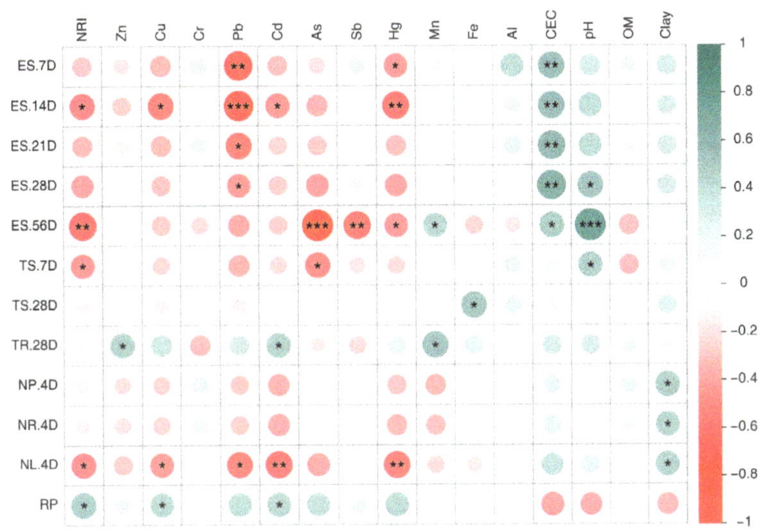

Figure 6. Spearman correlation analysis between toxic effects and soil composition. (*** $p < 0.001$, ** $0.001 < p < 0.01$, * $0.01 < p < 0.05$).

There is no correlation between *NRI* and *EI* for most single toxicity endpoints, which may be attributed to the following reasons. First, the stress to invertebrates in the combined contaminated soil was not only formed from the nine concerned pollutants, other components and the soil physicochemical properties may also have made an impact [40]. Second, although pollutants are the main source of soil toxicity, *NRI* neglects the interaction among pollutants and the bioavailability of pollutants, possessing great indeterminacy in the risk assessment of soil [41–43]. Spearman correlation analysis suggests that the survival of earthworms might be affected by soil CEC and pH ($p < 0.05$). The growth, pregnancy and reproduction of nematodes may be affected by clay content ($p < 0.05$). The survival of springtails may be affected by Fe content in soil ($p < 0.05$). In future toxicity studies of earthworms, nematodes and springtails, more attention needs to be paid to these soil indexes.

NRI and *RP* at four sites were further compared and analyzed (Figure 7). In site BY with high *NRI*, *RP* increases with the increase of *NRI*, which is consistent with the result of Spearman correlation analysis (Figure 6). However, the risk probability of BY−2 is higher than expected. In site DQ with low *NRI*, *RP* decreased with the increase of *NRI*, which was contrary to the result of Spearman correlation analysis (Figure 6). Possibly, lower concentrations of pollutants could be beneficial to soil invertebrates, or other soil factors may have exhibited greater effect on invertebrates than pollutants. In site QL with wide span of *NRI*, there was no monotone correlation between *NRI* and *RP*. The toxicity of QL−3 was greater than that expected by *NRI*, while the toxicity of QL−4 and QL−5 was much less than that expected by *NRI*. In site ZS, there is no monotone correlation between *NRI* and *RP*, and the toxicity of ZS−5 is much higher than that expected by *NRI*. The above results further indicate that the correlation between *NRI* and *RP* may be affected by soil components and physicochemical properties [44].

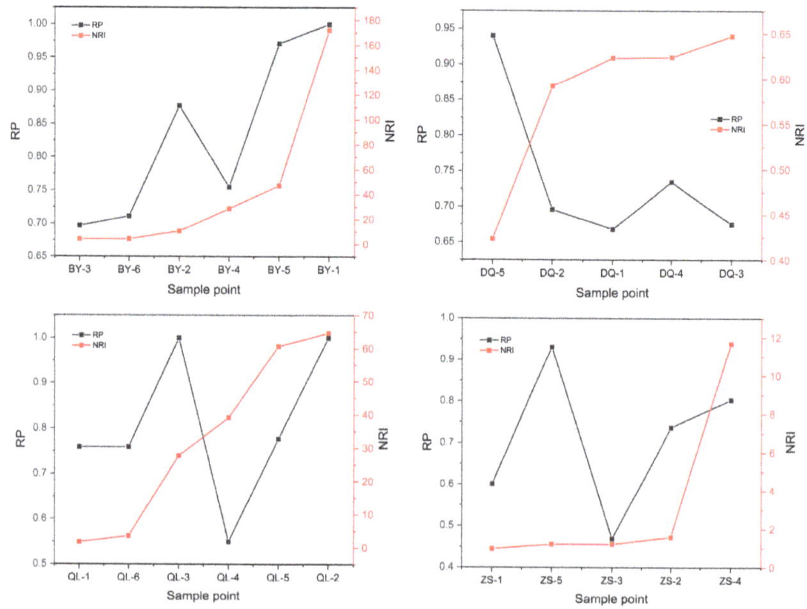

Figure 7. Comparison between *NRI* and *RP* of each soil sample.

Several studies have demonstrated that soils with higher pH and soil OM content exhibited less toxicity to earthworms and springtails [16,26,45–47]. Toxicity studies on nematodes also found that higher CEC and soil OM content were related to low toxicity [39]. The pH, OM and CEC of each soil sample collected from the four sites were significantly different (Table S4 and Figure 8), which was likely to have a great impact on the ecological risk of soil. For example, the high risk of BY−2 may be influenced by the low CEC and OM. The high risk of QL−2 and QL−3 may be affected by the lower pH. The lower risk of QL−4 and QL−5 may be related to the higher OM, pH, and CEC, and low CEC, OM and pH may have had an impact on the high risk of ZS−5. Therefore, the risk assessment based on the *EI* cumulative probability distribution method could essentially reflect the toxicity of soil to invertebrates, especially the actual combined contaminated soil. The complex interactions among soil pollutants and the effect of soil physicochemical properties were fully taken into consideration. Besides pollutant content, extreme soil physicochemical properties may also have a decisive influence on the life history of soil invertebrates. Taking these factors into account, the risk assessment result could be more convincing in the characterization of soil toxicity. Furthermore, accurate risk assessment could guide risk managers to formulate effective and customized remediation strategies.

3.4. Prospects for Risk Assessment Methods

Although the ecological risk assessment based on toxicity test is reliable, the cost of toxicity tests is high and the cycle is long. Therefore, in the long run, it is still the most economical and effective method to predict the ecological risk from pollutant content while taking the physicochemical properties of soil into account. However, it is very difficult to use traditional statistical methods to study the correlation between soil composition and toxic effects [48,49]. It has been shown that machine learning algorithms can make acceptable predictions about the relationship between responses and multiple factors. In order to reveal the superposition or confrontation of multiple factors, Yu et al. [50] established a tree−structure−based random forest feature importance and feature interaction network analysis framework (TBRFA), and successfully predicted the microbial composition of global soils under 18 environmental factors [51]. Machine learning methods also show great promise in the construction of prediction models of complex dose–effect relationships, if enough toxicity data of combined contamination are available. Therefore, the use of actual pollutant soil for future toxicity tests is recommended, as is the provision of original tabular data, so as to build an open database of composite pollution and toxic effects and to facilitate the prediction of dose–effect relationships by machine learning methods (earlier stage in Scheme 2). The following soil parameters should be considered: the total content of major pollutants, CEC, pH, and OM. In addition, accurate Latin names of species, trial duration, toxicity endpoint and toxicity effect index (*EI*) should be provided. Once the prediction model is established and validated, the toxicity effect index of a specific toxicity endpoint of soil can be predicted without further toxicity tests (later stage in Scheme 2). Finally, based on *EI* predicted by the model, the cumulative probability distribution method can be used to assess the ecological risk of any combined contaminated soil.

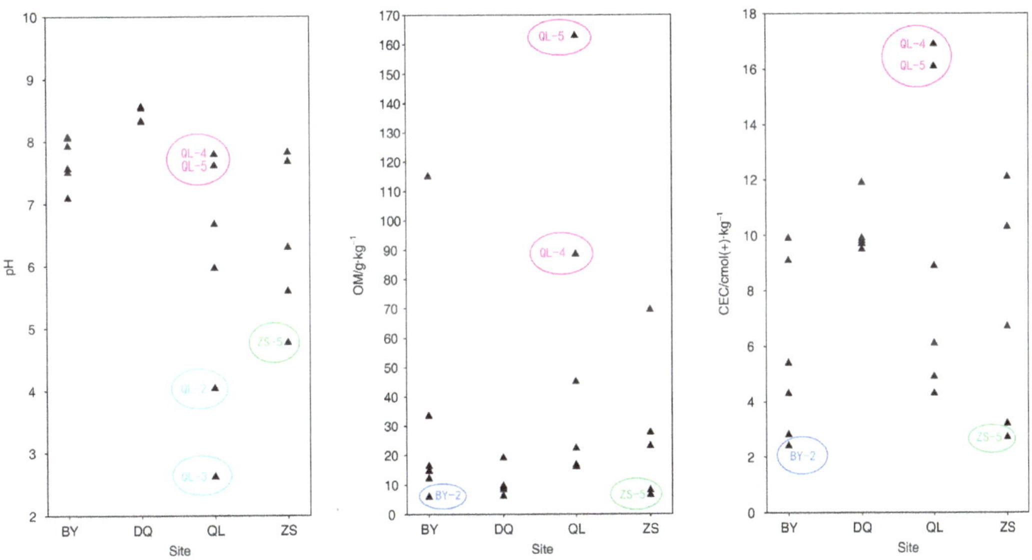

Figure 8. Comparison of physicochemical properties of soil samples.

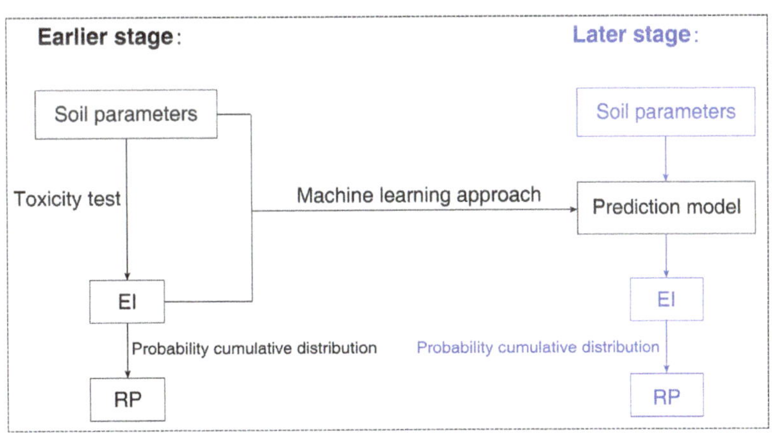

Scheme 2. The application of a machine learning model in probabilistic ecological risk assessment.

4. Conclusions

Toxicity effects risk index (*EI*) can normalize toxicity effects of different species at different toxicity endpoints ($0 \leq EI \leq 1$). Compared with *NRI*, the cumulative probability distribution curve of *EI* could present more information about toxicity endpoints, which is more practical for risk managers and decision makers. The applicability of the method was preliminarily validated using 11 toxicity endpoints of *E. fetida*, *F. candida*, and *C. elegans* in actual combined contaminated soil. Whether this method can be applied to ecological risk assessment of other species still needs to be verified by toxicity tests, including soil animals, plants, microorganisms and soil enzymes. It might make more sense to use native representative species. This new method is more practical for composite contaminated soils with pollution levels close to RSVs. For pollution levels where all pollutants are far above or far below RSVs, the risk of soil to organisms is foreseeable. Although this method is applicable, considering the cost of risk assessment, it may not be necessary to use this method.

Supplementary Materials: The following supporting information can be downloaded at https://www.mdpi.com/article/10.3390/toxics11050411/s1, Table S1: The concentration of heavy metals in soil samples; Table S2: Individual risk index (*RI*) and Nemerow risk index (*NRI*) of soil samples; Table S3: Toxicity effect index (*EI*) of soils; Table S4: Physicochemical properties of soils; Figure S1: Individual ecological risk index of soils based on risk screening values; Figure S2: *EI* cumulative probability distribution of soils collected at BY; Figure S3: *EI* cumulative probability distribution of soils collected at DQ; Figure S4: *EI* cumulative probability distribution of soils collected at QL; Figure S5: *EI* cumulative probability distribution of soils collected at ZS.

Author Contributions: Conceptualization, Q.W.; methodology, Q.W; formal analysis, Q.W.; investigation, J.C., Y.Z., J.G., X.W. and X.F.; resources, H.H.; writing—original draft preparation, Q.W. and J.W.; writing—review and editing, J.W. and H.H.; visualization, Q.W.; supervision, H.H.; project administration, H.H.; funding acquisition, H.H. All authors have read and agreed to the published version of the manuscript.

Funding: This research was funded by the National Key Research and Development Program of China, grant number 2019YFC1804603.

Institutional Review Board Statement: Not applicable.

Informed Consent Statement: Not applicable.

Data Availability Statement: The data presented in this study are available in this article and Supplementary Materials.

Conflicts of Interest: The authors declare no conflict of interest.

References

1. Cai, X.; Duan, Z.; Wang, J. Status Assessment, Spatial Distribution and HealthRisk of Heavy Metals in Agricultural Soils AroundMining-Impacted Communities in China. *Pol. J. Environ. Stud.* **2021**, *30*, 993–1002. [CrossRef]
2. Zhao, S.W.; Qin, L.Y.; Wang, L.F.; Sun, X.Y.; Yu, L.; Wang, M.; Chen, S.B. Ecological risk thresholds for Zn in Chinese soils. *Sci. Total Environ.* **2022**, *833*, 9. [CrossRef] [PubMed]
3. Song, W.E.; Chen, S.B.; Liu, J.F.; Chen, L.; Song, N.N.; Li, N.; Liu, B. Variation of Cd concentration in various rice cultivars and derivation of cadmium toxicity thresholds for paddy soil by species-sensitivity distribution. *J. Integr. Agric.* **2015**, *14*, 1845–1854. [CrossRef]
4. Qin, L.Y.; Wang, L.F.; Sun, X.Y.; Yu, L.; Wang, M.; Chen, S.B. Ecological toxicity (ECx) of Pb and its prediction models in Chinese soils withdifferent physiochemical properties. *Sci. Total Environ.* **2022**, *853*, 10. [CrossRef] [PubMed]
5. Qin, L.Y.; Sun, X.Y.; Yu, L.; Wang, J.; Modabberi, S.; Wang, M.; Chen, S.B. Ecological risk threshold for Pb in Chinese soils. *J. Hazard. Mater.* **2023**, *444*, 10. [CrossRef]
6. Jiang, B.; Ma, Y.B.; Zhu, G.Y.; Li, J. Prediction of soil copper phytotoxicity to barley root elongation by an EDTA extraction method. *J. Hazard. Mater.* **2020**, *389*, 8. [CrossRef]
7. Gao, J.T.; Ye, X.X.; Wang, X.Y.; Jiang, Y.J.; Li, D.C.; Ma, Y.B.; Sun, B. Derivation and validation of thresholds of cadmium, chromium, lead, mercury and arsenic for safe rice production in paddy soil. *Ecotoxicol. Environ. Saf.* **2021**, *220*, 10. [CrossRef]
8. Holtra, A.; Zamorska-Wojdyla, D. Application of individual and integrated pollution indices of trace elements to evaluate the noise barrier impact on the soil environment in Wroclaw (Poland). *Environ. Sci. Pollut. Res.* **2023**, *30*, 26858–26873. [CrossRef]

9. Nikolova, R.; Boteva, S.; Kenarova, A.; Dinev, N.; Radeva, G. Enzyme activities in soils under heavy metal pollution: A case study from the surroundings of a non−ferrous metal plant in Bulgaria. *Biotechnol. Biotechnol. Equip.* **2023**, *37*, 49–57. [CrossRef]
10. Bai, Z.; Wu, F.; He, Y.; Han, Z. Pollution and risk assessment of heavy metals in Zuoxiguo antimony mining area, southwest China. *Environ. Pollut. Bioavailab.* **2023**, *35*, 1–11. [CrossRef]
11. Shen, G.; Ru, X.; Gu, Y.; Liu, W.; Wang, K.; Li, B.; Guo, Y.; Han, J. Pollution Characteristics, Spatial Distribution, and Evaluation of Heavy Metal(loid)s in Farmland Soils in a Typical Mountainous Hilly Area in China. *Foods* **2023**, *12*, 681. [CrossRef] [PubMed]
12. Senoro, D.B.; Monjardin, C.E.F.; Fetalvero, E.G.; Benjamin, Z.E.C.; Gorospe, A.F.B.; de Jesus, K.L.M.; Ical, M.L.G.; Wong, J.P. Quantitative Assessment and Spatial Analysis of Metals and Metalloids in Soil Using the Geo-Accumulation Index in the Capital Town of Romblon Province, Philippines. *Toxics* **2022**, *10*, 633. [CrossRef] [PubMed]
13. Guo, Y.; Huang, M.; You, W.; Cai, L.; Hong, Y.; Xiao, Q.; Zheng, X.; Lin, R. Spatial analysis and risk assessment of heavy metal pollution in rice in Fujian Province, China. *Front. Environ. Sci.* **2022**, *10*, 2422. [CrossRef]
14. Shifaw, E. Review of Heavy Metals Pollution in China in Agricultural and Urban Soils. *J. Health Pollut.* **2020**, *8*, 180607. [CrossRef]
15. Adnan, M.; Xiao, B.; Xiao, P.; Zhao, P.; Li, R.; Bibi, S. Research Progress on Heavy Metals Pollution in the Soil of Smelting Sites in China. *Toxics* **2022**, *10*, 231. [CrossRef] [PubMed]
16. Wang, W.; Lin, X.; Zhao, L.; Zhang, J.; Fan, W.; Hou, H. Toxicity threshold and prediction model for zinc in soil-dwelling springtails in Chinese soils. *J. Agro−Environ. Sci.* **2021**, *40*, 766–773.
17. Zhang, J.; Liu, Z.; Tian, B.; Li, J.; Luo, J.; Wang, X.; Ai, S.; Wang, X. Assessment of soil heavy metal pollution in provinces of China based on different soil types: From normalization to soil quality criteria and ecological risk assessment. *J. Hazard. Mater.* **2023**, *441*, 129891. [CrossRef]
18. Han, D.; Zhao, L.; Zhang, N.; Hou, H.; Sun, Z. Classification of Cd Contaminated Paddy Soils in Carbonate Parent Material Area of Southwest China by Species Sensitivity Distribution Method (SSD). *Res. Environ. Sci.* **2021**, *34*, 409–418.
19. Sun, X.Y.; Qin, L.Y.; Wang, L.F.; Zhao, S.W.; Yu, L.; Wang, M.; Chen, S.B. Aging factor and its prediction models of chromium ecotoxicity in soils with various properties. *Sci. Total Environ.* **2022**, *847*, 8. [CrossRef] [PubMed]
20. Wan, Y.N.; Jiang, B.; Wei, D.P.; Ma, Y.B. Ecological criteria for zinc in Chinese soil as affected by soil properties. *Ecotoxicol. Environ. Saf.* **2020**, *194*, 7. [CrossRef] [PubMed]
21. Backhaus, T.; Faust, M. Predictive environmental risk assessment of chemical mixtures: A conceptual framework. *Environ. Sci. Technol.* **2012**, *46*, 2564–2573. [CrossRef]
22. Jiang, R.; Wang, M.; Chen, W.; Li, X. Ecological risk evaluation of combined pollution of herbicide siduron and heavy metals in soils. *Sci. Total Environ.* **2018**, *626*, 1047–1056. [CrossRef]
23. Chen, C.; Wang, Y.; Qian, Y.; Zhao, X.; Wang, Q. The synergistic toxicity of the multiple chemical mixtures: Implications for risk assessment in the terrestrial environment. *Environ. Int.* **2015**, *77*, 95–105. [CrossRef]
24. Rodriguez-Ruiz, A.; Etxebarria, J.; Boatti, L.; Marigomez, I. Scenario-targeted toxicity assessment through multiple endpoint bioassays in a soil posing unacceptable environmental risk according to regulatory screening values. *Environ. Sci. Pollut. Res. Int.* **2015**, *22*, 13344–13361. [CrossRef] [PubMed]
25. Oliveira Resende, A.P.; Santos, V.S.V.; Campos, C.F.; Morais, C.R.d.; de Campos Júnior, E.O.; Oliveira, A.M.M.d.; Pereira, B.B. Ecotoxicological risk assessment of contaminated soil from a complex of ceramic industries using earthworm Eisenia fetida. *J. Toxicol. Environ. Health Part A* **2018**, *81*, 1058–1065. [CrossRef]
26. Sujetoviene, G.; Cesynaite, J. Assessment of Toxicity to Earthworm Eisenia fetida of Lead Contaminated Shooting Range Soils with Different Properties. *Bull. Environ. Contam. Toxicol.* **2019**, *103*, 559–564. [CrossRef] [PubMed]
27. Marchand, C.; Jani, Y.; Kaczala, F.; Hijri, M.; Hogland, W. Physicochemical and Ecotoxicological Characterization of Petroleum Hydrocarbons and Trace Elements Contaminated Soil. *Polycycl. Aromat. Compd.* **2018**, *40*, 967–978. [CrossRef]
28. Gruss, I.; Stefanovska, T.; Twardowski, J.; Pidlisnyuk, V.; Shapoval, P. The ecological risk assessment of soil contamination with Ti and Fe at military sites in Ukraine: Avoidance and reproduction tests with Folsomia candida. *Rev. Environ. Health* **2019**, *34*, 303–307. [CrossRef]
29. *HJ 1068—2019*; Soil—Determination of Particle Size Distribution—Pipette Method and Hydrometer Method. Ministry of Ecology and Environment: Beijing, China, 2019.
30. OECD. *Test No. 207: Earthworm, Acute Toxicity Tests*; OECD: Paris, France, 1984.
31. OECD. *Test No. 222: Earthworm Reproduction Test (Eisenia fetida/Eisenia andrei)*; OECD: Paris, France, 2016.
32. EPA. *Ecological Soil Screening Levels for Zinc (Interim Final): OSWER Directive 9285.7−73*; EPA: Washington, DC, USA, 2007.
33. EPA. *Ecological Soil Screening Levels for Copper (Interim Final): OSWER Directive 9285.7−68*; EPA: Washington, DC, USA, 2007.
34. EPA. *Ecological Soil Screening Levels for Lead (Interim Final): OSWER Directive 9285.7−70*; EPA: Washington, DC, USA, 2005.
35. EPA. *Ecological Soil Screening Levels for Cadmium (Interim Final): OSWER Directive 9285.7−65*; EPA: Washington, DC, USA, 2005.
36. EPA. *Ecological Soil Screening Levels for Antimony (Interim Final): OSWER Directive 9285.7−61*; EPA: Washington, DC, USA, 2005.
37. EPA. *Ecological Soil Screening Levels for Manganese (Interim Final): OSWER Directive 9285.7−71*; EPA: Washington, DC, USA, 2007.
38. *GB 15618–2018*; Soil Environmental Quality: Risk Control Standard for Soil Contamination of Agricultural Land. Ministry of Ecology and Environment: Beijing, China, 2018.
39. Song, Z.; Dang, X.; Zhao, L.; Hou, H.; Wang, X.; Lu, H. Toxic effects of antimony on Caenorhabditis elegans in soils. *J. Agro-Environ. Sci.* **2022**, *41*, 1917–1925.

40. Lin, X.; Sun, Z.; Zhao, L.; Ma, J.; Li, X.; He, F.; Hou, H. The toxicity of exogenous arsenic to soil−dwelling springtail Folsomia candida in relation to soil properties and aging time. *Ecotoxicol. Environ. Saf.* **2019**, *171*, 530–538. [CrossRef]
41. Wang, Z.; Cui, Z.; Liu, L.; Ma, Q.; Xu, X. Toxicological and biochemical responses of the earthworm Eisenia fetida exposed to contaminated soil: Effects of arsenic species. *Chemosphere* **2016**, *154*, 161–170. [CrossRef]
42. Delistraty, D.; Yokel, J. Ecotoxicological study of arsenic and lead contaminated soils in former orchards at the Hanford Site, USA. *Environ. Toxicol.* **2014**, *29*, 10–20. [CrossRef]
43. Porfido, C.; Allegretta, I.; Panzarino, O.; Laforce, B.; Vekemans, B.; Vincze, L.; de Lillo, E.; Terzano, R.; Spagnuolo, M. Correlations between As in Earthworms' Coelomic Fluid and As Bioavailability in Highly Polluted Soils as Revealed by Combined Laboratory X−ray Techniques. *Environ. Sci. Technol.* **2019**, *53*, 10961–10968. [CrossRef]
44. Santorufo, L.; Van Gestel, C.A.M.; Maisto, G. Ecotoxicological assessment of metal−polluted urban soils using bioassays with three soil invertebrates. *Chemosphere* **2012**, *88*, 418–425. [CrossRef] [PubMed]
45. Liu, H.; Li, M.; Zhou, J.; Zhou, D.; Wang, Y. Effects of soil properties and aging process on the acute toxicity of cadmium to earthworm Eisenia fetida. *Environ. Sci. Pollut. Res. Int.* **2018**, *25*, 3708–3717. [CrossRef] [PubMed]
46. Nahmani, J.; Hodson, M.E.; Black, S. Effects of metals on life cycle parameters of the earthworm Eisenia fetida exposed to field-contaminated, metal−polluted soils. *Environ. Pollut.* **2007**, *149*, 44–58. [CrossRef]
47. Römbke, J.; Jänsch, S.; Junker, T.; Pohl, B.; Scheffczyk, A.; Schallnaß, H. Improvement of the applicability of ecotoxicological tests with earthworms, springtails, and plants for the assessment of metals in natural soils. *Environ. Toxicol. Chem.* **2006**, *25*, 776–787. [CrossRef]
48. Bonnard, M.; Eom, I.-C.; Morel, J.-L.; Vasseur, P. Genotoxic and reproductive effects of an industrially contaminated soil on the earthwormEisenia Fetida. *Environ. Mol. Mutagen.* **2009**, *50*, 60–67. [CrossRef]
49. Crouau, Y.; Pinelli, E. Comparative ecotoxicity of three polluted industrial soils for the Collembola Folsomia candida. *Ecotoxicol. Environ. Saf.* **2008**, *71*, 643–649. [CrossRef]
50. Yu, F.; Wei, C.; Deng, P.; Peng, T.; Hu, X. Deep exploration of random forest model boosts the interpretability of machine learning studies of complicated immune responses and lung burden of nanoparticles. *Sci. Adv.* **2021**, *7*, eabf4130. [CrossRef]
51. Hao, Y.; Yu, F.; Hu, X. Multiple factors drive imbalance in the global microbial assemblage in soil. *Sci. Total Environ.* **2022**, *831*, 154920. [CrossRef]

toxics

Article

Determinants Analysis Regarding Household Chemical Indoor Pollution

Paolo Montuori [1], Mariagiovanna Gioia [1], Michele Sorrentino [1], Fabiana Di Duca [1], Francesca Pennino [1,*], Giuseppe Messineo [1], Maria Luisa Maccauro [1], Simonetta Riello [1], Ugo Trama [2], Maria Triassi [1] and Antonio Nardone [1]

[1] Department of Public Health, University "Federico II", Via Sergio Pansini n° 5, 80131 Naples, Italy
[2] General Directorate of Health, Campania Region, Centro Direzionale Is. C3, 80143 Naples, Italy
* Correspondence: francesca.pennino@unina.it

Abstract: Indoor household pollution is not yet sufficiently studied in the general population. Over 4 million people die prematurely every year due to air pollution in households. This study aimed to propose quantitative data research through the administration of a KAP (Knowledge, Attitudes, and Practices) Survey Questionnaire. This cross-sectional study administered questionnaires to adults from the metropolitan city of Naples (Italy). Three Multiple Linear Regression Analyses (MLRA) were developed, including Knowledge, Attitudes, and Behavior regarding household chemical air pollution and the related risks. One thousand six hundred seventy subjects received a questionnaire to be filled out and collected anonymously. The mean age of the sample was 44.68 years, ranging from 21–78 years. Most of the people interviewed (76.13%) had good attitudes toward house cleaning, and 56.69% stated paying attention to cleaning products. Results of the regression analysis indicated that positive attitudes were significantly higher among subjects who graduated, with older age, male and non-smokers, but they were correlated with lower knowledge. In conclusion, a behavioral and attitudinal program targeted those with knowledge, such as younger subjects with high educational levels, but do not engage in correct practices towards household indoor chemical pollution.

Keywords: indoor air quality; chemical contaminants; knowledge; attitude; practice; cross-sectional survey

check for updates

Citation: Montuori, P.; Gioia, M.; Sorrentino, M.; Di Duca, F.; Pennino, F.; Messineo, G.; Maccauro, M.L.; Riello, S.; Trama, U.; Triassi, M.; et al. Determinants Analysis Regarding Household Chemical Indoor Pollution. *Toxics* **2023**, *11*, 264. https://doi.org/10.3390/toxics11030264

Academic Editors: Alina Barbulescu and Lucica Barbes

Received: 21 February 2023
Revised: 9 March 2023
Accepted: 12 March 2023
Published: 13 March 2023

1. Introduction

More than 4 million people die prematurely every year due to household air pollution [1,2]. Elevated concentrations of indoor pollutants are not only associated with increased mortality but also with a range of harmful health effects, such as adverse pregnancy outcomes [3], chronic obstructive pulmonary disease [4], severe pneumonia, especially in childhood [5], lung cancer [6], cardiovascular diseases [7,8]. The greatest risk comes from long-term exposure, as 80–90% of a lifetime is spent in confined spaces which may increase due to cumulative lifetime exposures [9–11].

Indoor air pollution is a significant public health issue, caused by various substances found in common household items and influenced by common indoor activities, such as heating, cooking, and the use of cleaning products, as well as behavioral practices like smoking, vaping, burning candles or incense [12–15]. Moreover, many of these pollutants can cause secondary reactions producing additional highly reactive and harmful substances [16].

Public policy is a crucial tool in reducing air pollution and improving air quality and people's health. Since 1990, measures designed to curb air pollution have prevented approximately 600,000 premature deaths annually [17]. The Control Action Plan introduced a decade ago has already prevented 15,822 associated morbidities in 2017 [18]. However, these policies have mostly focused on outdoor environments, ignoring indoor spaces where people spend most of their time [19].

In fact, individuals can play a crucial role in reducing their exposure to indoor air pollution, as their behavior significantly impacts the indoor environment [20]. By following specific yet reasonable behaviors, individuals can reduce the risks associated with indoor air pollution. Such practices include ensuring adequate ventilation, maintaining combustion appliances, limiting exposure to volatile organic compounds, and reducing smoking [21]. Improving ventilation rates in households by opening windows or using ventilation fans can lead to a reduction in emissions from human activities, thereby improving indoor air quality [22,23]. Additionally, higher ventilation rates have been linked to improved health outcomes [24].

A recent study assessed the dependence of community knowledge and attitude with socio-demographic factors and the dependence of the behaviors with knowledge, attitude, and socio-demographic factors using community KAB towards IAQ, revealing lower levels of knowledge and behaviors towards IAQ and moderate levels of attitude within the study population [25]. Daniel et al. described the perceptions, knowledge, and practices of adults concerning indoor environmental pollution, evidencing that well-integrated practices were not related to knowledge, level of education, or perceptions but rather to the responsibility of having a child and that implementation of less well-followed practices would be improved by better knowledge/information and a change in perceptions [26]. In 2018, Al-Khamees examined the knowledge, attitudes, and practices toward indoor pollution at Kuwait University, demonstrating poor knowledge regarding indoor pollution among university students and teachers [27]. Moreover, some papers focused their research only on certain types of pollutants, such as Adeolu et al., that conducted a study on knowledge and attitudes towards lead exposure in Nigeria [28], or over radon, a typical yet specific pollutant for some households, in a KAP model conducted in 2018 in a rural environment by Neri et al. [29].

For those reasons, the present study aims to propose quantitative research of data through the administration of a KAP (Knowledge, Attitudes, and Practices) Survey Questionnaire and the statistical analysis of the information collected towards household chemical air pollution in a population of a large metropolitan area to understand this phenomenon in order to collect data to develop specific and tailored educational programs.

2. Materials and Methods

2.1. Setting and Sample

This cross-sectional study was conducted by administering questionnaires to adults from the metropolitan city of Naples (Italy), with a population of 909,048 [30]. The study was conducted from the beginning of January 2022 to the end of September 2022. Subjects were selected to participate in the study using a snowballing sampling method among universities, working places, and community centers. The inclusion criteria in the study required that participants were 18 and older and residing in the metropolitan area of Naples. The required sample size was calculated using Slovin's formula to obtain a representative sample within a margin of error of 3%, and a confidence interval of 95%, determining a final number of subjects to be recruited of 1523. Finally, after accounting for a 30% non-response rate, the estimated total sample size was 1066.

2.2. Procedures

During the study period, experienced interviewers submitted to participants the questionnaire from Monday to Friday between 10:00 a.m. and 8:00 p.m. to avoid oversampling non-working individuals. The interviewers, at the beginning of the submission, stated that they were conducting a study on behalf of the University of the studies of Naples "Federico II", giving information to the participants about the nature and scope of the research, the methodology, that their participation was on a voluntary basis, that all the collected information would be processed anonymously and confidentially, and that they could end their participation at any time without disclosing a reason. Verbal informed consent was obtained prior to progressing with the interview. No incentive for participation or survey completion was provided. The present study conformed with the Declaration of Helsinki, and ethical clearance was obtained according to local legislation.

2.3. Data Collection

The questionnaire was developed through the meeting of a large commission of physicians, chemists, and biologists. Questions considered inappropriate or not useful for the study objectives were either removed or replaced. Before the commencement of the data collection, a pilot study was performed on 10 individuals in order to test the participants' understanding of the questionnaire items, the results of which were not taken into consideration for the study. The first section of the questionnaire assessed participants' socio-demographic characteristics and other health-related information, including gender, age, marital status, level of education, occupation, partner's occupation, and number of children. The second section investigated knowledge, attitudes, and behaviors concerning household chemical air pollution for a total of 36 questions. Knowledge and attitudes were assessed on a three-point Likert scale with options for "agree", "uncertain", and "disagree", while inquiries regarding behaviors were in a four-answer format of "never", "sometimes", "often", and "yes/always".

2.4. Statistical Analysis

Data reported by the study were analyzed using STATA MP v14.0 statistical software program (College Station, TX, USA). The analysis was carried out in two steps. First, a descriptive statistic was employed to sum up the basic information of the statistical units; then, a Multiple Linear Regression Analysis (MLRA) was performed, as previously extensively explained [31–33]. Briefly, three MLRA were developed, including the variables potentially associated with the following outcomes of interest:

(1) Knowledge regarding household chemical air pollution and the related risks (Model 1);
(2) Attitudes toward household chemical air pollution (Model 2);
(3) Behavior related to household chemical air pollution (Model 3).

Knowledge, Attitudes, and Behaviors, as dependent variables, were acquired by adding the results of the respective question scores (questions with inverse answers have been coded inversely). The independent variables were included in all models: sex (1 = male, 2 = female); age, in years; education level (1 = primary school, 2 = middle school, 3 = high school, 4 = university degree); marital status (1 = Single; 2 = In a relationship); smoking habits (1 = smoker, 2 = non-smoker); having children (1 = Yes; 2 = No). In Model 2, Knowledge was added to the independent variables, and in Model 3, both Knowledge and Attitudes were included in the independent variables. Attitudes and Knowledge were analyzed as indexes rather than scales; thus, each observed variable (A1, ..., A10 and K1, ..., K11) is presumed to cause the latent variables associated (Attitude and Knowledge). In other words, the relationship between observed and latent variables is formative. Therefore, inter-observed variables correlations are not required. On the contrary, the relationship between the observed variables (B1, ..., B11) and latent variable Behavior could be considered reflective (Cronbach's alpha = 0.825). All statistical tests were two-tailed, and results were statistically significant if the p-values were less than or equal to 0.05.

3. Results and Discussion

During the administration period, 1670 subjects were recruited to participate in the study and received a questionnaire to be filled out and collected anonymously. Among those, 1332 questionnaires were filled and returned with a response rate of 79.76%, slightly more than expected (70%) and calculated. Characteristics of the sample are described in Table 1. Regarding gender, 677 were male (50.83%) and 655 females (49.17%). The mean age of the study population was 44.68, ranging from 21–78 years. Educational levels were distributed as 51 subjects (3.83%) declaring an elementary school license, 332 (24.92%) middle school license, 506 (37.99%) responding to having a high school diploma, and

443 (33.26%) graduated with a university degree. Responding about their marital status, 379 respondents (28.45%) declared themselves to be single, and 953 were in a relationship; in addition, 675 of them stated to have at least a son, while 675 had none. Finally, more than half of the population surveyed (59.68%) said they did not smoke. Therefore, this sample can be considered representative of a standard European population in size and frequency of main demographic characteristics [34].

Table 1. Study population characteristics.

Study Population	N (1332)	Percentage
Sex		
Male	677	50.83
Female	655	49.17
Age		
<30	327	24.55
31–35	209	15.69
36–40	91	6.83
41–45	93	6.98
46–50	105	7.88
>51	507	38.06
Education		
Primary school	51	3.83
Middle school	332	24.92
High school	506	37.99
University Degree	443	33.26
Children		
Yes	675	50.68
No	657	49.32
Smoking habits		
Yes	537	40.32
No	795	59.68
Marital Status		
Single	379	28.45
In a relationship	953	71.55

Respondent's knowledge about household indoor pollution is presented in Table 2. Most of the people interviewed knew that the chemical pollution of the air in the household environment is more than that of the outdoors (74.25%). The 0.23% of the sample had not answered to K1 question. Only 38.29% of the population knew that gas stoves contribute to household pollution, while 36.94% did not. Half of the sample (50.00%) believed that plants at night release substances dangerous to health. Regarding smoke, most respondents (52.48%) knew where second-hand smoke comes from, while less than half (40.24%) disagreed that thirdhand smoke is less toxic than secondhand smoke. In addition, 65.24% of the population agreed that inadequate ventilation causes more than 50% of the chemical pollution of the air in a domestic environment. Moreover, a high percentage of the respondents knew that carbon monoxide is the main household pollutant (68.17%), and that formaldehyde is a household chemical pollutant (71.02%). However, only 46.16% of the population knew that formaldehyde is classified as a carcinogen. Half of the sample, 50.83%, knew what Sick Building Syndrome is, and only 21.40% were aware that currently, there are no laws governing the pollution of the domestic environment. Furthermore, with a mean score of 69.19%, the population showed a good knowledge regarding household chemical pollution. Similar results were evidenced in 2020 in France by Daniel et al., where a population of 554 adults totalized a mean score of 68.19% [26]. However, other studies revealed lower levels of knowledge, as evidenced in 2020 by Muro et al. in a study conducted in Nairobi County over a sample of 393 subjects, which indicated a low knowledge level on indoor air pollution with an average score of 38.5% and, previously, by Al-Khamees in 2018, in Kuwait, which demonstrated that the respondents had a low knowledge level on indoor air pollution at 41.47% [27,35]. These differences can be justified considering the diversity of the populations sampled for the study and the significant difference in the distribution of educational levels between them [36].

Table 2. Knowledge of respondents regarding household chemical indoor pollution.

N.	Statement (Variables)	Agree (%)	Uncertain (%)	Disagree (%)
K1	The chemical pollution of the air in the household environment is less than that of outdoors *	6.53	18.99	74.25
K2	Gas stoves contribute to household pollution	38.29	24.77	36.94
K3	Plants at night release substances dangerous to health	50.00	35.36	14.64
K4	Secondhand smoke comes from the smoke exhaled by a smoker	52.48	24.55	22.97
K5	Thirdhand smoke derives from the toxic substances of the smoke deposited in the environment	40.24	20.35	39.41
K6	Thirdhand smoke is less toxic than secondhand smoke	34.68	25.80	40.24
K7	Inadequate ventilation causes more than 50% of the chemical pollution of the air in the domestic environment	65.24	29.13	5.63
K8	The main household pollutant is carbon monoxide	68.17	24.17	7.66
K9	Formaldehyde is one of the household chemical pollutants	71.02	25.38	7.06
K10	Formaldehyde is a certain carcinogen	46.62	42.79	10.59
K11	The Sick Building Syndrome is a condition in which the occupants of a building show a series of symptoms and pathologies without specific causes	50.83	30.93	18.24
K12	There are laws governing the pollution of domestic environments	47.90	30.71	21.40

* 0.23% of the sample did not respond to the K1 question.

Table 3 describes attitudes towards household chemical indoor pollution. The vast majority of the respondents (90.90%) agreed with the good habit of opening the windows, in agreement with Amegah et al. [3], who observed that even with the air conditioner turned on (52.48%), spending time in home microenvironments may not offer sufficient protection from fine ambient aerosol particles ($PM_{2.5}$) [37] and that risk factors for fine particles ($PM_{2.5}$) are greater than for coarse particles (PM_{10}) [38]. Thus, 45.95% of the population deemed it necessary to ventilate the house in winter. The majority of the sample (76.13%) had a good attitude toward house cleaning, and 56.76% of the respondents stated that it was important to pay attention to the use of cleaning products. In addition, almost half of the sample (48.87%) believe it is convenient to use spray deodorant, and 41.67% think lighting candles at home is relaxing. Concerning the last two attitudes, an agreement was noted with the results obtained by Al-Khamees, who evidenced similar results [27]. More than half (56.53%) believed having plants in the house is nice. Unfortunately, only 33.56% thought that induction stoves are more comfortable than gas ones, and 64.19% believed that a fireplace improves the house. Roughly half of the sample (48.49%) deemed smoking on the sofa as not relaxing.

Table 3. The attitude of respondents toward household chemical indoor pollution.

N.	Statement (Variables)	Agree (%)	Uncertain (%)	Disagree (%)
A1	Opening windows is a good habit	90.90	8.78	1.13
A2	It is necessary to open the windows even with the air conditioner on	52.48	22.52	25.00
A3	In winter, it is still necessary to ventilate the house several times a day	45.95	20.05	34.00
A4	House cleaning is a waste of time	13.96	9.91	76.13
A5	A cleaning product is as good as another	20.72	22.52	56.76
A6	It is convenient to use spray deodorant	48.87	28.15	22.97
A7	It is nice to have plants in the house	56.53	19.82	23.65
A8	Lightning candles at home is relaxing	41.67	22.75	35.59
A9	Induction stoves are no more comfortable than the gas ones	44.59	21.85	33.56
A10	A fireplace graces the house	64.19	18.02	17.79
A11	Smoking on the sofa is relaxing	38.29	13.29	48.49

The behaviors of respondents are listed in Table 4. 55.18% of the sample replied that they were attentive to the ventilation of their own house, but only 10.81% claimed to use air purifiers. Regarding gas stoves, almost half of the population (43.47%) use them all the time. Despite the extensive use of gas stoves, only 29.28% operate the hood in the kitchen while cooking food. It has also been noted that there is still a large use of pellet or gas stoves (54.50%), while there is more focus on the use of filters for heating/conditioning systems (47.07%) and checking them (49.55%). Fortunately, the use of insecticides is not widespread, as well as that of air fresheners (23.42%). Of all the behaviors, the most comforting fact comes from smoking, which is never practiced at home, from 54.50% of the population about traditional cigarettes and 61.49% for heated tobacco cigarettes. A high percentage of incorrect behaviors were encountered in the sample, meanly 64.59% for men and 64.69% for women. Those scores were slightly higher than Al-Khamees et al., which were 51.0% for men and 53.5% for women [27]. Also, Daniel et al. found out that certain practices were not well followed by less than 60% of participants [26]. The reason why the results revealed high percentages of some incorrect behaviors rather than others is probably that these are actions performed repeatedly in daily life, and many of them become incorrect habits fueled by poor knowledge and understanding of household air pollution.

Table 4. Behaviors of respondents concerning household chemical indoor pollution.

N.	Questions	Yes/Always (%)	Often (%)	Sometimes (%)	Never (%)
B1	Do you ventilate your home?	55.18	27.70	6.98	10.14
B2	Do you use air purifiers?	10.81	18.69	30.63	39.86
B3	Do you use gas stoves?	43.47	15.09	12.16	29.28
B4	Do you operate the hood in the kitchen?	29.28	15.99	18.47	36.26
B5	Do you use gas and/or a pellet heater?	54.50	12.61	18.92	13.96
B6	Do you use filters for heating/air conditioning systems?	47.07	7.43	17.34	28.15
B7	Do you periodically check the heating, air conditioning, and ventilation systems?	49.55	10.36	12.84	27.25
B8	Do you use insecticides at home?	16.67	16.22	11.71	55.41
B9	Do you use air fresheners?	23.42	14.19	5.63	56.76
B10	Do you wash curtains and carpets?	18.24	10.81	10.59	60.36
B11	Do you decorate your home with plants?	45.95	18.47	18.24	17.34
B12	Do you smoke traditional cigarettes in your home?	22.75	9.46	13.29	54.50
B13	Do you smoke heated tobacco cigarettes in your home?	14.64	8.11	15.77	61.49

Table 5 illustrates the results of linear multiple regression in three models. Model I, Knowledge, as a dependent variable, was correlated with age and education, evidenced as younger subjects had a better overall consciousness of household chemical pollution. These findings agreed with a previous study by Unni in Singapore in 2022, which evidenced a decreasing level of knowledge within the elder population, and with another KAP study carried out over 1604 subjects resident in Ningbo, China, that showed a similar trend of declining levels of knowledge [25,39]. Furthermore, the findings of this investigation are consistent with Jin et al., who analyzed knowledge regarding Secondhand Smoke Exposure and assessed that about 60% of people aged between 15 and 34 had better knowledge of the harmful effects of smoking than people aged 60 [40]. Therefore, since, to the best of our knowledge in literature, no other paper has evidenced a higher knowledge regarding indoor pollution in the elder population, this result may suggest a more pronounced awareness of pollution in younger subjects, as clarified by Chin et al. in 2019 [41]. The second evidence of this MLRA was the statistically significant relation between knowledge regarding indoor air pollution and education. In particular, higher knowledge levels were found in subjects with higher education levels. This evidence is widely expected and confirmed by Kaur et al., who stated that, among a sample of urban homemakers in Ludhiana (India), urban respondents with a higher education level were more conscious of environmental concerns than their own rural counterparts [42]. In addition, Daniel et al. in 2020 found that a higher level of education was also associated with a higher knowledge score in a population of adults between 18 and 45 years in Brittany (France) [26]. These results are not surprising since educational level is widely reported in the literature as a predictor of pollution-related knowledge [43–45]. Moreover, in a cross-sectional study conducted in Italy over 15 universities, the perception of environmental health risks was positively associated with increasing years of attending classes, such as the interest in searching for different sources of information [46].

Table 5. Results of the linear multiple regression analysis (MLRA).

	Coefficients Not Standardized		**Coefficients Standardized**			
	b	**Standard Error**	**t**	**95% Conf. Interval**		***p*-Value**
Model I—Dependent variable: Knowledge						
Prob > F = 0.000		*R-squared = 0.0323*			*Root MSE = 3.929*	
Age	−0.022	0.009	−2.33	−0.041	−0.003	0.020
Sex	−0.175	0.216	−0.081	−0.599	0.025	0.416
Marital status	−0.218	0.274	0.80	−0.319	0.755	0.425
Children	0.082	0.306	0.27	−0.517	0.681	0.681
Education	0.544	0.140	3.88	0.269	0.818	0.000
Smoking habits	0.004	0.220	0.02	−0.427	0.435	0.099
Model II—Dependent variable: Attitudes						
Prob > F = 0.000		*R−squared = 0.0532*			*Root MSE = 3.386*	
Age	0.021	0.008	2.58	0.005	0.037	0.010
Sex	−0.396	0.186	−2.13	−0.761	−0.032	0.033
Marital status	−0.455	0.236	−1.93	−0.918	0.007	0.054
Children	0.238	0.263	−0.090	0.278	0.754	0.366
Education	0.597	0.121	4.92	0.359	0.835	0.000
Smoking habits	0.862	0.189	4.55	0.490	1.23	0.000
Knowledge	−0.1199	0.024	−5.08	−0.166	−0.074	0.000
Model III—Dependent variable: Behavior						
Prob > F = 0.000		*R−squared = 0.1617*			*Root MSE = 6.901*	
Age	0.004	0.017	0.22	−0.029	0.037	0.825
Sex	−0.013	0.379	−0.03	−0.757	0.731	0.973
Marital status	1.05	0.482	2.19	0.109	1.99	0.029
Children	−0.262	0.537	−0.49	−1.31	0.790	0.626
Education	1.39	0.249	5.59	0.905	1.88	0.000
Smoking habits	1.77	0.389	4.56	1.01	2.54	0.000
Knowledge	−0.415	−0.048	−8.55	−0.510	−0.319	0.000
Attitude	0.516	0.056	9.24	0.406	0.625	0.000

Model II uses Attitudes as a dependent variable assessing a positive correlation, statistically significant, with age, gender, education, smoking habits, and knowledge. In particular, the regression analysis results indicated that positive attitudes were significantly higher among subjects who graduated, with older age, male and non-smokers, but they were correlated with lower knowledge. Regarding the correlation between age and attitude, as found in the present study, the literature reports the study conducted by Unni et al., in 2022, on household residents in Singapore, which evidenced that older residents had a higher attitude score than newer counterparts [25]. This result is widely expected as it has been stated that younger persons have a significantly worse perception of air pollution [47] and of activities that may reduce related health harnesses, whereas elder subjects are more aware of the risks [48]. Also, with reference to the between attitudes and gender, the results of the MLRA highlighted that females had a better overall score in attitude, according to the study by Al-Khamees et al., which, in a sample of students and teachers at Kuwait University, found a significantly better attitude in females, stating that such correlation can be explained as women are more often involved in polluting activities and tend to be less on guard regarding the risks connected [27]. Moreover, the evidence related to positive attitudes and respondents with higher education was confirmed in the study by Unni et al., that assessed community levels of Knowledge, Attitude, and Behavior (KAB) towards indoor air quality in randomly selected adults in Singapore: those who were higher skilled had comparatively higher attitude scores [25]. This result also agreed with Egondi et al., who in 2013 reported the association between attitude and educational levels, and Liu et al., who stated that a lower consciousness of air pollution and health effects was associated with a low educational level [49,50]. Furthermore, a recent cross-sectional study carried out in Lebanon over 2623 participants assessed that attitude towards cumulative effects of smoking, therefore also related to indoor air pollution, was significantly higher in nonsmoker subjects [51]. In addition, Al-Haqwi reported that non-smokers among a population of students had more willingness to act against polluting activities and therefore had better attitudes [52]. Also, the surprising relationship between attitudes and lower knowledge scores was confirmed in the aforementioned study by Unni et al., who assessed the same correlation [25]. Since, as aforementioned, knowledge is negatively related to behaviors, another educational program has to be implemented, in this case, designed to improve knowledge targeted to categories of people who allegedly already have positive attitudes and correct behaviors with the aim to reinforce their habits and improve their already good practices such as subjects involved in a relationship, with high educational levels and, non-smokers.

In conclusion, a behavioral program targeted those with knowledge, such as younger subjects with high educational levels, but do not engage in the correct practice toward indoor chemical pollution.

Model III displays that practices regarding household air pollution were statistically significant and correlated to education, smoking habits, knowledge, and attitudes. It has also been noted that there was a positive correlation between correct behaviors and marital status. In relation to the latter, a cross-sectional study conducted in Nairobi (Africa) on over 5317 individuals aged 35+ showed that marital status was not associated with improved behavior leading to better air quality [49]. Moreover, Kim et al. indicated that married subjects had better attitudes toward air pollution, but an explanation was not provided [53]. Therefore, the literature suggests that people involved in a relationship are usually more concerned about environmental pollution because their partner synergically influences them [54]. The relation between positive practices and education level might appear obvious; however, some doubts arise from the review by Maung et al. about indoor air pollution, which highlighted how human activities, behaviors, and education level are associated with personal exposure to air pollutants [55]. Again, as expected, teachers had a higher level of knowledge than students, which was reflected in their use of less polluting behaviors [27].

The results of the MLRA also evidenced the relationship between behavior and non-smoking. However, although widely expected, the literature does not define this correlation well. So far, only one previous study, conducted on householders in the USA during 2010–2011, evidenced that subjects without smoking habits also had other behaviors correlated to reduced air pollution [56]. Therefore, indoor pollution-related behaviors and air pollution, in general, may be affected by having or not smoking habits. Besides, in this study, a relation between positive behavior and negative knowledge was found, also stated by Unni et al. [25].

Inherently, a review carried out by Barnes, comprehending data from several studies, defined the limited effectiveness of education in improving behaviors concerning indoor air pollution [57]. On the other hand, the study by Daniel et al. indicated an association between high knowledge levels and behavior scores [26]. This could explain the correlation found in our study related to positive behaviors and negative knowledge, unlike many other studies on indoor air pollution, which used the KAB model and found that higher knowledge levels for respondents towards IAQ were associated with significantly higher behavior scores [25,26]. Another important correlation found in Model III was between respondents with higher behavior scores and high attitude scores, in agreement with Unni et al. [25] and also consistent with previous literature, as pointed out by Pampel et al. in 2010, which demonstrated that subjects with better attitude also had better behaviors [58]. This relationship pointed out the dominant role of attitude in forming correct behaviors related to indoor pollution and led us to suggest that an educational program designed to improve attitude is mandatory to improve behaviors in the population. Moreover, it is necessary to organize a training program for those who demonstrate the worst behaviors, such as singles, smokers, and less-educated subjects, to improve their practices and reduce the quantity of indoor pollutants they are exposed to and, therefore, the risks associated with it.

4. Conclusions

In summary, as shown in this study, indoor household pollution is a phenomenon not yet sufficiently studied in the general population. Behaviors intended to reduce indoor pollution are difficult to practice, although the sample has a good knowledge of the harms resulting from some habits. Therefore, it is necessary to organize training programs for people with the worst behavior, such as singles, smokers, and less-educated people, to improve their practice and reduce the amount of pollutants they are exposed to within the house and, therefore, the risks associated with it. Since knowledge is negatively related to behavior and attitude, it is necessary to implement another educational program in this case to improve knowledge of a category of people who allegedly already have positive attitudes and correct behaviors in order to strengthen their habits and improve their good practices, such as subjects involved in high-level relationships and non-smokers. In conclusion, a behavior and attitude correction program is aimed at those with knowledge, such as young people with high education levels, but does not put proper practices for household indoor chemical pollution in practice.

Author Contributions: Conceptualization, P.M., M.G., M.S., F.D.D., F.P., G.M., M.L.M., S.R., M.T. and A.N.; Data curation, P.M.; Formal analysis, M.G., M.S., F.D.D., G.M., M.L.M. and S.R.; Investigation, M.G. and F.D.D.; Methodology, M.G., M.S., G.M., M.L.M. and S.R.; Resources, P.M., F.P., U.T., M.T. and A.N.; Software, M.S.; Supervision, P.M., F.P., U.T., M.T. and A.N.; Validation, M.S. and F.P.; Writing—original draft, M.G., F.D.D., G.M., M.L.M. and S.R.; Writing—review and editing, M.S., F.P. and U.T. All authors have read and agreed to the published version of the manuscript.

Funding: This research received no external funding.

Institutional Review Board Statement: Not applicable.

Informed Consent Statement: Not applicable.

Data Availability Statement: The data that support the findings of this study are available upon reasonable request from the corresponding author. The data are not publicly available due to privacy or ethical restrictions.

Conflicts of Interest: We declare no competing interest.

References

1. Gordon, S.B.; Bruce, N.G.; Grigg, J.; Hibberd, P.L.; Kurmi, O.P.; Lam, K.B.H.; Mortimer, K.; Asante, K.P.; Balakrishnan, K.; Balmes, J.; et al. Respiratory risks from household air pollution in low and middle income countries. *Lancet Respir. Med.* **2014**, *2*, 823–860. [CrossRef] [PubMed]
2. Raju, S.; Siddharthan, T.; McCormack, M.C. Indoor Air Pollution and Respiratory Health. *Clin. Chest Med.* **2020**, *41*, 825–843. [CrossRef] [PubMed]
3. Amegah, A.K.; Quansah, R.; Jaakkola, J.J.K. Household Air Pollution from Solid Fuel Use and Risk of Adverse Pregnancy Outcomes: A Systematic Review and Meta-Analysis of the Empirical Evidence. *PLoS ONE* **2014**, *9*, e113920. [CrossRef] [PubMed]
4. Kurmi, O.P.; Semple, S.; Simkhada, P.; Smith, W.C.S.; Ayres, J.G. COPD and chronic bronchitis risk of indoor air pollution from solid fuel: A systematic review and meta-analysis. *Thorax* **2010**, *65*, 221–228. [CrossRef]

5. Bruce, N. Indoor air pollution from unprocessed solid fuel use and pneumonia risk in children aged under five years: A systematic review and meta-analysis. *Bull. World Health Organ.* **2008**, *86*, 390–398. [CrossRef] [PubMed]
6. Hamra, G.B.; Guha, N.; Cohen, A.; Laden, F.; Raaschou-Nielsen, O.; Samet, J.M.; Vineis, P.; Forastiere, F.; Saldiva, P.; Yorifuji, T.; et al. Outdoor Particulate Matter Exposure and Lung Cancer: A Systematic Review and Meta-Analysis. *Environ. Health Perspect.* **2014**, *122*, 906–911. [CrossRef]
7. Kantipudi, N.; Patel, V.; Jones, G.; Kamath, M.V.; Upton, A.R.M. Air Pollution's Effects on the Human Respiratory System. *Crit. Rev. Biomed. Eng.* **2016**, *44*, 383–395. [CrossRef]
8. Vardoulakis, S.; Giagloglou, E.; Steinle, S.; Davis, A.; Sleeuwenhoek, A.; Galea, K.S.; Dixon, K.; Crawford, J.O. Indoor Exposure to Selected Air Pollutants in the Home Environment: A Systematic Review. *Int. J. Environ. Res. Public Health* **2020**, *17*, 8972. [CrossRef]
9. Balmes, J.R. Household air pollution from domestic combustion of solid fuels and health. *J. Allergy Clin. Immunol.* **2019**, *143*, 1979–1987. [CrossRef]
10. National Academies of Sciences, Engineering, and Medicine. *Microbiomes of the Built Environment: A Research Agenda for Indoor Microbiology, Human Health, and Buildings*; The National Academies Press: Washington, DC, USA, 2017. [CrossRef]
11. Ni, Y.; Shi, G.; Qu, J. Indoor PM2.5, tobacco smoking and chronic lung diseases: A narrative review. *Environ. Res.* **2019**, *181*, 108910. [CrossRef]
12. González-Martín, J.; Kraakman, N.J.R.; Pérez, C.; Lebrero, R.; Muñoz, R. A state–of–the-art review on indoor air pollution and strategies for indoor air pollution control. *Chemosphere* **2021**, *262*, 128376. [CrossRef] [PubMed]
13. Shen, H.; Luo, Z.; Xiong, R.; Liu, X.; Zhang, L.; Li, Y.; Du, W.; Chen, Y.; Cheng, H.; Shen, G.; et al. A critical review of pollutant emission factors from fuel combustion in home stoves. *Environ. Int.* **2021**, *157*, 106841. [CrossRef] [PubMed]
14. Van Tran, V.; Park, D.; Lee, Y.-C. Indoor Air Pollution, Related Human Diseases, and Recent Trends in the Control and Improvement of Indoor Air Quality. *Int. J. Environ. Res. Public Health* **2020**, *17*, 2927. [CrossRef] [PubMed]
15. Wickliffe, J.K.; Stock, T.H.; Howard, J.L.; Frahm, E.; Simon-Friedt, B.R.; Montgomery, K.; Wilson, M.J.; Lichtveld, M.Y.; Harville, E. Increased long-term health risks attributable to select volatile organic compounds in residential indoor air in southeast Louisiana. *Sci. Rep.* **2020**, *10*, 21649. [CrossRef] [PubMed]
16. Wong, J.P.S.; Carslaw, N.; Zhao, R.; Zhou, S.; Abbatt, J.P.D. Observations and impacts of bleach washing on indoor chlorine chemistry. *Indoor Air* **2017**, *27*, 1082–1090. [CrossRef] [PubMed]
17. UNECE. United Nations Economic Commmission For Europe. *Protecting the Air We Breathe. 40 Years of Cooperation under the Convention on Long-Range Transboundary Air Pollution.* 2019. Available online: https://unece.org/environment-policy/publications/protecting-air-we-breathe (accessed on 21 September 2022).
18. Huang, J.; Pan, X.; Guo, X.; Li, G. Health impact of China's Air Pollution Prevention and Control Action Plan: An analysis of national air quality monitoring and mortality data. *Lancet Planet. Health* **2018**, *2*, e313–e323. [CrossRef]
19. Mazaheri, M.; Clifford, S.; Yeganeh, B.; Viana, M.; Rizza, V.; Flament, R.; Buonanno, G.; Morawska, L. Investigations into factors affecting personal exposure to particles in urban microenvironments using low-cost sensors. *Environ. Int.* **2018**, *120*, 496–504. [CrossRef]
20. Sierra-Vargas, M.P.; Teran, L.M. Air pollution: Impact and prevention. *Respirology* **2012**, *17*, 1031–1038. [CrossRef]
21. Cooper, E.; Wang, Y.; Stamp, S.; Burman, E.; Mumovic, D. Use of portable air purifiers in homes: Operating behaviour, effect on indoor PM2.5 and perceived indoor air quality. *Build. Environ.* **2021**, *191*, 107621. [CrossRef]
22. National Aeronautics and Space Administration. The Effects of Climate Change. 2019. Available online: https://climate.nasa.gov/effects/ (accessed on 2 February 2023).
23. Pamonpol, K.; Areerob, T.; Prueksakorn, K. Indoor Air Quality Improvement by Simple Ventilated Practice and Sansevieria Trifasciata. *Atmosphere* **2020**, *11*, 271. [CrossRef]
24. Sundell, J.; Levin, H.; Nazaroff, W.W.; Cain, W.S.; Fisk, W.J.; Grimsrud, D.T.; Gyntelberg, F.; Li, Y.; Persily, A.K.; Pickering, A.C.; et al. Ventilation rates and health: Multidisciplinary review of the scientific literature. *Indoor Air* **2011**, *21*, 191–204. [CrossRef]
25. Unni, B.; Tang, N.; Cheng, Y.M.; Gan, D.; Aik, J. Community knowledge, attitude and behaviour towards indoor air quality: A national cross-sectional study in Singapore. *Environ. Sci. Policy* **2022**, *136*, 348–356. [CrossRef]
26. Daniel, L.; Michot, M.; Esvan, M.; Guérin, P.; Chauvet, G.; Pelé, F. Perceptions, Knowledge, and Practices Concerning Indoor Environmental Pollution of Parents or Future Parents. *Int. J. Environ. Res. Public Health* **2020**, *17*, 7669. [CrossRef]
27. Al-Khamees, N.A. Knowledge of, Attitudes toward, and Practices regarding Indoor Pollution at Kuwait University. *J. Geosci. Environ. Prot.* **2018**, *6*, 146–157. [CrossRef]
28. Adeolu, A.T.; Odipe, O.E.; Raimi, M.O. Practices and Knowledge of Household Residents to Lead Exposure in Indoor Environment in Ibadan, Oyo State, Nigeria. *J. Sci. Res. Rep.* **2018**, *19*, 1–10. [CrossRef]
29. Neri, A.; McNaughton, C.; Momin, B.; Puckett, M.; Gallaway, M.S. Measuring public knowledge, attitudes, and behaviors related to radon to inform cancer control activities and practices. *Indoor Air* **2018**, *28*, 604–610. [CrossRef]
30. ISTAT. Bilancio Demografico Mensile e Popolazione Residente per Sesso. 2022. Available online: https://demo.istat.it/app/?i=POS&l=it (accessed on 13 June 2022).

31. Montuori, P.; Loperto, I.; Paolo, C.; Castrianni, D.; Nubi, R.; De Rosa, E.; Palladino, R.; Triassi, M. Bodybuilding, dietary supplements and hormones use: Behaviour and determinant analysis in young bodybuilders. *BMC Sports Sci. Med. Rehabil.* **2021**, *13*, 147. [CrossRef]
32. Montuori, P.; Sarnacchiaro, P.; Nubi, R.; Di Ruocco, D.; Belpiede, A.; Sacco, A.; De Rosa, E.; Triassi, M. The use of mobile phone while driving: Behavior and determinant analysis in one of the largest metropolitan area of Italy. *Accid. Anal. Prev.* **2021**, *157*, 106161. [CrossRef]
33. Montuori, P.; Sorrentino, M.; Sarnacchiaro, P.; Di Duca, F.; Nardo, A.; Ferrante, B.; D'Angelo, D.; Di Sarno, S.; Pennino, F.; Masucci, A.; et al. Job Satisfaction: Knowledge, Attitudes, and Practices Analysis in a Well-Educated Population. *Int. J. Environ. Res. Public Health* **2022**, *19*, 14214. [CrossRef]
34. Educational Attainment Statistics, 2020—European Commission. Available online: https://ec.europa.eu/eurostat/index (accessed on 17 November 2022).
35. Muro, M.B.; Njogu, E.; Orinda, G. Caregivers'level of Knowledge on Indoor Air Pollution and Acute Respiratory Infections Among Under-Fives In Informal Settlement: Makadara, Nairobi County. *Int. Acad. J. Health Med. Nurs.* **2020**, *5*, 3. Available online: https://www.iprjb.org/journals/index.php/JHMN/article/view/1106 (accessed on 15 November 2022).
36. Rosengren, A.; Smyth, A.; Rangarajan, S.; Ramasundarahettige, C.; Bangdiwala, S.I.; Alhabib, K.F.; Avezum, A.; Boström, K.B.; Chifamba, J.; Gulec, S.; et al. Socioeconomic status and risk of cardiovascular disease in 20 low-income, middle-income, and high-income countries: The Prospective Urban Rural Epidemiologic (PURE) study. *Lancet Glob. Health* **2019**, *7*, e748–e760. [CrossRef]
37. Gall, E.T.; Chen, A.; Chang, V.W.-C.; Nazaroff, W.W. Exposure to particulate matter and ozone of outdoor origin in Singapore. *Build. Environ.* **2015**, *93*, 3–13. [CrossRef]
38. Pope, C.A.; Turner, M.C.; Burnett, R.T.; Jerrett, M.; Gapstur, S.M.; Diver, W.R.; Krewski, D.; Brook, R.D. Relationships between Fine Particulate Air Pollution, Cardiometabolic Disorders, and Cardiovascular Mortality. *Circ. Res.* **2015**, *116*, 108–115. [CrossRef]
39. Qian, X.; Xu, G.; Li, L.; Shen, Y.; He, T.; Liang, Y.; Yang, Z.; Zhou, W.W.; Xu, J. Knowledge and perceptions of air pollution in Ningbo, China. *BMC Public Health* **2016**, *16*, 1138. [CrossRef]
40. Jin, Y.; Wang, L.; Lu, B.; Ferketich, A.K. Secondhand Smoke Exposure, Indoor Smoking Bans and Smoking-Related Knowledge in China. *Int. J. Environ. Res. Public Health* **2014**, *11*, 12835–12847. [CrossRef]
41. Chin, Y.S.J.; De Pretto, L.; Thuppil, V.; Ashfold, M.J. Public awareness and support for environmental protection—A focus on air pollution in peninsular Malaysia. *PLoS ONE* **2019**, *14*, e0212206. [CrossRef]
42. Kaur, D.; Sidhu, M.; Bal, S. A study on subjective assessment of indoor pollution among rural and urban homemakers of Ludhiana city. *Asian J. Environ. Sci.* **2015**, *10*, 95–99. [CrossRef]
43. Jansen, T.; Rademakers, J.; Waverijn, G.; Verheij, R.; Osborne, R.; Heijmans, M. The role of health literacy in explaining the association between educational attainment and the use of out-of-hours primary care services in chronically ill people: A survey study. *BMC Health Serv. Res.* **2018**, *18*, 394. [CrossRef]
44. Lee, T.M.; Markowitz, E.M.; Howe, P.D.; Ko, C.-Y.; Leiserowitz, A.A. Predictors of public climate change awareness and risk perception around the world. *Nat. Clim. Chang.* **2015**, *5*, 1014–1020. [CrossRef]
45. Zsóka, Á.; Szerényi, Z.M.; Széchy, A.; Kocsis, T. Greening due to environmental education? Environmental knowledge, attitudes, consumer behavior and everyday pro-environmental activities of Hungarian high school and university students. *J. Clean. Prod.* **2013**, *48*, 126–138. [CrossRef]
46. Carducci, A.; Fiore, M.; Azara, A.; Bonaccorsi, G.; Bortoletto, M.; Caggiano, G.; Calamusa, A.; De Donno, A.; De Giglio, O.; Dettori, M.; et al. Environment and health: Risk perception and its determinants among Italian university students. *Sci. Total Environ.* **2019**, *691*, 1162–1172. [CrossRef]
47. Skov, T.; Cordtz, T.; Jensen, L.K.; Saugman, P.; Schmidt, K.; Theilade, P. Modifications of health behaviour in response to air pollution notifications in Copenhagen. *Soc. Sci. Med.* **1991**, *33*, 621–626. [CrossRef]
48. Al-Shidi, H.K.; Ambusaidi, A.K.; Sulaiman, H. Public awareness, perceptions and attitudes on air pollution and its health effects in Muscat, Oman. *J. Air Waste Manag. Assoc.* **2021**, *71*, 1159–1174. [CrossRef]
49. Egondi, T.; Kyobutungi, C.; Ng, N.; Muindi, K.; Oti, S.; Van De Vijver, S.; Ettarh, R.; Rocklöv, J. Community Perceptions of Air Pollution and Related Health Risks in Nairobi Slums. *Int. J. Environ. Res. Public Health* **2013**, *10*, 4851–4868. [CrossRef]
50. Liu, H.; Kobernus, M.; Liu, H. Public Perception Survey Study on Air Quality Issues in Wuhan, China. *J. Environ. Prot.* **2017**, *8*, 1194–1218. [CrossRef]
51. Haddad, C.; Sacre, H.; Hajj, A.; Lahoud, N.; Akiki, Z.; Akel, M.; Saade, D.; Zeidan, R.K.; Farah, R.; Hallit, S.; et al. Comparing cigarette smoking knowledge and attitudes among smokers and non-smokers. *Environ. Sci. Pollut. Res.* **2020**, *27*, 19352–19362. [CrossRef]
52. Al-Haqwi, A.I.; Tamim, H.; Asery, A. Knowledge, attitude and practice of tobacco smoking by medical students in Riyadh, Saudi Arabia. *Ann. Thorac. Med.* **2010**, *5*, 145–148. [CrossRef]
53. Kim, H.; Cho, J.; Isehunwa, O.; Noh, J.; Noh, Y.; Oh, S.S.; Koh, S.-B.; Kim, C. Marriage as a social tie in the relation of depressive symptoms attributable to air pollution exposure among the elderly. *J. Affect. Disord.* **2020**, *272*, 125–131. [CrossRef]
54. Brown, M.A.; Macey, S.M. Understanding Residential Energy Conservation through Attitudes and Beliefs. *Environ. Plan. A Econ. Space* **1983**, *15*, 405–416. [CrossRef]

55. Maung, T.Z.; Bishop, J.E.; Holt, E.; Turner, A.M.; Pfrang, C. Indoor Air Pollution and the Health of Vulnerable Groups: A Systematic Review Focused on Particulate Matter (PM), Volatile Organic Compounds (VOCs) and Their Effects on Children and People with Pre-Existing Lung Disease. *Int. J. Environ. Res. Public Health* **2022**, *19*, 8752. [CrossRef]
56. Zhang, X.; Martinez-Donate, A.; Rhoads, N. Parental Practices and Attitudes Related to Smoke-Free Rules in Homes, Cars, and Outdoor Playgrounds in US Households with Underage Children and Smokers, 2010–2011. *Prev. Chronic Dis.* **2015**, *12*, e96. [CrossRef]
57. Barnes, B.R. Behavioural Change, Indoor Air Pollution and Child Respiratory Health in Developing Countries: A Review. *Int. J. Environ. Res. Public Health* **2014**, *11*, 4607–4618. [CrossRef]
58. Pampel, F.C.; Krueger, P.M.; Denney, J.T. Socioeconomic Disparities in Health Behaviors. *Annu. Rev. Sociol.* **2010**, *36*, 349–370. [CrossRef]

toxics

MDPI

Article

Health Risk Assessment of PAHs from Estuarine Sediments in the South of Italy

Fabiana Di Duca [1], Paolo Montuori [1,*], Ugo Trama [2], Armando Masucci [1], Gennaro Maria Borrelli [1] and Maria Triassi [1]

1 Department of Public Health, University "Federico II", Via Sergio Pansini n° 5, 80131 Naples, Italy
2 General Directorate of Health, Campania Region, Centro Direzionale Is. C3, 80143 Naples, Italy
* Correspondence: pmontuor@unina.it

Abstract: Increased concerns about the toxicities of Polycyclic Aromatic Hydrocarbons (PAHs), ubiquitous and persistent compounds, as well as the associated ecotoxicology issue in estuarine sediments, have drawn attention worldwide in the last few years. The levels of PAHs in the Sele, Sarno, and Volturno Rivers sediments were evaluated. Moreover, the cancerogenic risk resulting from dermal and ingestion exposure to PAHs was estimated using the incremental lifetime cancer risk (ILCR) assessment and the toxic equivalent concentration (TEQ_{BaP}). For Sele River, the results showed that the total PAH concentration ranged from 632.42 to 844.93 ng g^{-1} dw, with an average value of 738.68 ng g^{-1} dw. ∑PAHs were in the range of 5.2–678.6 ng g^{-1} dw and 434.8–872.1 ng g^{-1} dw for the Sarno and Volturno River sediments, respectively. The cancerogenic risk from the accidental ingestion of PAHs in estuarine sediments was low at all sampling sites. However, based on the $ILCR_{dermal}$ values obtained, the risk of cancer associated with exposure by dermal contact with the PAHs present in the sediments was moderate, with a mean $ILCR_{dermal}$ value of 2.77×10^{-6}. This study revealed the pollution levels of PAHs across the South of Italy and provided a scientific basis for PAH pollution control and environmental protection.

Keywords: polycyclic aromatic hydrocarbons (PAHs); river sediment; occurrence; incremental lifetime cancer risk; carcinogenic risk

check for updates

Citation: Di Duca, F.; Montuori, P.; Trama, U.; Masucci, A.; Borrelli, G.M.; Triassi, M. Health Risk Assessment of PAHs from Estuarine Sediments in the South of Italy. *Toxics* **2023**, *11*, 172. https://doi.org/10.3390/toxics11020172

Academic Editor: Crispin J. Halsall

Received: 5 January 2023
Revised: 1 February 2023
Accepted: 11 February 2023
Published: 13 February 2023

1. Introduction

Estuaries are the main deposits for the disposal of industrial and domestic effluents, sewage sludge, and dredged material with a significant load of contaminants, including PAHs, from pipeline discharges, vehicular emissions, atmospheric deposition, surface runoff, as well as oil spills in aquatic environments [1]. Due to their low water solubility and high lipophilicity, PAHs tend to accumulate in the sediments of aquatic systems for long periods due to their high degradation resistance and high organic carbon content [2]. In soil and sediment compartments, PAHs can undergo biodegradation processes by microorganisms. However, due to their stable physic–chemical characteristics, hydrophobicity, and a strong tendency to absorb into the soil matrix, their biodegradation rate is low. Consequently, PAHs do not degrade easily, and they can accumulate in the solid phase of the terrestrial and aquatic environment, where they persist for a long time [3,4]. Thus, sediments constitute a natural reserve of PAHs in the aquatic system [5,6]. Moreover, they can be released into the surrounding environment by means of resuspension phenomena, thus giving rise to "secondary pollution". Consequently, since high concentrations of PAHs in sediments may reveal a potential pollution risk for the environment and human health, it is essential to monitor these compounds in sediment to protect and preserve the aquatic environment and human health [7].

PAHs, as persistent organic pollutants (POPs), are widely present in the ecosystem [8–10]. Their spread to the environment has raised many concerns for human health, as some of them have been identified as carcinogens, mutagens, and teratogens. Indeed, PAHs released into the environment may enter the food chain, and exposure to them may result in a risk of cancer or other adverse effects on human health [11–13]. The International Agency for Research on Cancer (IARC) has classified PAHs according to their carcinogenicity as carcinogenic (Group 1), probable carcinogenic (Group 2A), possible carcinogenic (Group 2B), and non-carcinogenic (Group 3) [14]. In general, high molecular weight PAHs (4–6 rings) are more toxic than low molecular weight PAHs (2–3 rings) [15]. The greater toxicity of the former is due to the greater number of aromatic rings from which dihydrodioloepoxides are formed [16]. In particular, the fat solubility of PAHs makes them dangerous because they can cross cell membranes, penetrate, and deposit in tissues. In tissues, PAHs can be oxidized to epoxide (where an oxygen atom replaces one of the double bonds C=C) by a monooxygenase associated with cytochrome P 450 present in the endoplasmic reticulum of cells. The epoxide thus formed can attack macromolecules such as DNA, hence the mutagenic and carcinogenic action of the PAHs, or be transformed into diol by enzymatic systems such as epoxide hydrolase (EH). This reaction of detoxification allows the formed diol, with two hydrophilic alcohol groups, to be more soluble than the starting compound and then to be expelled from the body more easily [17]. In fact, Lee et al. reported that cytochrome P450 enzymes can metabolize BaP and activate it into a carcinogenic reactive intermediate or metabolite. Consequently, these substances can bind to DNA, resulting in DNA adducts that interfere with DNA replication, causing cytotoxicity, teratogenicity, genotoxicity, immunotoxicity, mutagenesis, and carcinogenesis [18].

Mainly, Benzo[a]pyrene (BaP) has been classified as genotoxic using in vitro tests and in vivo studies. In laboratory animals, oral administration of BaP induced tumors of the stomach and mammary gland and skin cancer [19,20]. According to the WHO, BaP is the compound with the greatest negative consequences for human health and has been included in Group 1, which includes all those substances for which there is sufficient evidence of carcinogenicity in humans [20]. BaP has been classified as genotoxic [10], and it has been involved in tumor development in all test animal species tested, regardless of the route of exposure (oral, cutaneous, subcutaneous, inhalator, intratracheal, intrabronchial, intraperitoneal, or intravenous) [21]. Furthermore, dibenz[a,h]anthracene (DahA) has been classified as a probable carcinogen and/or mutagen for humans, and it is included in Group 2A unlike Benzo[a]anthracene (BaA), Benzo[b]fluoranthrene (BbF), Benzo[k]fluoranthrene (BkF), Chrysene (Chr), and Indeno [123-cd]pyrene (IcdP), which have been included in Group 2B as possible carcinogens in humans [22].

The main routes of exposure to PAHs in the general population are inhalation from breathing ambient and indoor air or smoking cigarettes, ingesting food containing PAHs, and breathing smoke from open fireplaces [10,23,24]. According to Adeniji et al., exposure via inhalation, ingestion, or skin contact may lead to human health problems resulting from short- and long-term effects, including some serious respiratory and cardiovascular diseases [13].

The acute effects of PAHs on human health mainly depend on the duration of exposure, PAH concentration during exposure, and the toxicity of the compounds to which one is exposed, as well as the route of exposure. Short-term exposure to PAHs has been reported to cause impaired lung function such as asthma and thrombotic effects in people with coronary heart disease [24]. However, there is currently no full understanding of the effects of short-term exposure to PAHs, but it is well-known that occupational exposure to high levels of PAH-containing mixtures causes symptoms such as eye irritation, nausea, and vomiting [25]. Moreover, PAHs mixtures are also known to cause skin irritation and inflammation, as Anthracene (Ant), Benzo[a]Pyrene (BaP), and Naphthalene (NaP) are skin irritants [26]. In addition, PAHs interfere with hormonal systems and, as a result, can have harmful effects on reproduction and immune function [27]. The adverse effects of exposure to PAHs have been extensively investigated, but the information currently available on human exposure to individual PAHs is scattered and incomplete, except for some accidental contact with NaP and BaP [28–30]. Srogi et al. stated that prolonged dermal contact with NaP may cause redness and inflammation of the skin [31]. In addition, Diggs et al. reported that long-term exposure to low levels of Pyr and BaP has been identified as the cause of cancer in laboratory animals [32]. Animal studies have also shown adverse effects on reproduction and development due to exposure to PAHs, whereas these effects were not commonly detected in humans [33,34]. Moreover, Anyahara stated that exposure to PAHs can induce cataracts and cause kidney and liver damage and jaundice [35].

Estuaries are important aquatic systems largely affected by PAH pollution. In fact, due to their chemical properties, PAHs can persist in water where they are readily adsorbed onto particulate matter, settling in river sediments and soils. Thus, river sediments behave as the primary sink and reservoir for PAHs in the aquatic environment [36,37]. Consequently, sediments are important indicators since they can reflect the pollution status of the environment [38].

To date, no previous studies have evaluated the carcinogenic risk to human health associated with dermal and accidental ingestion exposure to PAHs from surface sediments in the South of Italy. Therefore, the main aim of this study was to assess the risk to human health from exposure to PAHs present in the sediments of surface waters in a large coastal area of the Campania Region, in the South of Italy. Specifically, the purpose of this study was to evaluate the distribution patterns of PAHs and to assess the carcinogenic risk to human health from dermal and ingestion exposure to these contaminants from the estuarine sediments of the Sarno, Volturno, and Sele Rivers, which are the main surface water streams of the Campania Region, in the South of Italy.

2. Materials and Methods

2.1. Study Area

This assessment of the human health risk from exposure to PAHs was carried out in a study area of approximately 3100 km^2 and included the three largest plains in the Campania Region, in the South of Italy. Particularly, the research area was close to the estuaries of the Sele, Volturno, and Sarno Rivers, which traverse the same-named plains. Figure 1 shows the three plains of interest and the respective estuaries of the rivers that cross them.

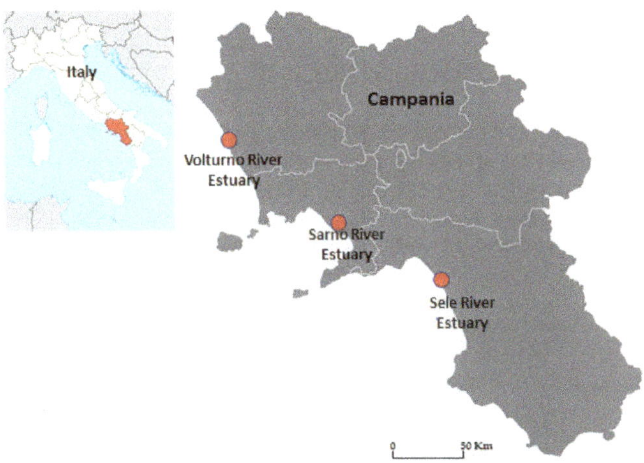

Figure 1. Study area.

2.2. Sampling

A sampling campaign was carried out during the spring season (April 2021) at 10 sampling sites near the Sele River Estuary. In detail, sediment samples were taken from the mouth of the river (Site 1) at different distances, 500 m, 1000 m, and 1500 m from the mouth, and directions, to the north, west, and south of the estuary (Figure 2). During sampling, a global positioning system (GPS) was used to locate all sampling sites.

Information on the identification number (ID), characteristics, and coordinates of each sampling location are shown in Table 1. The samples were collected at a depth of 0 to 5 cm using a scraping sampler (Van Veen Grab) and placed in aluminum containers. Then, they were transferred under refrigeration to the laboratory and stored at −20 °C until analysis.

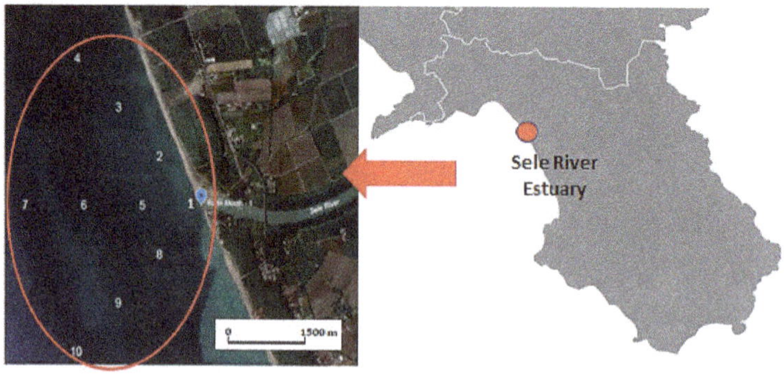

Figure 2. Hydrographic network and sampling sites near the Sele River.

Table 1. Sampling sites with their identification number (ID), location name, and coordinates from the Sele River.

ID	Location	Coordinates	ID	Location	Coordinates
1	Sele River mouth	40°28′55″ N 14°56′33″ E	6	1000 m west	40°28′55″ N 14°55′50″ E
2	500 m north	40°29′04″ N 14°56′14″ E	7	1000 m south	40°28′39″ N 14°55′56″ E
3	500 m west	40°28′55″ N 14°56′12″ E	8	1500 m north	40°29′20″ N 14°55′38″ E
4	500 m south	40°28′47″ N 14°56′16″ E	9	1500 m west	40°28′55″ N 14°55′28″ E
5	1000 m north	40°29′12″ N 14°55′56″ E	10	1500 m south	40°28′30″ N 14°55′38″ E

The PAHs levels in sediment samples from the Sarno and Volturno Rivers were evaluated previously [39,40]. Briefly, for the Sarno River, a sampling campaign was carried out during the spring of 2008 at the source of the river (site 1), just before and after the junction with Alveo Comune, at the river mouth (site 4), and in 9 sites located at different distances from the estuary (Figure 3). More detailed information about the sediment sampling in the Sarno River is given in Table 2.

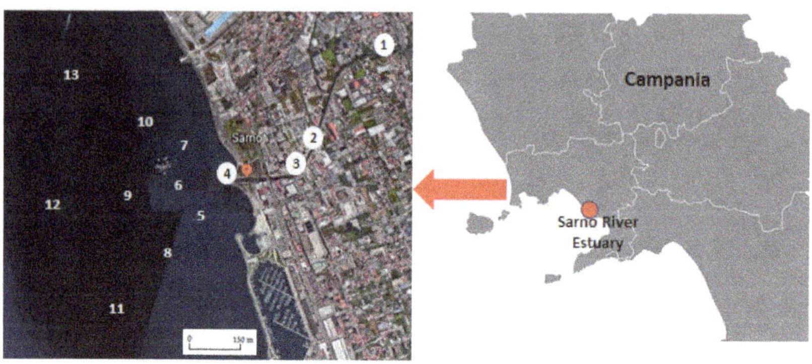

Figure 3. Hydrographic network and sampling sites near the Sarno River.

Table 2. Sampling sites with their identification number (ID), location name, and coordinates from the Sarno River.

ID	Location	Coordinates	ID	Location	Coordinates
1	Source of Sarno River	40°48′54.03″ N 14°36′45.36″ E	8	150 m south	40°43′35.68″ N 14°28′02.94″ E
2	Before junction with Alveo Comune	40°46′42.73″ N 14°34′00.48″ E	9	150 m west	40°43′42.25″ N 14°27′59.97″ E
3	After junction with Alveo Comune	40°46′00.34″ N 14°33′10.68″ E	10	150 m north	40°43′49.26″ N 14°27′30.31″ E
4	Sarno River mouth	40°46′10.68″ N 14°28′07.89″ E	11	500 m south	40°43′30.31″ N 14°27′58.94″ E
5	50 m south	40°43′40.11″ N 14°28′06.45″ E	12	500 m west	40°43′42.29″ N 14°27′46.41″ E
6	50 m west	40°43′42.46″ N 14°28′05.03″ E	13	500 m north	40°43′57.85″ N 14°27′48.68″ E
7	50 m north	40°43′45.09″ N 14°28′05.17″ E			

For the Volturno River, the sampling campaign was carried out in April 2018 near the mouth of the Volturno River and in 9 sites located at different distances from it (Figure 4). The specific details are provided in Table 3.

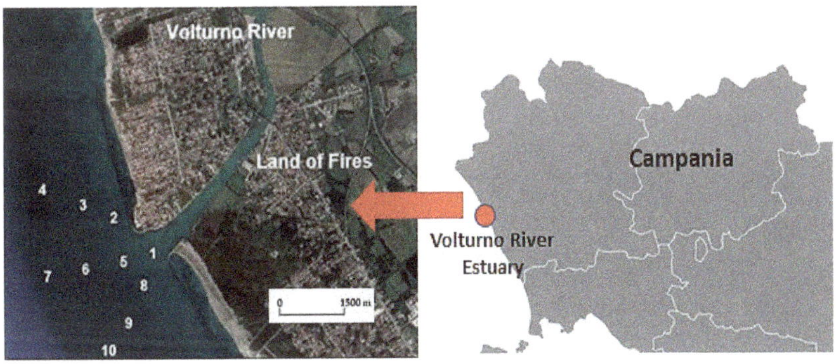

Figure 4. Hydrographic network and sampling sites near the Volturno River.

Table 3. Sampling sites with their identification number (ID), location name, and coordinates from the Volturno River.

ID	Location	Coordinates	ID	Location	Coordinates
1	Volturno River mouth	40°48′54.03″ N 14°36′45.36″ E	6	1000 m west	40°43′42.46″ N 14°28′05.03″ E
2	500 m north	40°46′42.73″ N 14°34′00.48″ E	7	1000 m south	40°43′45.09″ N 14°28′05.17″ E
3	500 m west	40°46′00.34″ N 14°33′10.68″ E	8	1500 m north	40°43′35.68″ N 14°28′02.94″ E
4	500 m south	40°43′42.62″ N 14°28′07.89″ E	9	1500 m west	40°43′42.25″ N 14°27′59.97″ E
5	1000 m north	40°43′40.11″ N 14°28′06.45″ E	10	1500 m south	40°43′49.26″ N 14°27′59.82″ E

2.3. Extraction Procedure and Clean-Up

The analyses were performed as described previously [41]. Briefly, for PAH extraction, the sediment samples were air-dried, crushed, sieved in 250 μm particles, and then divided into portions of 5 g. The PAH concentrations were indicated as dry weight (ng/g dw) [42,43].

The PAH extraction was performed with a Soxhlet extractor using methylene chloride as solvent. Subsequently, the extracts, first purified using a column composed of sodium sulfate/silica gel and then eluted with 70 mL of a hexane:methylene chloride (7:3, v/v) solution, were evaporated to dryness and reduced to a final volume (500 µL) with the aid of a weak current of nitrogen. Finally, the extracts were analyzed using gas chromatography coupled with mass spectrometry (GC-MS). A TOC analyzer was used to evaluate the total organic carbon (TOC) content in the sediment samples (TOC-VCPH, Shimadzu Corp., Kyoto, Japan).

2.4. Instrumental Analysis, Quality Assurance, and Quality Control

A TRACE^{TM}1310 gas chromatograph coupled to an ISQ^{TM}7000 single quadrupole mass spectrometer (GC-MS, Thermo Scientific, Waltham, MA, USA) was used, equipped with a capillary column TG-5MS (length 30 mm, inner diameter 0.25 mm, film thickness 0.25 µm) and helium as a gas carrier (constant flow of 1 mL/min), operating in the electronic ionization mode (EI) set to 70 eV. The injector operated at 280 °C, and the temperature of the detector was set to 300 °C. The acquisition was performed with the Selected Ion Monitoring (SIM) mode using two characteristic fragments for each selected analyte. A splitless injection mode was adopted with an injection volume of 1 µL. The quantification of PAHs was carried out using response factors related to the respective internal standards based on a six-point calibration curve for individual PAHs (Dr. Ehrenstorfer GmbH, Augsburg, Germany) ($R^2 > 0.97$). Chrysene-d_{12} was used as an internal standard for sample quantification. Before the analysis, all the glassware to be used was thoroughly washed with methanol, acetone, and dichloromethane and placed in the oven at 200 °C to minimize possible sources of contamination. The column temperature was set with different gradients: from 60 °C to 200 °C with an increase of 25 °C/min (kept for 2 min), to 270 °C increasing at 10 °C min^{-1} (kept for 6 min), and to 310 °C with a rise of 25 °C min^{-1} (kept for 10 min). The single ion monitoring mode (SIM) was used for the acquisition using characteristic ions for each target analyte. The 16 priority IPA, according to the WHO and USEPA, were evaluated (Table 4) [20,44].

Six-point calibration curves (5–10–50–250–500–1000 ng/L), procedural blanks, and sample triplicates were carried out for every set of samples. The PAH concentrations were calculated as dry weight (ng/g dw). The limits of detection (LOD) and quantification (LOQ) were evaluated as three and ten times the noise in blank samples, respectively. They were in the range of 1.5–1.9 ng g^{-1} and 5.1–6.3 ng g^{-1}, respectively. In the procedural blanks, analyzed as the samples, the PAHs showed a concentration below the LOD. Moreover, for individual PAHs, the recovery test values ranged from 80% to 97%, meeting the quality control criteria (70–130%). For the effective and reproducible detection and quantification of low concentrations of PAHs in sediments, the linear range, precision, limits of detection, and limits of quantification were performed. The precision of the method was determined using repeatability tests and was expressed as standard deviation (SD) (Tables S1–S3). The average of the results was used to estimate the precision of the method.

2.5. Human Health Risk Assessment

The human health risk assessment is useful in determining whether exposure to a chemical in a specific dose may cause an increase in the frequency of adverse effects on human health [39,42,45–47]. For the population, PAH exposure represents a health and hygiene risk, which is assessed as a carcinogenic risk. Therefore, the USEPA defined carcinogenic risk as the probability that an individual may develop cancer over a lifetime from exposure to a specific substance classified as mutagenic or carcinogenic. Thus, this risk assessment consists of two phases, which are the estimation of the probability of an event occurring and the study of the likely magnitude of its adverse effect over a specific time frame [48].

Table 4. The 16 Priority PAHs according to the United States Environmental Protection Agency (US EPA) and the World Health Organization (WHO) [20,44].

Name	Abbreviation	Molecular Formula	Number of Rings	Chemical Structure	Molecular Weight
Naphthalene	Nap	$C_{10}H_8$	2	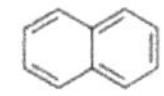	128.2 g mol^{-1}
Acenaphthylene	Acy	$C_{12}H_8$	3	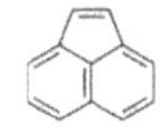	152.2 g mol^{-1}
Acenaphthalene	Ace	$C_{12}H_{10}$	3		154.2 g mol^{-1}
Fluorene	Flu	$C_{13}H_{10}$	3	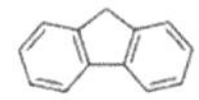	166.2 g mol^{-1}
Phenanthrene	Phe	$C_{14}H_{10}$	3	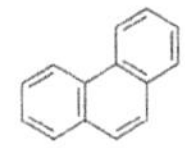	178.2 g mol^{-1}
Anthracene	Ant	$C_{14}H_{10}$	3	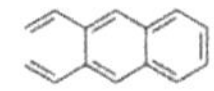	178.2 g mol^{-1}
Fluoranthene	Fla	$C_{16}H_{10}$	4	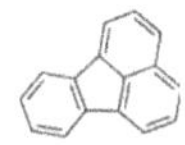	202.3 g mol^{-1}
Pyrene	Pyr	$C_{16}H_{10}$	4		202.3 g mol^{-1}
Benzo[a]anthracene	BaA	$C_{18}H_{12}$	4	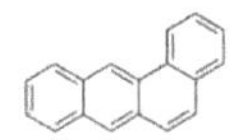	228.3 g mol^{-1}
Crysene	Chr	$C_{18}H_{12}$	4	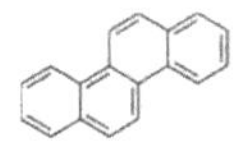	228.3 g mol^{-1}
Benzo[b]fluoranthene	BbF	$C_{20}H_{12}$	5	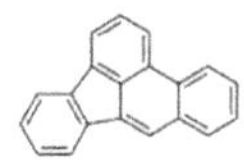	252.3 g mol^{-1}
Benzo[k]fluoranthene	BkF	$C_{20}H_{12}$	5	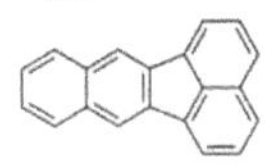	252.3 g mol^{-1}
Benzo[a]pyrene	BaP	$C_{20}H_{12}$	5	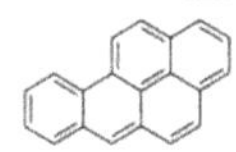	252.3 g mol^{-1}
Indeno [123-cd]pyrene	IcdP	$C_{22}H_{12}$	6	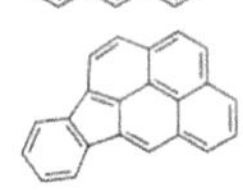	276.3 g mol^{-1}
Benzo[ghi]perylene	BghiP	$C_{22}H_{12}$	6		276.3 g mol^{-1}
Dibenzo[a,h]anthracene	DahA	$C_{22}H_{14}$	5	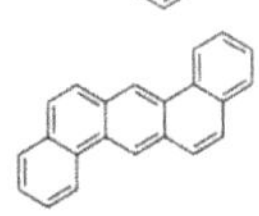	278.3 g mol^{-1}

The incremental lifetime cancer risk (ILCR) due to exposure by direct ingestion and skin contact to PAHs present in the sediment was evaluated [49]. First, the doses of contaminants taken up by human receptors through the two different exposure pathways considered were calculated according to Equations (1) and (2) [50]:

$$Dose_{ing} = \frac{C_s \times IR_s \times RAF_{oral} \times D_{hours} \times D_{days} \times D_{weeks} \times ED_{years}}{BW \times LE} \quad (1)$$

$$Dose_{derm} = \frac{C_s \times SA_h \times SL_h \times RAF_{derm} \times EF \times D_{days} \times D_{weeks} \times ED_{years}}{BW \times LE} \quad (2)$$

where:

$Dose_{ingestion}$ (mg/kg-day) indicates the dose from accidental sediment ingestion; $Dose_{dermal}$ (mg/kg-day) is the dose from skin contact with sediment; C_s (mg/kg) represents the concentration of the contaminant in the sediment; IR_s (kg/day) is the rate of accidental sediment ingestion; RAF_{oral} indicates the relative absorption factor for the gastrointestinal tract; RAF_{derm} (dimensionless) expresses the relative absorption factor for the skin. Moreover, the dose was evaluated based on the hours per day with exposure: 0–16/16 h for accidental ingestion of sediment (D_{hours}); days in a week with exposure [(0–7)/7 days] (D_{days}); weeks in a year with exposure [(0–52)/52 weeks] (D_{weeks}); total years with exposure (ED_{years}); surface of hands (assuming only hands are exposed) (SA_h (cm^2); SL_h (kg/cm^2-event) = Sediment load rate on exposed skin; EF (event/day) = Number of skin exposures per day; BW (kg) = receptor body weight; LE = life expectancy/average life expectancy expressed in years; CF (conversion coefficient) = 1×10^{-6} kg/mg.

Additionally, since Benzo(a)pyrene (BaP) is the most carcinogenic compound among the PAHs considered [10], all individual analyte concentrations were converted to the corresponding toxic equivalent concentrations of BaP. These concentrations are referred to as TEQ_{BaP} or BaP_{eq} and were obtained using the concentration product for toxic equivalence factor (TEF). The TEF factors for the 16 US EPA priority PAHs are shown in Table 5 [51].

Table 5. Equivalent toxicity factors (TEF) of the 16 PAHs [51].

Compound	TEF
Acenaphthalene (Ace)	0.001
Acenaphthylene (Acy)	0.001
Anthracene (Ant)	0.01
Benzo[a]anthracene (BaA)	0.1
Benzo[a]pyrene (BaP)	1
Benzo[b]fluoranthene (BbF)	0.1
Benzo[g,h,i]perylene (BghiP)	0.01
Benzo[k]fluoranthene (BkF)	0.1
Crysene (Chr)	0.01
Dibenzo[a,h]anthracene (DahA)	1
Fluoranthene (Fla)	0.001
Fluorene (Flu)	0.001
Indeno [1,2,3-cd] pyrene (IcdP)	0.1
Naphtalene (Nap)	0.001
Fenanthrene (Phe)	0.001
Pyrene (Pyr)	0.001

The total concentrations of PAHs were obtained using the sum of the calculated toxic equivalents for each compound in relation to BaP. BaPeq were calculated according to Equation (3) [52]:

$$BaP_{eq} = \Sigma C_s \times TEF_i \quad (3)$$

where C_s represents the average concentration of an individual PAH.

Humans can encounter the PAHs present in estuarine sediments by oral ingestion and dermal contact. As a result, in addition to calculating the doses of contaminants taken up by human receptors, the incremental lifetime cancer risk by oral ingestion ($ILCR_{ingestion}$) and dermal contact ($ILCR_{dermal}$) was evaluated according to Equations (4) and (5) [49]:

$$ILCR_{ingestion} = \frac{C_s \times SF_{ingestion} \times \sqrt[3]{\frac{BW}{70}} \times IR_{ingestion} \times EF \times Dyears}{BW \times AT \times 10^6} \quad (4)$$

$$ILCR_{dermal} = \frac{C_s \times SF_{dermal} \times \sqrt[3]{\frac{BW}{70}} \times AS \times AF \times ABS \times EF \times Dyears}{BW \times AT \times 10^6} \quad (5)$$

where:

$SF_{ingestion}$ (Kg-day/mg) indicates the oral slope factor.
SF_{dermal} (Kg-day/mg) is the dermal slope factor.
SA (cm^2/kg) represents the area of dermal contact with the sediment.
AF (mg/cm^2) is the skin absorption coefficient for the sediment.
ABS is the skin absorption coefficient for contaminants.
AT (years) is the average lifespan.

The parameters used for the calculation of the carcinogenic risk are shown in Table 6.

Table 6. Parameters used for the calculation of doses taken up by human receptors through different routes of exposure.

Parameter	Unit of Measure	Value	References
BW	Kg	70.7	[53]
$IR_{ingestion}$	Kg/days	2.00×10^{-5}	[54,55]
AT	Years	80	[56,57]
SA_h	cm^2	890	[53]
SL_h	Kg/cm^2-event	1.00×10^{-7}	[58]
D_{hours}	Hours	0–16/16 h	[59]
D_{days}	Days	0–7/7 days	[59]
D_{weeks}	Weeks	0–52/52 weeks	[59]
RAF_{oral}	-	1	[60]
ED_{years}	Years	60	[60]
RAF_{derm} [a]	-	0.148	[61]
$SF_{ingestion}$ [a]	Kg-day/mg	2.3	[60]
SF_{dermal} [a]	Kg-day/mg	25	[62]
EF	(event/day)	1	[63]
SA	cm^2/kg	5000	[64]
AF	mg/cm^2	0.04	[64]
ABS	/	0.1	[64]

[a] expressed in relation to BaP.

3. Results

3.1. PAH Concentrations in Sediment from the Sele River

The concentrations of the 16 USEPA priority PAHs obtained from instrumental analyses of sediment samples taken near the mouth of the Sele River are given in Table S1. In particular, the total concentration of the PAHs ranged from 632.42 ng g^{-1} dw (site 10) to 844.93 ng g^{-1} dw (at site 1), with an average value of 738.68 ng g^{-1} dw. Specifically, the concentrations ranged from 2.23 to 70.64 ng g^{-1} dw with an average value of 36.43 ng g^{-1} dw for PAHs with 2 rings (NaP), from 5.45 to 51.03 ng g^{-1} dw for 3-ring PAHs (Acy, Ace, Flu, Phe, Ant), from 0.70 to 74.6 ng g^{-1} dw for 4-ring PAHs (Flu, Pyr, BaA, Chr), from 39.12 to 154.99 ng g^{-1} dw for 5-ring PAHs (BbF, BkF, BaP, DahA), and from 3.01 to 75.13 ng g^{-1} dw for 6-ring PAHs (BghiP, IcdP). Figure 5 shows the individual concentrations of PAHs detected in sediment samples from different sampling sites. The figure reveals that the highest concentrations of PAHs were found at the mouth of the Sele River (site 1) and 500 m from the mouth in the southerly direction (site 8). The composition profile of PAHs in the sediment is shown in Figure 6. PAHs with 5 rings were found in most test sites at 57.4% of the total PAHs in the sediment.

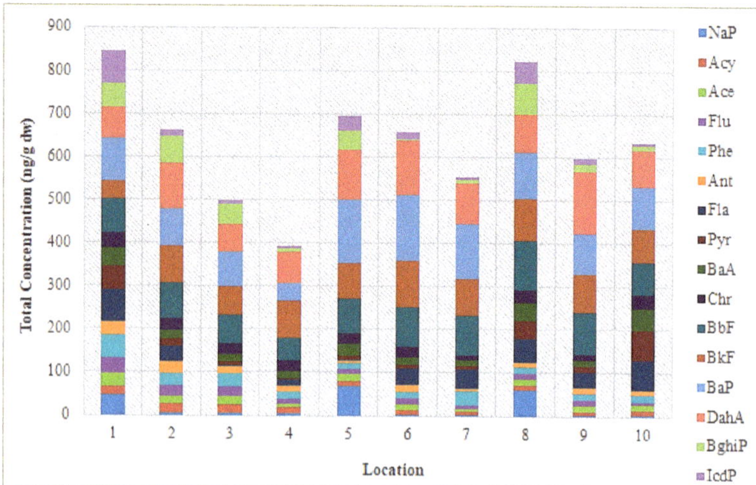

Figure 5. Concentrations of the 16 priority PAHs (ng/g dw) found in sediment samples from the Sele River at 10 sampling locations (April 2021).

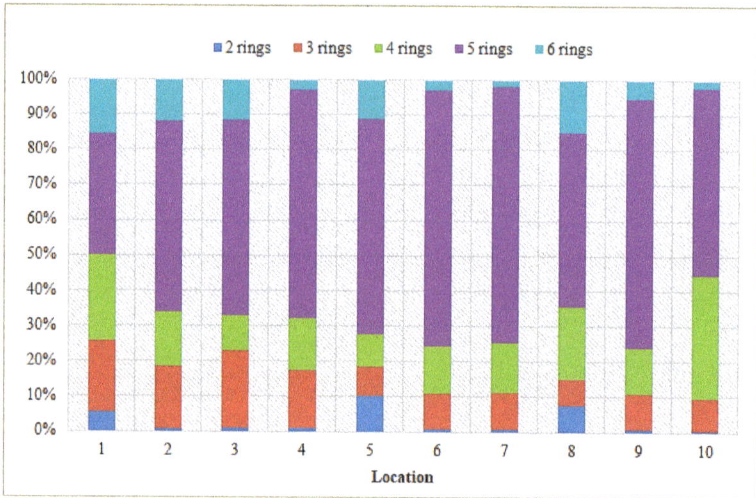

Figure 6. Composition profile of total PAHs in sediment samples from the Sele River.

3.2. *PAH Concentrations in Sediment from the Sarno River*

Data on PAH concentrations found in the Sarno River are indicated in Table S2 [39]. The total concentration of PAHs in the sediment ranged from 5.2 ng g^{-1} dw at the source of the river (site 1) to 678.6 ng g^{-1} dw at the point 150 m to the west of the mouth (site 9), with an average value of 266.9 ng g^{-1} dw. The measured PAH concentrations ranged from 0.2 to 31.6 ng g^{-1} dw with an average of 9.7 ng g^{-1} dw for 2-ring PAHs (Nap), from 0.2 to 46.3 ng g^{-1} dw for 3-ring PAHs (Acy, Ace, Flu, Phe, Ant), from 0.3 to 47.2 ng g^{-1} dw for 4-ring PAHs (Fla, Pyr, BaA, Chr), from 0.2 to 46.6 ng g^{-1} dw for 5-ring PAHs (BbF, BkF, BaP, DahA), and from 0.5 to 46.7 ng g^{-1} dw for 6-ring PAHs (BghiP, IcdP). Figure 7 shows the individual and total concentrations of PAHs found in the sediment samples taken from the different sampling sites located near the mouth of the Sarno River. The figure indicates an increase in total PAH levels at the sampling point 150 m west of the river mouth (site 8). The composition profile of PAHs in the sediment is shown in Figure 8. Three-ring PAHs were found in most test sites, at a percentage of 47.3% of the total PAH amount in the sediment, followed by 5-ring PAHs at a percentage of 20.6%.

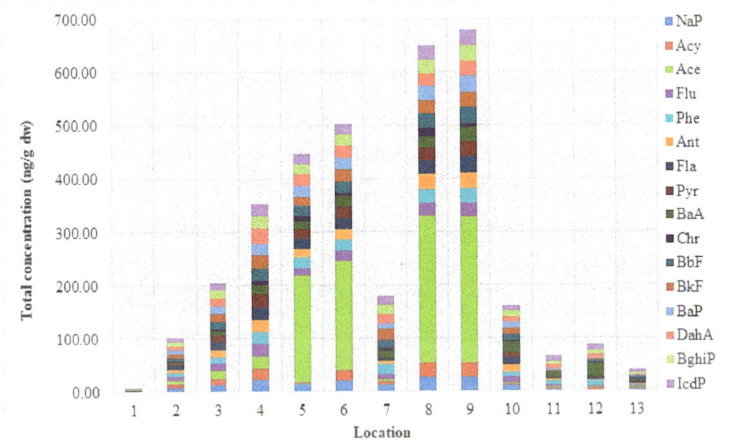

Figure 7. Concentrations of the 16 priority PAHs (ng/g dw) found in sediment samples from the Sarno River at the 13 sampling sites (April 2008).

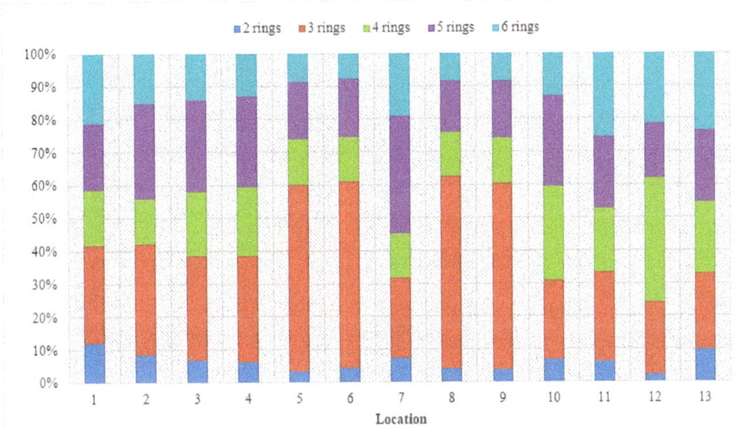

Figure 8. Composition profile of total PAHs in sediment samples from the Sarno River.

3.3. *PAH Concentrations in Sediment from the Volturno River*

Data on individual PAH concentrations found in the Volturno River are given in Table S3, while total concentrations were previously reported [40]. In detail, total concentrations were between 434.8 ng g^{-1} dw (site 8) and 872.1 ng g^{-1} dw (site 1), with an average value of 659.1 ng g^{-1} dw. For 2-ring PAHs (NaP), the levels ranged from 5.3 to 73.8 ng g^{-1} dw with an average value of 24.1 ng g^{-1} dw; for 3-ring PAHs (Acy, Ace, Flu, Phe, Ant), from 42.9 to 186.3 ng g^{-1} dw; for 4-ring PAHs (Fla, Pyr, BaA, Chr), from 61.7 ng g^{-1} dw to 199.7 ng g^{-1} dw; for 5-ring PAHs (BbF, BkF, BaP, DahA), from 262.7 to 507.1 ng g^{-1} dw; and for 6-ring PAHs (BghiP, IcdP), from 17.5 to 133.2 ng g^{-1} dw. Figure 9 shows the individual and total concentrations of PAHs found in the sediment samples taken at the different sampling sites near the mouth of the Volturno River. The figure shows that the highest concentrations of PAHs were found at the mouth of the river (site 1) and 500 m from the mouth in the southerly direction (site 4). The composition profile of PAHs in sediment is shown in Figure 10. Five-ring PAHs were found in most of the test sites at 57,4% of the total PAHs in the sediment.

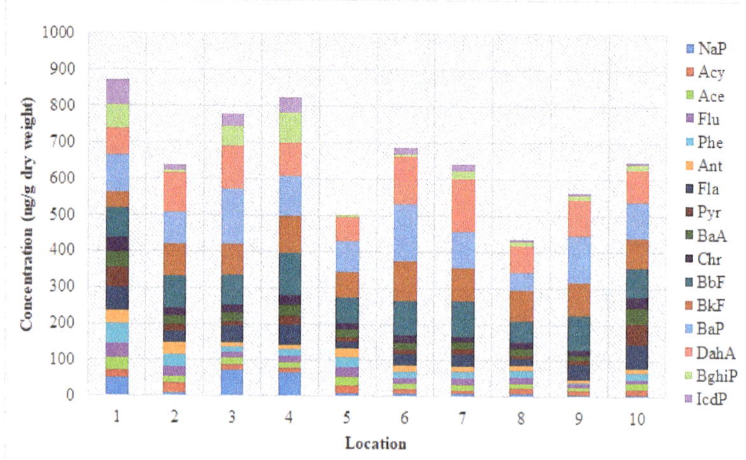

Figure 9. Concentrations of the 16 priority PAHs (ng/g dry weight) found in sediment samples from the Volturno River at 10 sampling locations (April 2018).

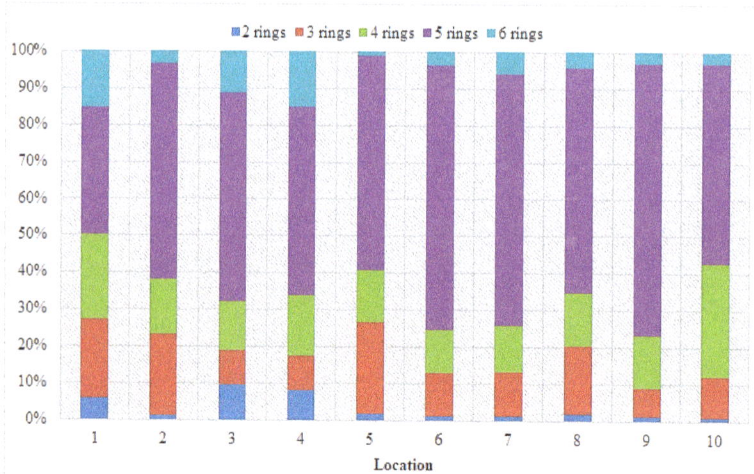

Figure 10. Composition profile of total PAHs in sediment samples from the Volturno River.

3.4. *Evaluation of the Carcinogenic Risk for Human Health from Dermal and Accidental Ingestion Exposure to the PAHs Present in the Sediments of the Surface Waters*

To assess the Carcinogenic Risk from exposure to PAHs present in estuarine sediments of the Sele, Volturno, and Sarno Rivers, doses of contaminants taken up by human receptors through the different routes of exposure were evaluated, and the results obtained are reported in Table 7. Particularly, for all three rivers, the doses of each individual PAH and the total doses relating to the entire class of compounds taken up by human receptors through dermal and oral exposure were calculated. The doses from accidental ingestion of PAHs from sediments ($Dose_{ingestion}$) were 7.86×10^{-4} mg Kg^{-1}/day for the Sele River, 8.13×10^{-4} mg Kg^{-1}/day for the Volturno River, and 3.31×10^{-4} mg Kg^{-1}/day for the Sarno River. On the other hand, in relation to the doses of PAHs taken up by skin contact with sediment ($Dose_{dermal}$), the results were 3.23×10^{-5}, 1.36×10^{-5}, and 3.35×10^{-5} mg Kg^{-1}/day for the Sele, Sarno, and Volturno Rivers, respectively.

Table 7. Average total PAH concentrations (C_s), intakes by accidental ingestion ($Dose_{ingestion}$), and dermal contact ($Dose_{dermal}$) of PAH present in estuarine sediments of the Sele, Volturno, and Sarno Rivers and equivalent toxic concentrations (BaP_{eq}).

River	C_s (mg Kg^{-1} dw)	$Dose_{ingestion}$ (mg Kg^{-1}/day)	$Dose_{dermal}$ (mg Kg^{-1}/day)	BaP_{eq} (mg Kg^{-1} dw)
Sele	0.6360	7.86×10^{-4}	3.23×10^{-5}	2.23×10^{-1}
Sarno	0.2677	3.31×10^{-4}	1.36×10^{-5}	3.38×10^{-2}
Volturno	0.6577	8.13×10^{-4}	3.35×10^{-5}	2.30×10^{-1}

Moreover, to assess the carcinogenic and mutagenic potencies of PAHs in relation to BaP, the most carcinogenic compound among the PAH considered [10], the average concentrations of individual analytes were converted to the corresponding toxic equivalent concentrations (BaP_{eq}). The equivalent toxic concentrations (BaP_{eq}) obtained for the Sele, Volturno, and Sarno Rivers are given in Table 7.

Furthermore, the incremental risk of developing lifelong cancer expressed as ILCR was assessed for the exposure to PAHs by ingestion ($ILCR_{ingestion}$) and dermal contact ($ILCR_{dermal}$) [65]. The $ILCR_{ingestion}$ and $ILCR_{dermal}$ values obtained for Sele, Volturno, and Sarno Rivers are shown in Table 8.

Table 8. Incremental lifetime cancer risk values ($ILCR_{ingestion}$ and $ILCR_{dermal}$) due to exposure by ingestion and dermal contact from PAHs present in the estuarine sediments of the Sele, Volturno, and Sarno Rivers.

River	$ILCR_{ingestion}$ (mg Kg^{-1} dw)	$ILCR_{dermal}$ (mg Kg^{-1} dw)	$ILCR_{ingestion/dermal}$ [66] Cancerogenic Risk
Sele	3.11×10^{-13}	3.38×10^{-6}	If ILCR < 1×10^{-6} Low or Zero Risk
Sarno	1.31×10^{-13}	1.42×10^{-6}	If 1×10^{-6} < ILCR < 1×10^{-4} Medium Risk
Volturno	3.22×10^{-13}	3.50×10^{-6}	If ILCR >1×10^{-4} High Risk

The values obtained for carcinogenic risk due to exposure to PAHs by ingestion ($ILCR_{ingestion}$) and dermal contact ($ILCR_{dermal}$) were found to be comparable for the three rivers. According to the USEPA, the ILCR values were interpreted by reference to three ranges, each of which is associated with a risk of carcinogenicity: an ILCR value < 1×10^{-6} is associated with a low or zero carcinogenic risk; ILCR values between 1×10^{-4} and 1×10^{-6} are indicators of a moderate carcinogenic risk; and an ILCR value higher than 1×10^{-4} corresponds to a high carcinogenic risk associated with exposure to PAHs in sediment [66,67]. The $ILCR_{ingestion}$ values obtained for the three rivers were found to be much lower than the $ILCR_{dermal}$, indicating that the risk of cancer associated with dermal contact exposure to PAHs present in estuarine sediments may be higher than that associated with accidental ingestion exposure. Specifically, the $ILCR_{ingestion}$ and $ILCR_{dermal}$ values obtained for all sediment samples taken near the mouth of the Sele River ranged between 6.69×10^{-15} and 3.11×10^{-13} and 7.27×10^{-8} and 3.38×10^{-6}, respectively. The $ILCR_{ingestion}$ and $ILCR_{dermal}$ values obtained for the Sarno River ranged between 2.45×10^{-15} and 1.31×10^{-13} and 2.66×10^{-8} and 1.42×10^{-6}, respectively. For the Volturno River, the $ILCR_{ingestion}$ values were between 8.23×10^{-15} and 3.22×10^{-13}, while those of $ILCR_{dermal}$ were between 8.95×10^{-8} and 3.50×10^{-6}. Thus, the risk of cancer associated with exposure to PAHs by ingestion of estuarine sediments of the Sele, Sarno, and Volturno Rivers was found to be low at all sampling sites. However, based on the $ILCR_{dermal}$ values obtained, the risk of cancer associated with exposure by dermal contact with the PAHs present in the sediments was found to be moderate (average $ILCR_{dermal}$ for the three rivers of 2.77×10^{-6}). In addition, the values of the $ILCR_{ingestion}$ and $ILCR_{dermal}$ indices for sediment samples taken at sites with the highest concentrations of PAHs were evaluated. For the Sele and Volturno Rivers, the assessment was carried out at the mouth (site 1), for which total PAH concentrations of 0.8449 and 0.8721 mg Kg^{-1} dw were found, respectively. For the Sarno River, the assessment was carried out at the sampling site 150 m to the west of the estuary (site 9), where a total concentration of PAHs of 0.6792 mg Kg^{-1} dw was found. The results obtained for the three rivers are shown in Table 9.

Table 9. Incremental lifetime cancer risk values ($ILCR_{ingestion}$ and $ILCR_{dermal}$) due to exposure by ingestion and dermal contact to the highest PAHs levels found in the sediment samples of the Sele, Volturno, and Sarno Rivers.

River	$ILCR_{ingestion}$ (mg Kg^{-1} dw)	$ILCR_{dermal}$ (mg Kg^{-1} dw)	$ILCR_{ingestion/dermal}$ [66] Cancerogenic Risk
Sele	4.14×10^{-13}	4.50×10^{-6}	If ILCR < 1×10^{-6} Low or Zero Risk
Sarno	3.33×10^{-13}	3.61×10^{-6}	If 1×10^{-6} < ILCR < 1×10^{-4} Medium Risk
Volturno	4.27×10^{-13}	4.64×10^{-6}	If ILCR >1×10^{-4} High Risk

The $ILCR_{ingestion}$ values obtained for the three rivers at the sampling sites with the highest PAH concentrations were $<1 \times 10^{-6}$ (order of 10^{-13}), suggesting that the risk of cancer associated with exposure by ingestion of PAHs present in estuarine sediments was low or zero. On the other hand, the $ILCR_{dermal}$ values obtained for the three rivers at the sampling sites with the highest PAH concentrations were in the order of 10^{-6}, suggesting that the risk of cancer associated with dermal contact exposure to PAHs present in estuarine sediments was moderate. In fact, as stated also by Cheng et al., ILCR values between 1×10^{-4} and 1×10^{-6} are associated with a moderate carcinogenic risk [66,67]. Thus, on the basis of the ILCR values obtained by taking into account total PAH concentrations at all sites or considering only the total concentrations recorded at the most polluted sites, the risk of cancer associated with exposure by ingestion was found to be low or zero, but the risk associated with dermal contact exposure of PAHs present in estuarine sediments of the Sele, Volturno, and Sarno Rivers is moderate. Thus, since these areas were previously considered potentially contaminated according to Italian environmental law (D. Lgs. 152/2006), and as stated by Albanese et al., who assessed an incremental lifetime cancer risk higher than 1×10^{-5} for the city of Naples [9], a continuous monitoring of potentially hazardous substances is necessary to ensure the protection of public health.

4. Conclusions

This paper presents for the first time an assessment of the carcinogenic risk to human health from dermal and ingestion exposure to PAHs present in sediments of the main surface water streams of the Campania Region, southern Italy. The paper also provides information on the concentrations, spatial distribution, and composition profiles of the PAHs detected in sediments collected near the Sele, Sarno, and Volturno River estuaries. The results obtained indicate that the risk of cancer following oral exposure to PAHs in estuarine sediments, expressed as incremental lifetime cancer risk ($ILCR_{ingestion}$), is low, unlike the risk from accidental skin exposure, which was moderate with $ILCR_{dermal}$ values between 1×10^{-4} and 1×10^{-6}. This calls for ongoing assessment of the carcinogenic risk to human health posed by cutaneous and oral exposure to PAHs, as well as constant monitoring of PAH concentrations in surface water sediments in the Campania Region. In conclusion, this study represents a starting point for future studies aimed at assessing the risk of carcinogenicity to human health due to exposure to the PAHs in order to provide support for pollution prevention measures and ecological restoration strategies for rivers, as well as for the preservation of the general well-being.

Supplementary Materials: The following supporting information can be downloaded at: https://www.mdpi.com/article/10.3390/toxics11020172/s1, Table S1: PAH levels (ng g-1 dw) with SD (Standard Deviation) detected in sediment samples from Sele River; Table S2. PAH concentrations (ng g-1 dw) with SD (Standard Deviation) found in sediment samples from the Sarno River (April 2008). Table S3. PAH levels (ng g-1 dry weight) with SD (Standard Deviation) detected in sediment samples from Volturno River.

Author Contributions: Conceptualization, F.D.D., P.M., U.T., A.M., G.M.B. and M.T.; data curation, U.T.; formal analysis, F.D.D. and G.M.B.; investigation, F.D.D. and G.M.B.; resources, P.M., U.T. and A.M.; supervision, P.M. and M.T.; validation, M.T.; visualization, M.T.; writing—original draft, F.D.D. and G.M.B.; writing—review and editing, P.M. All authors have read and agreed to the published version of the manuscript.

Funding: This research received no external funding.

Institutional Review Board Statement: Not applicable.

Informed Consent Statement: Not applicable.

Data Availability Statement: Not applicable.

Conflicts of Interest: The authors declare no conflict of interest.

References

1. Huang, Y.; Zhang, W.; Pang, S.; Chen, J.; Bhatt, P.; Mishra, S.; Chen, S. Insights into the microbial degradation and catalytic mechanisms of chlorpyrifos. *Environ. Res.* **2021**, *194*, 110660. [CrossRef]
2. Zhang, A.; Zhao, S.; Wang, L.; Yang, X.; Zhao, Q.; Fan, J.; Yuan, X. Polycyclic aromatic hydrocarbons (PAHs) in seawater and sediments from the northern Liaodong Bay, China. *Mar. Pollut. Bull.* **2016**, *113*, 592–599. [CrossRef] [PubMed]
3. Brion, D.; Pelletier, É. Modelling PAHs adsorption and sequestration in freshwater and marine sediments. *Chemosphere* **2005**, *61*, 867–876. [CrossRef]
4. Shahsavari, E.; Schwarz, A.; Aburto-Medina, A.; Ball, A.S. Biological degradation of polycyclic aromatic compounds (PAHs) in soil: A current perspective. *Curr. Pollut. Rep.* **2019**, *5*, 84–92. [CrossRef]
5. Guo, W.; He, M.; Yang, Z.; Lin, C.; Quan, X.; Wang, H. Distribution of polycyclic aromatic hydrocarbons in water, suspended particulate matter and sediment from Daliao River watershed, China. *Chemosphere* **2007**, *68*, 93–104. [CrossRef]
6. Awe, A.A.; Opeolu, B.O.; Olatunji, O.S.; Fatoki, O.S.; Jackson, V.A.; Snyman, R. Occurrence and probabilistic risk assessment of PAHs in water and sediment samples of the Diep River, South Africa. *Heliyon* **2020**, *6*, e04306. [CrossRef]
7. Lin, Y.; Ma, Y.; Qiu, X.; Li, R.; Fang, Y.; Wang, J.; Hu, D. Sources, transformation, and health implications of PAHs and their nitrated, hydroxylated, and oxygenated derivatives in PM2.5 in Beijing. *J. Geophys. Res. Atmos.* **2015**, *120*, 7219–7228. [CrossRef]
8. CCME (Canadian Council of Ministers of the Environment). *Canadian Soil Quality Guidelines for Carcinogenic and Other Polycyclic Aromatic Hydrocarbons (Environmental and Human Health Effects)*; Canadian Council of Ministers of the Environment: Winnipeg, MB, Canada, 2008.
9. Albanese, S.; Fontaine, B.; Chen, W.; Lima, A.; Cannatelli, C.; Piccolo, A.; De Vivo, B. Polycyclic aromatic hydrocarbons in the soils of a densely populated region and associated human health risks: The Campania Plain (Southern Italy) case study. *Environ. Geochem. Health* **2015**, *37*, 1–20. [CrossRef] [PubMed]
10. Wang, Y.; Liu, M.; Dai, Y.; Luo, Y.; Zhang, S. Health and ecotoxicological risk assessment for human and aquatic organism exposure to polycyclic aromatic hydrocarbons in the Baiyangdian Lake. *Environ. Sci. Pollut. Res.* **2021**, *28*, 574–586. [CrossRef]
11. Omar, W.A.M.; Mahmoud, H.M. Risk assessment of polycyclic aromatic hydrocarbons (PAHs) in River Nile up-and downstream of a densely populated area. *J. Environ. Sci. Health A* **2017**, *52*, 166–173. [CrossRef]
12. Zhu, Y.; Liang, B.; Xia, W.; Gao, M.; Zheng, H.; Chen, J.; Tian, M. Assessing potential risks of aquatic polycyclic aromatic compounds via multiple approaches: A case study in Jialing and Yangtze Rivers in downtown Chongqing, China. *Environ. Pollut.* **2022**, *294*, 118620. [CrossRef] [PubMed]
13. Adeniji, A.O.; Okoh, O.O.; Okoh, A.I. Levels of polycyclic aromatic hydrocarbons in the water and sediment of Buffalo River Estuary, South Africa and their health risk assessment. *Arch. Environ. Contam. Toxicol.* **2019**, *76*, 657–669. [CrossRef]
14. Regional Office for Europe. *Air Quality Guidelines for Europe*, 2nd ed.; World Health Organization, Regional Office for Europe: Geneva, Switzerland, 2000; Available online: https://apps.who.int/iris/handle/10665/107335 (accessed on 16 September 2022).
15. Rocha, M.J.; Dores-Sousa, J.L.; Cruzeiro, C.; Rocha, E. PAHs in water and surface sediments from Douro River estuary and Porto Atlantic coast (Portugal)-impacts on human health. *Environ. Monit. Assess.* **2017**, *189*, 1–14. [CrossRef] [PubMed]
16. Khan, A.; Ahsan, A.; Farooq, M.A.; Naveed, M.; Li, H. Role of polycyclic aromatic hydrocarbons as EDCs in metabolic disorders. In *Endocrine Disrupting Chemicals-induced Metabolic Disorders and Treatment Strategies*; Springer: Cham, Switzerland, 2021; pp. 323–341.

17. Gaber, M.; Sequely, A.A.; Monem, N.A.; Balbaa, M. Effect of polyaromatic hydrocarbons on cellular cytochrome P450 1A induction. *Ocean Coast Manag.* **2021**, *69*, 21026. [CrossRef]
18. Lee, T.Y.; Tseng, Y.H. The potential of phytochemicals in oral cancer prevention and therapy: A review of the evidence. *Biomolecules* **2020**, *10*, 1150. [CrossRef]
19. Kobets, T.; Smith, B.P.; Williams, G.M. Food-Borne Chemical Carcinogens and the Evidence for Human Cancer Risk. *Foods* **2022**, *11*, 2828. [CrossRef] [PubMed]
20. Joint WHO/Convention Task Force on the Health Aspects of Air Pollution; World Health Organization. *Health Risks of Persistent Organic Pollutants from Long-Range Transboundary Air Pollution*; WHO Regional Office for Europe: Copenhagen, Denmark, 2003; Available online: https://apps.who.int/iris/handle/10665/107471 (accessed on 16 September 2022).
21. IARC Working Group on the Evaluation of Carcinogenic Risks to Humans. Air pollution, Part 1, some non-heterocyclic polycyclic aromatic hydrocarbons and some related industrial exposure. *IARC Monogr. Eval. Carcinog. Risks Hum.* **2010**, *92*, 1–853. Available online: https://pubmed.ncbi.nlm.nih.gov/21141735/ (accessed on 3 October 2022).
22. Tay, C.K.; Doamekpor, L.K.; Mohammed, S.; Dartey, G.; Kuddy, R.; Fianyaglo, E.; Mawuena, M. Health risk assessment and source identification of Polycyclic Aromatic Hydrocarbons (PAHs) in commercially available singed cowhide within the Greater Accra Region, Ghana. *West Afr. J. Appl. Ecol.* **2022**, *30*, 13–34.
23. Ravindra, K.; Sokhi, R.; Van Grieken, R. Atmospheric polycyclic aromatic hydrocarbons: Source attribution, emission factors and regulation. *Atmos. Environ.* **2008**, *42*, 2895–2921. [CrossRef]
24. Campling, B.G.; El-Deiry, W.S. Clinical implications of p53 mutations in lung cancer. *Lung Cancer* **2003**, *75*, 53–78.
25. Unwin, J.; Cocker, J.; Scobbie, E.; Chambers, H. An assessment of occupational exposure to polycyclic aromatic hydrocarbons in the UK. *Ann. Occup. Hyg.* **2006**, *50*, 395–403. [CrossRef]
26. Kim, K.H.; Jahan, S.A.; Kabir, E.; Brown, R.J. A review of airborne polycyclic aromatic hydrocarbons (PAHs) and their human health effects. *Environ. Int.* **2013**, *60*, 71–80. [CrossRef] [PubMed]
27. Sun, K.; Song, Y.; He, F.; Jing, M.; Tang, J.; Liu, R. A review of human and animals exposure to polycyclic aromatic hydrocarbons: Health risk and adverse effects, photo-induced toxicity and regulating effect of microplastics. *Sci. Total Environ.* **2021**, *773*, 145403. [CrossRef] [PubMed]
28. Gunter, M.J.; Divi, R.L.; Kulldorff, M.; Vermeulen, R.; Haverkos, K.J.; Kuo, M.M.; Sinha, R. Leukocyte polycyclic aromatic hydrocarbon–DNA adduct formation and colorectal adenoma. *Carcinogenesis* **2007**, *28*, 1426–1429. [CrossRef] [PubMed]
29. John, K.; Ragavan, N.; Pratt, M.M.; Singh, P.B.; Al-Buheissi, S.; Matanhelia, S.S.; Martin, F.L. Quantification of phase I/II metabolizing enzyme gene expression and polycyclic aromatic hydrocarbon–DNA adduct levels in human prostate. *The Prostate* **2009**, *69*, 505–519. [CrossRef]
30. Garcia-Suastegui, W.A.; Huerta-Chagoya, A.; Carrasco-Colín, K.L.; Pratt, M.M.; John, K.; Petrosyan, P.; Gonsebatt, M.E. Seasonal variations in the levels of PAH–DNA adducts in young adults living in Mexico City. *Mutagenesis* **2010**, *26*, 385–391. [CrossRef] [PubMed]
31. Srogi, K. Monitoring of environmental exposure to polycyclic aromatic hydrocarbons: A review. *Environ. Chem. Lett.* **2007**, *5*, 169–195. [CrossRef]
32. Diggs, D.L.; Harris, K.L.; Rekhadevi, P.V.; Ramesh, A. *Tumor.* microsomal metabolism of the food toxicant, benzo (a) pyrene, in Apc Min mouse model of colon cancer. *Tumor. Biol.* **2012**, *33*, 1255–1260. [CrossRef]
33. Wells, P.G.; Lee, C.J.; McCallum, G.P.; Perstin, J.; Harper, P.A. Receptor-and reactive intermediate-mediated mechanisms of teratogenesis. In *Adverse Drug Reactions*; Springer: Berlin/Heidelberg, Germany, 2010; Volume 43, pp. 221–242. [CrossRef]
34. Drwal, E.; Rak, A.; Gregoraszczuk, E.L. Polycyclic aromatic hydrocarbons (PAHs)—Action on placental function and health risks in future life of newborns. *Toxicology* **2019**, *411*, 133–142. [CrossRef]
35. Anyahara, J.N. Effects of Polycyclic Aromatic Hydrocarbons (PAHs) on the environment: A systematic review. *Int. J. Adv. Acad. Res.* **2021**, *7*, e7303. [CrossRef]
36. Ambade, B.; Sethi, S.S.; Giri, B.; Biswas, J.K.; Bauddh, K. Characterization, behavior, and risk assessment of polycyclic aromatic hydrocarbons (PAHs) in the estuary sediments. *Bull. Environ. Contam. Toxicol.* **2022**, *108*, 243–252. [CrossRef]
37. Li, Y.; Liu, M.; Hou, L.; Li, X.; Yin, G.; Sun, P.; Zheng, D. Geographical distribution of polycyclic aromatic hydrocarbons in estuarine sediments over China: Human impacts and source apportionment. *Sci. Total Environ.* **2021**, *768*, 145279. [CrossRef]
38. Du, J.; Jing, C. Anthropogenic PAHs in lake sediments: A literature review (2002–2018). *Environ. Sci. Process. Impacts* **2018**, *20*, 1649–1666. [CrossRef] [PubMed]
39. Montuori, P.; Triassi, M. Polycyclic aromatic hydrocarbons loads into the Mediterranean Sea: Estimate of Sarno River inputs. *Mar. Pollut. Bull.* **2012**, *64*, 512–520. [CrossRef]
40. Montuori, P.; De Rosa, E.; Di Duca, F.; Provvisiero, D.P.; Sarnacchiaro, P.; Nardone, A.; Triassi, M. Estimation of Polycyclic Aromatic Hydrocarbons Pollution in Mediterranean Sea from Volturno River, Southern Italy: Distribution, Risk Assessment and Loads. *Int. J. Environ. Res. Public Health.* **2021**, *18*, 1383. [CrossRef]
41. Montuori, P.; De Rosa, E.; Di Duca, F.; De Simone, B.; Scippa, S.; Russo, I.; Triassi, M. Polycyclic Aromatic Hydrocarbons (PAHs) in the Dissolved Phase, Particulate Matter, and Sediment of the Sele River, Southern Italy: A Focus on Distribution, Risk Assessment, and Sources. *Toxics* **2022**, *10*, 401. [CrossRef] [PubMed]

42. Lin, L.; Dong, L.; Meng, X.; Li, Q.; Huang, Z.; Li, C.; Li, R.; Yang, W.; Crittenden, J. Distribution and sources of polycyclic aromatic hydrocarbons and phthalic acid esters in water and surface sediment from the Three Gorges Reservoir. *J. Environ. Sci.* **2018**, *69*, 271–280. [CrossRef] [PubMed]
43. USEPA (United States Environmental Protection Agency). *Method 3540C. Soxhlet Extraction*; USEPA: Washington, DC, USA, 1996.
44. US Environmental Protection Agency. *Protocol for Equipment Leak Emission Estimates*; EPA-453/R-95—17; US Environmental Protection Agency: Washington, DC, USA, 1995.
45. Wcisło, E.; Ioven, D.; Kucharski, R.; Szdzuj, J. Human health risk assessment case study: An abandoned metal smelter site in Poland. *Chemosphere* **2002**, *47*, 507–515. [CrossRef] [PubMed]
46. Li, Y.L.; Liu, Y.G.; Liu, J.L.; Zeng, G.M.; Li, X. Effects of EDTA on lead uptake by Typha oreentalis Presl: A new lead-accumulating species in southern China. *Bull. Environ. Contam. Toxicol.* **2008**, *81*, 36–41. [CrossRef]
47. Li, Y.; Liu, J.; Cao, Z.; Lin, C.; Yang, Z. Spatial distribution and health risk of heavy metals and polycyclic aromatic hydrocarbons (PAHs) in the water of the Luanhe River basin, China. *Environ. Monit. Assess.* **2010**, *163*, 1–13. [CrossRef]
48. Gerba, C.P.; Pepper, I.L.; Gerb, C.P.; Brusseau, M. *Risk Assessment. Environmental and Pollution Science*; Academic Press, Elsevier: San Diego, CA, California, 2006.
49. Han, J.; Liang, Y.; Zhao, B.; Wang, Y.; Xing, F.; Qin, L. Polycyclic aromatic hydrocarbon (PAHs) geographical distribution in China and their source, risk assessment analysis. *Environ. Pollut.* **2019**, *251*, 312–327. [CrossRef]
50. Akpan, A.D.; Okori, B.S.U.; Ekpechi, D.C. Human Health Risk Assessment of Polycyclic Aromatic Hydrocarbons in Water Samples around Eket Metropolis, Akwa Ibom State, Nigeria. *Asian J. Environ. Sci.* **2022**, *19*, 58–71. [CrossRef]
51. Nisbet, I.C.T.; LaGoy, P.K. Toxic equivalency factors (TEFs) for polycyclic aromatic hydrocarbons (PAHs). *Regul. Toxicol. Pharmacol.* **1992**, *16*, 290–300. [CrossRef]
52. Deelaman, W.; Choochuay, C.; Pongpiachan, S.; Han, Y. Ecological and health risks of polycyclic aromatic hydrocarbons in the sediment core of Phayao Lake, Thailand. *J. Exp. Sci. Environ. Epidemiol.* **2023**, *2*, 3. [CrossRef]
53. Richardson, G.M. *Compendium of Canadian Human Exposure Factors for Risk Assessment*; O'Connor Associates Environmental Incorporated: Ottawa, ON, Canada, 1997.
54. Khiari, N.; Charef, A.; Atoui, A.; Azouzi, R.; Khalil, N.; Khadhar, S. Southern Mediterranean coast pollution: Long-term assessment and evolution of PAH pollutants in Monastir Bay (Tunisia). *Mar. Pollut. Bull.* **2021**, *167*, 112268. [CrossRef] [PubMed]
55. MassDEP. *Technical Update: Calculation of Enhanced Soil Ingestion Rate*; MassDEP: Boston, MA, USA, 2002.
56. Health Canada. *Federal Contaminated Site Risk Assessment in Canada, Part V: Guidance on Human Health Detailed Quantitative Risk Assessment for Chemicals (DQRAChem.)*; Health Canada: Ottawa, ON, Canada, 2010.
57. Haney Jr, J.T.; Forsberg, N.D.; Hoeger, G.C.; Magee, B.H.; Meyer, A.K. Risk assessment implications of site-specific oral relative bioavailability factors and dermal absorption fractions for polycyclic aromatic hydrocarbons in surface soils impacted by clay skeet target fragments. *Regul. Toxicol. Pharmacol.* **2020**, *113*, 104649. [CrossRef]
58. Kissel, J.C.; Richter, K.Y.; Fenske, R.A. Field measurement of dermal soil loading attributable to various activities: Implications for exposure assessment. *Risk Anal.* **1996**, *16*, 115–125. [CrossRef]
59. Albanese, S.; Taiani, M.V.E.; De Vivo, B.; Lima, A. An environmental epidemiological study based on the stream sediment geochemistry of the Salerno province (Campania region, Southern Italy). *J. Geochem. Explor.* **2013**, *131*, 59–66. [CrossRef]
60. Alberico, I.; Amato, V.; Aucelli, P.P.C.; D'Argenio, B.; Di Paola, G.; Pappone, G. Historical shoreline change of the Sele Plain (Southern Italy): The 1870–2009 time window. *J. Coast. Res.* **2012**, *28*, 1638–1647. [CrossRef]
61. Moody, R.P.; Joncasa, J.; Richardsonb, M.; Chua, I. Contaminated soils (I): In vitro dermal absorption of benzo[a]pyrene in human skin. *J. Toxicol. Environ. Health Part A* **2007**, *70*, 1858–1865. [CrossRef]
62. Knafla, A.; Phillipps, K.A.; Brecher, R.W.; Petrovic, S.; Richardson, M. Development of a dermal cancer slope factor for benzo[a]pyrene. *Regul. Toxicol. Pharmacol.* **2006**, *45*, 159–168. [CrossRef]
63. Grmasha, R.A.; Al-sareji, O.J.; Salman, J.M.; Hashim, K.S. Polycyclic aromatic hydrocarbons (PAHs) in urban street dust within three land-uses of Babylon governorate, Iraq: Distribution, sources, and health risk assessment. *J. King Saud Univ. Eng. Sci.* **2020**, *34*, 231–239. [CrossRef]
64. Huang, B.; Liu, M.; Bi, X.; Chaemfa, C.; Ren, Z.; Wang, X.; Fu, J. Phase distribution, sources and risk assessment of PAHs, NPAHs and OPAHs in a rural site of Pearl River Delta region, China. *Atmos. Pollut. Res.* **2014**, *5*, 210–218. [CrossRef]
65. Srivastava, P.; Sreekrishnan, T.R.; Nema, A.K. Human health risk assessment and PAHs in a stretch of river Ganges near Kanpur. *Environ. Monit. Assess.* **2017**, *189*, 445. [CrossRef] [PubMed]
66. USEPA (US Environmental Protection Agency). *Risk Assessment Guidance for Superfund Volume I: Human Health Evaluation Manual (Part E, Supplemental Guidance for Dermal Risk Assessment) Final*; EPA/540/R/99/005 OSWER 9285.7-02EP PB99-963312 July 2004; Office of Superfund Remediation and Technology Innovation: Washington, DC, USA, 2004.
67. Cheng, J.; Wang, X.; Zheng, B.; Zhang, X.; Han, J. Evaluation of distribution characteristics and health risk of polycyclic aromatic hydrocarbons in sediments: From the perspective of land-ocean coordination. *J. Hydrol* **2022**, *607*, 127514. [CrossRef]

toxics MDPI

Article

Occurrence and Distribution of Persistent Organic Pollutants (POPs) from Sele River, Southern Italy: Analysis of Polychlorinated Biphenyls and Organochlorine Pesticides in a Water–Sediment System

Elvira De Rosa, Paolo Montuori *, Maria Triassi, Armando Masucci and Antonio Nardone

Department of Public Health, University "Federico II", Via Sergio Pansini n° 5, 80131 Naples, Italy
* Correspondence: pmontuor@unina.it

Abstract: The concentrations, possible sources, and ecological risk of polychlorinated biphenyls (PCBs) and organochlorine pesticides (OCPs) were studied by analyzing water column (DP), suspended particulate matter (SPM) and sediment samples from 10 sites on the Sele River. Total PCBs concentration ranged from 2.94 to 54.4 ng/L and 5.01 to 79.3 ng/g in the seawater and sediment samples, with OCPs concentration in the range of 0.51 to 8.76 ng/L and 0.50 to 10.2 ng/g, respectively. Pollutants loads in the seaside were measured in approximately 89.7 kg/year (73.2 kg/year of PCBs and 16.5 kg/year of OCPs), indicating that the watercourse could be an important cause of contamination to the Tyrrhenian Sea. Statistical analysis indicates that all polychlorinated biphenyls analytes are more probable to derive from surface runoff than an atmospheric deposition. The results explain that higher concentrations of these pollutants were built in sediment samples rather than in the other two phases, which are evidence of historical loads of PCBs and OCPs contaminants. The Sediment Quality Guidelines (SQGs), the Ecological Risk Index (ERI) and the Risk Quotient (RQ) show that the Sele river and its estuary would reputedly be a zone possibly at risk.

Keywords: persistent organic pollutants; Sele river; toxicity equivalent; risk assessment; Principal Component Analysis

Citation: De Rosa, E.; Montuori, P.; Triassi, M.; Masucci, A.; Nardone, A. Occurrence and Distribution of Persistent Organic Pollutants (POPs) from Sele River, Southern Italy: Analysis of Polychlorinated Biphenyls and Organochlorine Pesticides in a Water–Sediment System. *Toxics* **2022**, *10*, 662. https://doi.org/10.3390/toxics10110662

Academic Editor: Luis Alberto Henríquez-Hernández

Received: 10 October 2022
Accepted: 2 November 2022
Published: 4 November 2022

1. Introduction

The importance of riverine ecosystems for human living has attracted the interest of authorities and researchers, especially after the development of cities and the increase in industrial and agricultural activities, which have released significant amounts of contaminants into these ecosystems [1–3]. Among these, the persistent organic pollutants (POPs) such as Polychlorinated biphenyls (PCBs) and Organochlorine pesticides (OCPs) [4], have raised concern due to their physico-chemical properties and high toxicity [5].

POPs are a set of toxic chemicals that are persistent in the environment and able to last for several years before breaking down. Several were concluded regionally and globally to develop better risk management so as to reduce the impact of these toxic substances on humans' health and the environment [6]. Among these treaties, the Stockholm Convention on POPs is the most important. Accordingly, it has been necessary to introduce a set of rules for the forbidden and restricted worldwide use of POPs that are harmful to human health and the environment, because these are very stable compounds that resist photolytic, biological and chemical degradation and that thus persist in the environment with long half-lives [7,8]. These compounds can be transferred from air to surface soil and water by dry and wet deposition, from soil to aquatic bodies by rainfall runoff, and from soil and aquatic bodies back to air by volatilization. Due to the long-range atmospheric transport, they have been found in most areas of the world [9,10]. They greatly affect the quality of environmental ecosystems and human health.

PCBs are man-made organic compounds composed of a biphenyl with different numbers of chlorine atoms replaced with two six-carbon benzene rings [11]. They are composed of more than 200 individual chemical compounds produced by industrial mixtures via introducing elementary chlorine into biphenyl. Therefore, the primary source of PCBs is industrial production, including industrial wastewaters and slag discharged into the receiving environment. PCBs could have 10 homologs and 209 distinct congeners counting on the number and location of chlorine atoms. Given their property of low electrical conductivity and high resistance to heat and thermal degradation, PCBs are applied as heat exchange fluids in transformers and capacitors. Furthermore, PCBs were ideal additives in paints, dyed paper and plastics [12].

OCPs have been extensively applied in agriculture worldwide for several decades and they mainly originate from improperly treated industrial wastewaters originating from pesticide manufacturing plants. Different species of OCPs, including hexachlorocyclohexanes (HCHs) and dichlorodiphenyltrichloroethanes (DDTs), are still extensively present in water, sediments, atmosphere, fish and even food, due to their persistence, even though the production and application of these contaminants were banned in evolved countries in the 1970s and 1980s [13,14]. Because of their high refractiveness and hydrophobicity, most OCPs firmly adhere to the surface of suspended particles and eventually to sediments at the bottom of water bodies when entering the water environment. They can be subsequently released into the water column under certain conditions such as water turbulence, posing a serious threat to aquatic organisms and human health [15–17].

Many studies have confirmed that the marine environment appears to be one of the primary locations for the accumulation of PCBs and OCPs [18–20].

This study investigates the concentrations of PCBs and OCPs found from the Sele river, one of the main rivers of the Campania plain. Campania is one of the most populated regions of Italy, in which are developed numerous industrial activity and rich agricultural practices such as livestock farming (buffalo farms); the large-scale production of vegetables and fruits feeds the local food industry. These activities include a vast use of pesticides and fertilizers, which can damage water quality [21,22].

Hence, this study is intended to evaluate the concentrations of PCBs and OCPs from the Sele river estuary, southern Italy, and their environmental impact on the Mediterranean Sea. In particular, this paper aims to (i) estimate the PCBs and OCPs levels from the Sele river estuary; (ii) evaluate their distribution between the phases analyzed; (iii) define their spatial distribution and temporal trends in the study area; (iv) assess the potential environmental impact of PCBs and OCPs from the Sele river on the Mediterranean Sea. To the best of our knowledge, there are no previous studies that have evaluated the loads of PCBs and OCPs from the Sele river and the environmental impact on the Mediterranean Sea.

2. Materials and Methods

2.1. Study Area

The Sele river is the second river of the Campania region in the South of Italy, after the Volturno river, and it is a tributary of the Tyrrhenian Sea. It is one of the most important watercourses of the region with a drainage basin of 3235 km^2, a length of 64 km and an annual mean flow rate of 69 m^3/s (Figure 1) [21,23]. The basin is located on the western (i.e., Tyrrhenian) side of southern Italy and includes a large alluvial plain. The plain has a triangular surface area of about 400 km^2 and it is flanked versus the sea by a straight sandy coast between the towns of Salerno and Agropoli. In the Campania region (CP), the city of Salerno is amongst the most tourist-oriented areas around the Mediterranean Sea; furthermore, it has one of the largest transportation networks in south Italy, including railway, highway and various road connections into and around the region. The Sele plain is characterized by agriculture and agro-industries that still provides the major economic income and, from an environmental point of view, the stream network system in the Sele plain is responsible for carrying fertilizers and related products into the Mediterranean Sea. Instead, in the last decade, another source of pollution has been represented by a large number of illegal waste dumps, uncontrolled burning sites (especially in the north of Campania) and industrial wastes from manufacturing enterprises operating in the textile and leather goods sector, which contribute to an increase in the concentrations of the main pollutants [24,25].

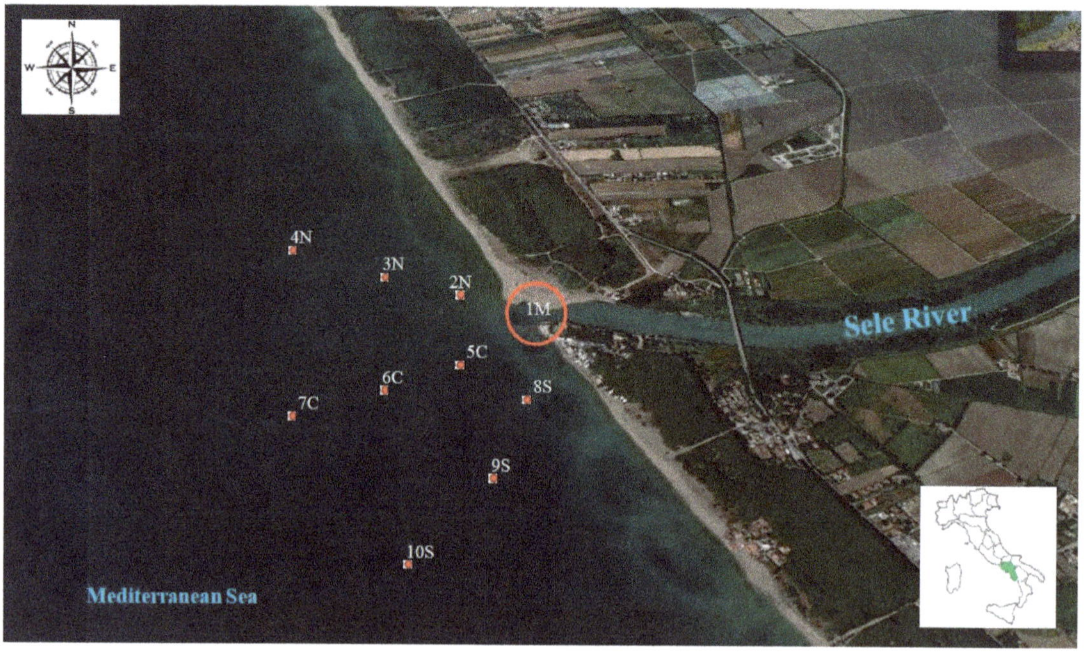

Figure 1. Study area in the Mediterranean Central Sea: solid dots show sampling stations from the Sele river and estuary, southern Italy.

The Sele river basin is characterized by a Mediterranean climate with a particularly dry climate in summer and mild temperatures in winter. The sea contributes to determining the climate, which is warm temperate, with modest daily and annual temperature ranges (less than 21 °C); in fact, the sea maintains the summer heat, accumulating and then releasing it during the winter. The dry summers and rainy winters are a typical characteristic of the Mediterranean climate [26,27].

2.2. Sample Collection

To assess temporal trends of pollutants, between 2020 and 2021, four sampling campaigns were conducted in the summer, autumn, winter and spring from 10 sampling points along the Sele river: the first sampling point was the Sele mouth and the other nine were at diverse distances from the mouth, i.e., 500 m, 1000 m and 1500 m to the north, south and west (Table 1). Three aliquots were sampled at each chosen point and for each season. Once collected, the samples were carried out to the laboratory and analyzed in triplicate, in order to assess the repeatability of the method. For any locations 2.5 L of water (approximately a depth of 0–50 cm from the sampling points) were collected from the surface layer with amber bottles using a portable water collector. All water samples were sent to the laboratory and placed in a 4 °C refrigerator. Sediment samples were obtained at a depth of 0–5 cm in a 0.04 m^2 range area with a Van Veen Grab sampler, and the overall weight of the sediment samples was not less than 500 g. The samples were quickly wrapped in polyethylene bags, shipped to the laboratory and placed in a refrigerator at −20 °C.

Table 1. Total PCBs concentrations in the three phases (DP, SPM, SED) analyzed from the samples collected from the Sele river, southern Italy.

Sampling Location			ΣPCBs								
Site Number Identificatin	Site	Sampling Point	DP (ng L^{-1})				SPM (ng L^{-1}) (ng g^{-1} Dry wt)				SED (ng g^{-1} Dry wt)
			Apr	Jul	Nov	Feb	Apr	Jul	Nov	Feb	Apr
1 (river water)	Sele River Source	40°28′55″ N 14°56′33″ E	6.80	12.1	7.01	4.20	14.0 (1758.6)	9.21 (1026.2)	26.0 (2622.7)	35.1 (1895.3)	79.3
2 (sea water)	River Mouth at 500 mt North	40°29′04″ N 14°56′14″ E	5.71	6.70	6.56	4.68	2.11 (223.2)	2.85 (1236.0)	10.7 (402.2)	22.2 (179.0)	51.2
3 (sea water)	River Mouth at 500 mt Central	40°29′12″ N 14°55′56″ E	6.51	7.29	6.84	4.76	4.2 (1126.7)	5.04 (2514.3)	8.81 (2589.2)	6.85 (1674.7)	36.4
4 (sea water)	River Mouth at 500 mt South	40°29′20″ N 14°55′38″ E	8.72	10.2	7.77	5.02	7.00 (952.1)	6.18 (2698.2)	24.9 (589.5)	30.6 (212.5)	62.1
5 (sea water)	River Mouth at 1000 mt North	40°28′55″ N 14°56′12″ E	6.21	6.66	6.32	3.94	1.52 (118.0)	1.17 (263.1)	6.50 (374.3)	8.08 (125.3)	34.2
6 (sea water)	River Mouth at 1000 mt Central	40°28′55″ N 14°55′50″ E	6.35	5.90	6.71	3.74	2.90 (1569.3)	2.52 (1524.0)	3.40 (1348.7)	2.66 (910.3)	12.3
7 (sea water)	River Mouth at 1000 mt South	40°28′55″ N 14°55′28″ E	6.90	8.20	6.74	4.73	4.10 (460.5)	3.10 (325.3)	15.3 (548.6)	11.5 (84.2)	35.4
8 (sea water)	River Mouth at 1500 mt North	40°28′47″ N 14°56′16″ E	4.90	5.55	4.84	2.41	1.10 (106.9)	0.35 (36.4)	2.18 (774.0)	3.76 (1048.6)	19.2
9 (sea water)	River Mouth at 1500 mt Central	40°28′39″ N 14°55′56″ E	5.30	5.89	5.00	1.98	3.22 (582.2)	1.00 (614.3)	1.50 (486.1)	1.78 (547.2)	5.0
10 (sea water)	River Mouth at 1500 mt South	40°28′30″ N 14°55′38″ E	7.00	7.22	4.32	3.16	4.21 (1986.5)	2.12 (156.1)	4.10 (120.3)	7.54 (1486.4)	10.1

2.3. Sample Processing and Chemical Analysis

The method used for extraction and analytical determination has been published previously [28]. Briefly, water samples were filtered through a previously kiln-fired (400 °C overnight) GF/F glass fiber filter (47 mm × 0.7 µm; Whatman, Maidstone, UK). Filters (suspended particulate matter, SPM) were kept in the dark at −20 °C until analysis. Dissolved phases (fraction of contaminants passing through the filter) were kept in the dark at 4 °C and extracted within the same day of sampling (3–6 h from sampling). Filters were fortified with 2 ng of PCB #65 and PCB #166 as a recovery standard, respectively. After, they were extracted three times by sonication and concentrated to 0.5 mL [29]. The dissolved phase (DP) was fortified with PCB #65 and PCB #166 as a recovery standard, in order to obtain a final concentration of 5 ng L^{-1}. Two liters of sample (DP) were preconcentrated and analyzed using SPE for solid phase extraction; subsequently, they were eluted and concentrated at 0.5 mL.

Sediments were oven desiccated at 60 °C and sifted at 250 µm. A samples rate was fortified with the same surrogate standards used previously, extracted three times and concentrated as the water samples [29]. In each sample analyzed of DP, SPM and sediment, the amount of the following 32 chosen PCBs were quantified (PCBs 8, 28, 37, 44, 49, 52, 60, 66, 70, 74, 77, 82, 87, 99, 101, 105, 114, 118, 126, 128, 138, 153, 156, 158, 166, 169, 170, 179, 180, 183, 187 and 189) (C-SCA-06 PCB Congeners Mix #6; AccuStandard, Inc., New Haven, CT 06513, USA). Instead, the mixed OCPs standard solution included: aldrin, α-BHC, βBHC, δ-BHC, γ-BHC (lindane), p,p′-DDD, p,p′-DDE, p,p′-DDT, dieldrin, endosulfan I, endosulfan II, endosulfan sulfate, endrin, endrin aldehyde, heptachlor, heptachlor epoxide (isomer B) and methoxychlor (M-8080 Organochlorine Pesticides; AccuStandard, Inc., CT 06513, USA). Analysis of sample extracts and standards was performed using a GC17A Shimadzu (Kyoto, Japan), equipped with an electron capture detector (ECD) and an AOC-

20i Shimadzu (Kyoto, Japan) autosampler. Identification of the compounds was achieved by comparing the retention times of the samples with those of the individual PCBs, while quantitative analysis was based on multilevel calibration curves. To confirm the presence of OCPs, GC–MS using a GC–MS 2010 Plus Shimadzu (Kyoto, Japan) was used, working in the electron impact mode and operating at 70 eV.

The mass spectrometer was operated in Single-Ion Monitoring (SIM) mode with the molecular ions of the studied pollutants. PCBs and OCPs are quantified using the response factors of internal standards.

2.4. Quality Assurance and Quality Control

All results were subject to precise quality control process. For each set of 10 samples, a procedural blank and a spiked sample consisting of all reagents were used to check interferences and cross-contaminations. Surrogate standards in DP, SPM and SED samples were analyzed carefully. The mean recovery of a surrogate for the DP sample was 80.5 $\pm$ 8.2%, for SPM samples was 79.3 $\pm$ 6.2%, and for sediment samples was 83.7 $\pm$ 3.1%. Spiked samples in each set of 10 samples were analyzed with mean recoveries ranging from 78.8 to 102.7%. Each extract was evaluated in two copies, in addition, the errors involved in sampling were assessed by carrying out triplicate sampling of water and sediment at the same site and the analysis of sample extracts. Results showed good reproducibility of the sampling process.

The Metod Detection Limit (MDL) was calculated as the average blank values plus three times the standard deviation and it ranged from 0.006 to 0.100 ng L^{-1} in the dissolved phase and in the particulate phase and ranged from 0.0005 to 0.0050 ng g^{-1} in the sediment. IDL was calculated as three times the noise in a blank sample chromatogram. If the amount of any compound in a sample was under its MDL/IDL, this analyte was reputed as not detected in the sample (under the limit of detection, <LOD). Data obtained for PCBs and OCPs were rectified for surrogate recoveries.

2.5. Analysis and Contaminants Load

All statistical analyses were performed with the SPSS 22.0 statistical package (IBM-SPSS Inc., Chicago, IL, USA). The significance level was $p < 0.05$ unless otherwise stated.

According to the UNEP guidelines [30], the method to evaluate the annual pollutants loads has been used (F_{annual}): The mean of the total concentrations was multiplied by the annual average flow rate (m^3/year) of the Sele river for each sampling event and corrected by the total water load for the sampling period. The average flow considered is 69 m^3/s and this information was found in the database of the Autorità di Bacino Distrettuale dell'Appennino Meridionale Sede Basilicata.

Principal Component Analysis (PCA) is a statistical process that purposes an orthogonal transformation to change a group of observations of potentially associated variables into a group of values of linearly uncorrelated variables called principal components. It is one of the oldest and most widely technique used. It reduces the dimensionality of a dataset, while preserving as much variability as possible [31]. In this study, PCA was performed to determine the possible sources of PCBs.

2.6. Toxicity and Dioxin-like PCBs

Dioxin-like PCBs (dl-PCBs) are compounds containing four to eight chlorine atoms. They are very toxic contaminants, bioaccumulative and pose a major health risk due to certain molecular characteristics. In fact, dl-PCBs have a comparable chemical conformation to dioxins and furans. For the combined risk assessment of these substances, the toxic equivalent (TEQ) concentrations for dioxin-like PCBs were calculated according to toxic equivalency factors (TEFs) adopted by the World Health Organization (WHO) in 2005 [32]. TEFs are a fundamental element of TEQ and have developed in the last few decades for dioxins/dioxin-like compounds.

TEF values used in this study are indicated by WHO 2005 for human and mammals [32]: 0.0001 for PCB 77; 0.0003 for PCB 81; 0.00003 for PCB 105, 114, 118, 123, 156, 157, 167 and 189; 0.03 for PCB 169 and 0.1 for PCB 126.

The maximum tolerable value established by US EPA is 0.7 pg WHO-TEQ/kg body weight, and the Equation (1) used to calculate the TEQ is the following:

$$\Sigma TEQ = \Sigma C_i \times TEF_i \quad (1)$$

C_i represents the amount of dl-PCBs (expressed in ng/g). In this study, the TEQ values were calculated in sediment samples to evaluate the presence of humans and environmental risks.

2.7. Risk Assessment

Sediment quality guidelines (SQGs) are generally employed as the effective tool for the estimation of ecological pollution of PCBs in the sediments samples, and have been used in many applications, including monitoring programs, ecological risk assessments and preventing additional pollution.

There are two set of SQGs identified as: (ERL) effect range low and (ERM) effect range median, which evaluate the probably negative effects on organisms concerning individual PCBs as well as the cumulative toxic effects due to the sum of total PCBs [33]; (TEL) threshold effect level and (PEL) probable effect level, which constitute the chemical amount under which the probability of toxicity and other effect are rare [28]. To evaluate the ecological risk related to PCBs and OCPs in the water environment, two indices have been estimated: The Ecological Risk Index (ERI) suggested by Hakanson [34], to evaluate the level of PCBs contamination in the watercourse environment; and Risk Quotient (RQ) method [35], for OCPs pollution. The ERI can be calculated using the following equations:

$$RI = \sum E^i{}_r \quad (2)$$

$$E^i{}_r = T^i{}_r\, C^i{}_f \quad (3)$$

$$C^i{}_f = C^i{}_0 / C^i \quad (4)$$

where ERI is the sum of potential ecological risk for all trace PCBs in the sediments, ERI was equal to $E^i{}_r$, $E^i{}_r$ and $T^i{}_r$ are the toxicity coefficient and individual potential ecological risk for PCBs, which for these pollutants was equal to 40, in line with the standardization elaborated by Hakanson [34]. $C^i{}_f$ was the contamination factor, $C^i{}_0$ was the PCBs amount in the sediment and $C^i{}_n$ was an established value equal to 10 μg/kg. The interpretation and significance of ERI is given as follows: low potential ecological risk, ERI < 40; moderate potential ecological risk, ERI = 40–79; considerable potential ecological risk, ERI = 80–159; high potential ecological risk, ERI =160–319; and very high potential ecological risk, ERI > 320 [34]. Regarding OCPs, the risk quotient (RQ) was conducted via calculation of RQ using Equation (5):

$$RQ = C/PNEC \quad (5)$$

where C was the concentration and PNEC was the predicted no-effect concentrations for particular OCPs. The PNEC results were procured from the ECOTOX database [36]. When $RQ < 0.01$, the OCP has a very low risk to aquatic organisms, and when $0.01 \leq RQ < 0.1$, the ecological risk level is low. When $0.1 \leq RQ < 1$, the OCP has a moderate risk to aquatic organisms. When $1 \leq RQ < 10$, the OCP has a high risk to aquatic organisms, and when $RQ \geq 10$, the ecological risk level is very high [37,38].

3. Results and Discussions

3.1. PCBs Distribution in DP, SPM and Sediment Samples

PCBs were identified in all sampling sites. This result shows that PCBs are extensively spread in the study area. The sum of amounts of PCBs, as demonstrated in (Tables 1 and S8), found in DP, extended from 1.98 ng L^{-1} (site 9) to 12.1 ng L^{-1} (site 1) with a mean value of 6.30 ± 2.10 ng L^{-1}. In Tables S1–S3 (percentage values), the data show that, as reported in (Figure 2a), the main PCBs detected in collected samples were tetra, penta and hexa-CBs, suggesting an average over 82% of ΣPCBs. The abundant presence of this class of PCBs is probably due to the fact that these compounds have stronger hydrophilicity than PCBs, with more chlorine atom substitutions [39]; in fact, when the number of chlorine atoms increases, the solubility decreases [40,41]. In DP samples, hepta-CB were present only for 9% of total PCBs.

In the SPM phase, the PCBs concentrations varied from 0.35 ng L^{-1} (36.4 ng g^{-1}) in site 8 to 35.1 ng L^{-1} (1895.3 ng g^{-1}) in site 1 on dry weight (Tables 1 and S9).

The PCBs most present are those with more chlorine atoms; in fact, in this phase, there is an increase in the percentage of hepta PCBs compared to the DP. This event can be explained through the chemical properties of the higher chlorinated PCBs, which are low hydrophilic and therefore, tend to bind more with the particulate (Figure 2a).

Regarding the sediment samples, the total PCBs values ranged from 5.0 ng g^{-1} (site 9) to 79.3 ng g^{-1} (site 1) (Tables 1 and S10). Data show that the amount of hepta-PCBs increased to 10%. Moreover, the amount of di- + tri-PCBs decreased in sediments samples compared to SPM and DP samples. It can therefore be said that the percentage of highly chlorinated PCBs in the sediments samples was higher than that in the DP and SPM phases, and the percentage of less chlorinated PCBs was lower than that in the DP and SPM phases; furthermore, in the sediment have been found the highest concentrations of PCBs. The characteristic of PCBs depends on the degree of chlorination, i.e., the higher the degree of chlorination, the lower the water solubility and vapor pressure [39]. In the Sele river, sediments turn out to be a sink for these contaminants and are a measurement of their amount during the years [42–44]. PCBs being hydrophobic organic compounds, they are characterized by extraordinary stability, high toxicity, extremely high long-range atmospheric transportability [45,46]. In the aquatic environment, PCBs are removed from the water column and adsorbed onto suspended particulate matter; they can subsequently bio-accumulate in sediment and thereby, transfer to higher trophic levels through the food chain. Due to their persistent and hydrophobic nature, the fate and transport of PCBs in a water environment are highly affected by their adsorption behavior on the sediment [47,48]. Many factors influence the adsorption behavior of PCBs. In this study, among them, pH, temperature and salinity were considered. Salinity, for example, can alter the water solubility of hydrophobic compounds and the physicochemical properties of sediment, through which it influences the adsorption capacity of hydrophobic compounds on the sediment. Table S4 shows the data of the factors that may have contributed to a higher concentration of PCBs in the sediment and may have influenced the distribution of these contaminants analysed in this study characterized by a predominantly mineral sediment.

Figure 2. Mean concentrations of PCBs (**A**) and OCPs (**B**) in the water samples (DP), suspended particulate matter (SPM) and sediment (SED) from the Sele River, southern Italy.

3.2. OCPs Distribution in DP, SPM and Sediment Samples

Data showed that samples raised from the Sele river included rests of HCH (sum of a-HCH, b-HCH, g-HCH, and d-HCH), DDT (p,p′-DDE, p,p′-DDD, p,p′-DDT isomers and methoxychlor) and cyclodienes (aldrin, dieldrin, endosulfan I, endosulfan II, endosulfan sulphate, endrin, heptachlor and heptachlor epoxide). In Tables 2 and S11 were reported the results of the DP, SPM and sediment sample analyses. In the DP phase, the total concentrations varied from 0.36 ng L^{-1} (site 9) to 5.71 ng L^{-1} (site 1) (mean value of 1.22 $\pm$ 0.23 ng L^{-1}). Particularly, as indicated in Figure 2b and in Tables S5–S7 (percentage values), they varied from ND to 0.75 ng L^{-1} for HCH, from ND to 1.0 ng L^{-1} for DDT and its degradates, and from ND to 3.20 ng L^{-1} for cyclodienes. In SPM, the amounts acquired for total OCPs extended from 0.05 ng L^{-1} (65.3 ng g^{-1} dw) in site 9 to 4.82 ng L^{-1} (201.4 ng g^{-1} dw) in site 1 (Tables 2 and S12). The HCHs extended from ND to 0.89 ng L^{-1}, the DDTs from ND to 0.96 ng L^{-1}, and the cyclodienes from ND to 2.62 ng L^{-1}, as shown in Figure 2b. In sediment samples, instead, the total OCPs concentration (Tables 2 and S13) extended from 1.1 ng g^{-1} (site 9) to 15.0 ng g^{-1} (site 1). The HCHs ranged from 0.10 to 1.24 ng g^{-1}, the DDTs from 0.10 to 6.12 and the cyclodienes from 0.15 to 3.10 ng g^{-1} (Figure 2b). The results show that in the Sele river, a higher percentage of cyclodienes and DDT was found compared to HCH; in fact, the results of the ratio indicate that the DDTs/cyclodienes ratio was <1 at most sites (mean, 0.70), such as the HCHs/DDTs and HCHs/cyclodienes ratios (means, 0.40 and 0.20, respectively). The dominant HCH was b-HCH (1.90 $\pm$ 1.00), followed by a-HCH (1.65 $\pm$ 0.80). This pesticide had a lower solubility in water, and dissolved organic matter can assimilate on this compound, which may raise the amount in water. The ratio of b-HCH in the HCHs was high and indicates that these contaminants maybe represent a historical input rather than a fresh input [49]. A similar trend for b-HCH has also been reported by Dong et al. [50] and Salem et al. [51].

In this study, it was also significant to assess the biodegradation of DDT in its metabolites in the aquatic system. DDT not only controls crop pests and malaria but is also used as an active ingredient in antifouling coatings on fishing boats in several developing countries [52]; in Italy this pesticide has been prohibited from rural application and limited for public health [28]. DDT is composed of p,p′-DDT, p,p′-DDD, p,p′-DDE. DDT will dechlorinate to DDD under anaerobic conditions and degrade to DDE under aerobic conditions [52]. To determine the indicated levels of DDT in this study, the ratio of p,p′-DDT to its metabolites was estimated. When the ratio < 0.5, the DDT input was recent while when the ratio > 0.5 the DDT present in the environment is attributable to the historical input [53]. In the Sele river, the ratio in DP, SPM and sediment was 15.1, 16.6 and 18.7, respectively, so these data indicate that most of the DDTs in the Sele river were obtained from historical input (Figure 3).

Among the cyclodiene compounds and their metabolites, endosulfan sulfate was in abundance with the highest grades in water (DP + SPM), justifying 9% of total OCPs. Heptachlor epoxide is the metabolite of heptachlor and the ratio of heptachlor/heptachlor epoxide in the water system of the Sele river was 0.17. According to Kuranchie Mensah et al. [54], when the trend of metabolites was higher than the parent compound present, there were no fresh inputs of this contaminant in the water stream.

Table 2. Total OCPs concentrations in the three phases (DP, SPM, SED) analyzed for the samples collected from the Sele river, southern Italy.

Sampling Location			ΣOCPs								
Site Number Identification	Site	Sampling Point	DP (ng L^{-1})				SPM (ng L^{-1}) (ng g^{-1} Dry wt)				SED (ng g^{-1} Dry wt)
			Apr	Jul	Nov	Feb	Apr	Jul	Nov	Feb	Apr
1 (river water)	Sele River Source	40°48′54.03″ N 14°36′45.36″ E	4.01	5.71	3.75	1.95	2.08 (198.5)	1.56 (154.1)	3.98 (243.0)	4.82 (201.4)	15.2
2 (sea water)	River Mouth at 500 mt North	40°46′42.73″ N 14°34′00.48″ E	1.70	2.98	2.12	1.10	1.06 (70.2	0.55 (284.1)	1.22 (51.8)	1.80 (174.3)	1.39
3 (sea water)	River Mouth at 500 mt Central	40°46′00.34″ N 14°33′10.68″ E	2.03	2.01	1.98	0.80	1.20 (185.4)	0.50 (154.2)	1.26 (274.6)	1.86 (119.4)	1.54
4 (sea water)	River Mouth at 500 mt South	40°43′42.62″ N 14°28′07.89″ E	3.24	4.38	2.18	1.82	1.48 (150.2)	0.68 (298.4)	1.54 (97.5)	2.20 (241.2)	3.85
5 (sea water)	River Mouth at 1000 mt North	40°43′40.11″ N 14°28′06.45″ E	1.00	2.00	1.78	0.75	1.00 (94.1)	0.48 (102.3)	0.98 (95.4)	1.03 (100.1)	1.20
6 (sea water)	River Mouth at 1000 mt Central	40°43′42.46″ N 14°28′05.03″ E	0.98	1.32	1.20	0.49	1.10 (254.3)	0.32 (36.8)	0.99 (198.4)	1.23 (155.2)	1.32
7 (sea water)	River Mouth at 1000 mt South	40°43′45.09″ N 14°28′05.17″ E	2.12	2.85	1.60	1.10	1.24 (110.4)	0.38 (89.2)	1.26 (114.7)	1.30 (10.2)	2.84
8 (sea water)	River Mouth at 1500 mt North	40°43′35.68″ N 14°28′02.94″ E	0.84	1.20	0.90	0.70	0.50 (71.4)	0.39 (96.8)	0.84 (195.7)	1.00 (180.3)	1.21
9 (sea water)	River Mouth at 1500 mt Central	40°43′42.25″ N 14°27′59.97″ E	0.84	0.90	0.81	0.36	0.47 (112.2)	0.05 (65.3)	0.91 (117.8)	0.89 (148.3)	1.10
10 (sea water)	River Mouth at 1500 mt South	40°43′49.26″ N 14°27′59.82″ E	1.45	1.85	0.73	0.79	0.60 (196.5)	0.10 (52.7)	0.74 (17.4)	0.89 (185.6)	1.02

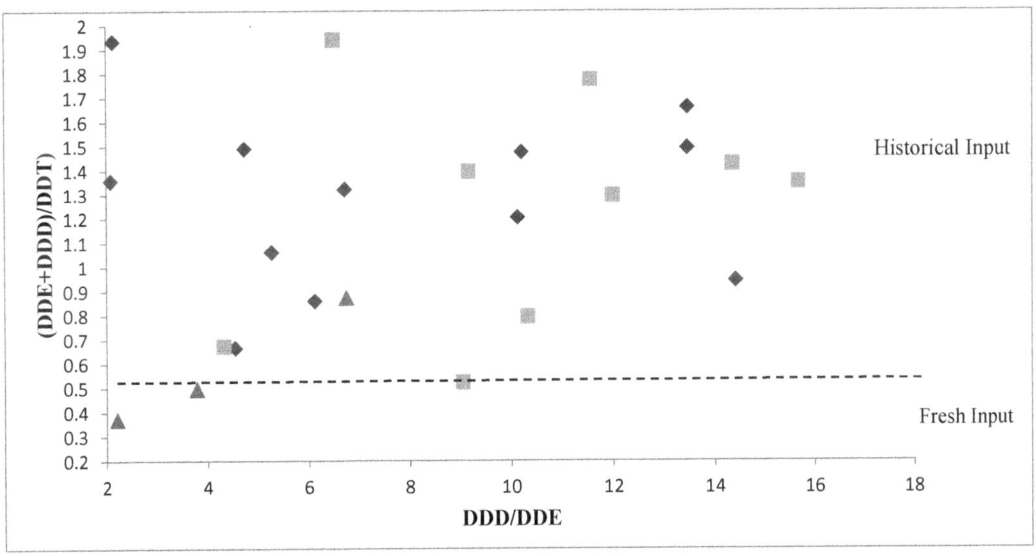

Figure 3. Isomeric ratios of DDT and its metabolites: DDD/DDE vs (DDE + DDD)/DDTs in the samples from Sele River. In the Figure the dotted line represents the point (0.5) where the fresh input becomes historical input. ◆ DP samples. ■ SPM samples. ▲ SED samples.

3.3. Spatiotemporal Diffusion

The spatial diffusion designs of ∑ PCBs, ∑ OCPs and isomers concentrations in water and sediments of the Sele river are illustrated in Figure 4a,b, respectively. The results shown were obtained by studying and comparing the concentrations in the different sites in the dry and rainy seasons. The data showed a similar trend for both classes of compounds.

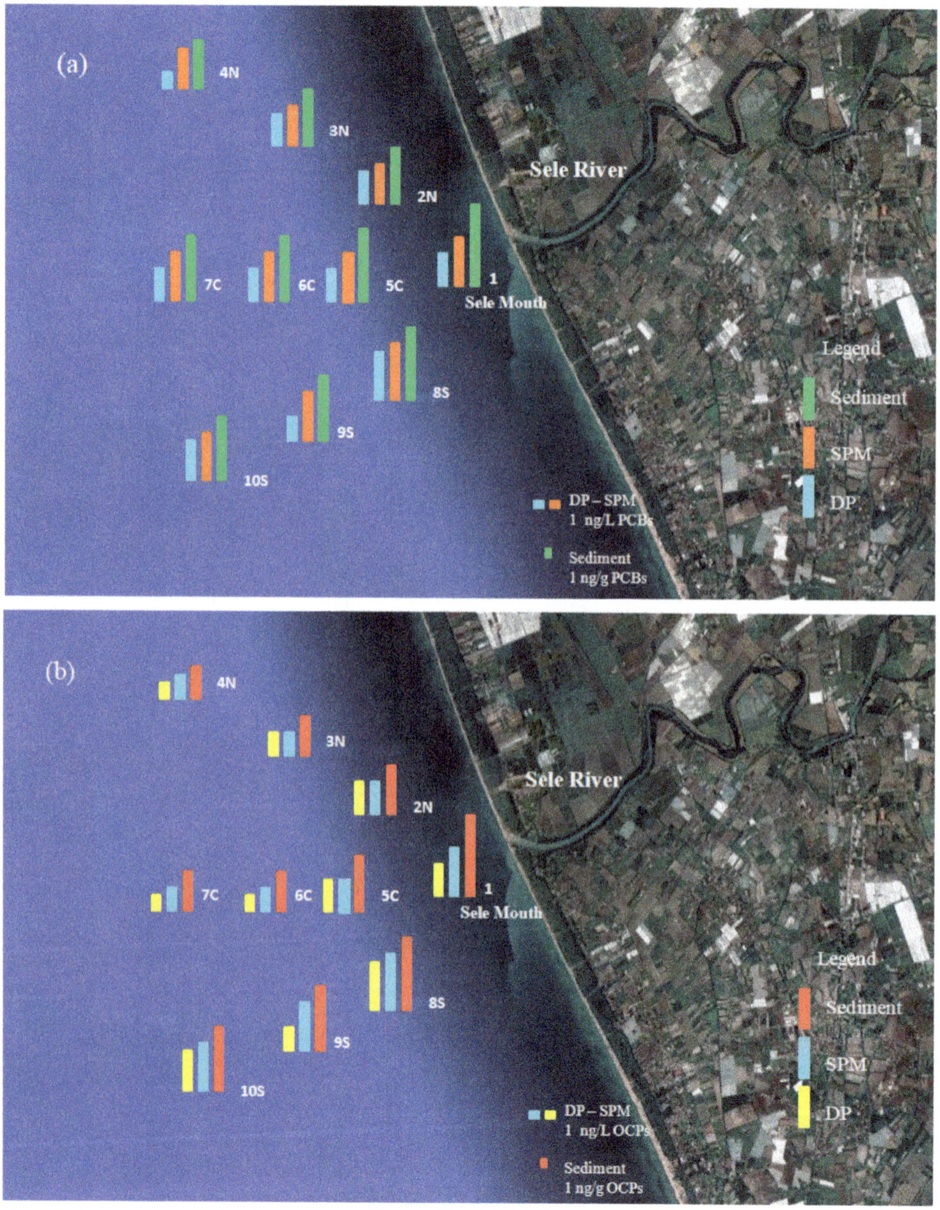

Figure 4. (**a**,**b**) show the spatial distribution of PCBs and OCPs in the water phase (DP ng/L), suspended particulate matter (SPM ng/L) and sediments (ng/g) from the Sele river.

The Mouth of the Sele river is the most contaminated with a more elevated total concentration of PCBs and OCPs. Concentrations decrease as you move away from the mouth up to 1500 m from the coast, where the concentrations of PCBs and OCPs are significantly lower. Figure 4a,b show that the highest concentrations have been obtained around the Sele river mouth, as the contaminants present in the aqueous phase are diluted as one moves away. In particular, the contaminants load from the Sele river mouth has been shown to move southward into the Tyrrhenian Sea. In this study, the pollutant load drained into the Tyrrhenian Sea by the Sele river was also calculated. The results show that the total estimated value is equal to 89.7 kg year^{-1} (73.2 kg year^{-1} of PCBs and 16.5 kg year^{-1} of OCPs) In the water samples (DP phase), the total amount of pollutants was considerably lower mainly during the wet season (February), due to the abundant rains which caused water dilution effects. On the other hand, in SPM samples, the amounts were lowest in all sampling sites during the dry season. The results showed that the contaminants concentrations in DP decreased from July to February, in parallel with the increase in rainfall, which could cause dilution ratio variations. Therefore, the decrease of the pollutants amount moving from the Sele river mouth to the Mediterranean Sea is also affected by the high flow in the rainfall season, which results in an even higher dilution ratio. The lowest concentrations in SPM were recorded in the dry season (July), due to the decrease in flow and a greater stagnation of SPM, which led to the shift of contaminants from SPM to DP.

3.4. Potential Sources of PCBs

For the purpose of more accurately controlling the emission and release of PCBs, it is deemed necessity to define their contamination sources as much as possible. Principal Component Analysis (PCA) has been executed on the different sediment datasets. Six groups of PCBs were identified in this study (Di- PCB, Tri-PCB, Tetra-PCB, Penta-PCB and Hepta-PCB). The obtained results from PCA manifested that the first three principal components show 57.1% (PC1), 15% (PC2) and 10% (PC3) of the total variance, respectively (Figure 5). Considering the three PCA axes individually, PC1 was principally composed of tetra-PCB, penta-PCB and hexa-PCB (high chlorinated congeners), PC2 was composed of Di-PCB and Tri-PCB, and the third component PC3 was composed of Hepta-PCBs.

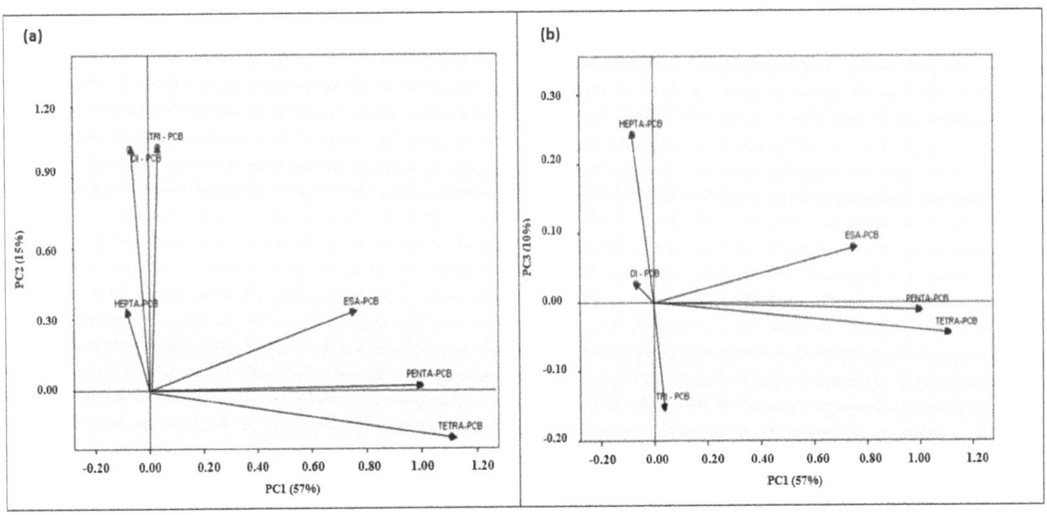

Figure 5. Principal Component Analysis (PCA) of the sediments PCBs results: (**a**) Score plot for the first and second principal component. (**b**) Loading plot for the first and third principal component.

The first component dominated by highly chlorinated PCBs could be unintentionally formed by anthropogenic activities such as industrial processes, waste incineration and vehicle exhaust [55,56]. Many studies [57,58] have demonstrated that PCB amount levels in the lower atmosphere near the water are confirmed 4Cl PCBs evaporated from the surface layer. In addition, the loss of a chlorine atom of highly chlorinated compounds with an anaerobic microbe can manifest in the sediments [26], which provides a good availability of molecules with few chlorine atoms. Therefore, PC1 represented PCBs originated from unintentionally formed local sources directly discharged into coastal water. The second component, dominated by 2Cl and 3Cl PCBs, suggests that these compounds could be transferred to the watercourse by surface runoff after rain cases, and cumulate in the estuary. The third component, composed of Hepta-PCB, suggests a point source deposition industrial loads along the Sele river: for example, discharge pipes from factories, sewage treatment plants and various organizations could be responsible for point source pollution in the Sele river. The existence can be assumed of a single major source in the watercourse related to the point source [59].

3.5. Dioxin Toxicity Equivalency

TEQs were calculated for eight PCBs (PCB 77, 105, 114, 118, 126, 156, 169 and 189) having dioxin-like properties by TEF, described in detail by Van den Berg et al. (2006) for all sediment samples. The TEQ concentrations of dioxin-like PCBs (DL-PCBs) detected at all sampling sites ranged from 0.004 to 0.270 ng/g with an average level of 0.050 ng/g. The highest $\sum$ TEQ_{PCB} concentrations were found at the Sele mouth (site 1). Despite PCB-114 indicating an amount higher than others PCB-DL, PCB-126 and PCB-169 contributed for 95.7% to TEQ_{PCB}, because of their higher TEF.

The data indicated that TEQ_{PCB} values of the Sele river and its estuary were in a low level, suggesting that the toxicity of the PCBs in the watercourse could negatively cause a great threat to organisms and ecosystem, and endanger human health through bioconcentration and the food chain [38].

3.6. Risk Assessment of PCBs and OCPs

The SQGs guidelines can estimate the level of the possible negative effects and toxicity thresholds of specific organic contaminants in sediment for the ecological environment [60,61]. In this study, the total PCBs amount in sediment samples of the Sele river were considerably lower than PEL and ERM (Table 3), while 40% and 40% of analyzed samples indicated concentrations above TEL and ERL values, respectively, in the Sele river. and the risk factor of analyzed samples indicated concentrations above TEL and ERL values, respectively.

Table 3. A comparison of the TEL, PEL, ERL and ERM guideline values (μg Kg^{-1}) for PCBs and OCPs and data from the Sele river and estuary, southern Italy.

	TEL	Percentage over the TEL	PEL	Percentage over the PEL	ERL	Percentage over the ERL	ERM	Percentage over the ERM
PCBs								
Total PCBs	21.6	40	189	0	22.7	40	180	0
OCPs								
γ-HCH (lindane)	0.32	0	0.99	0	-		-	
Dieldrin	0.72	0	4.3	0	0.02	50	8	0
4,4-DDD	1.22	20	7.81	0	2	0	20	0
4,4-DDE	2.07	0	374	0	2.2	0	27	0
4,4-DDT	1.19	10	4.77	0	1	10	7	0
Total DDT	3.89	0	51.7	0	1.58	10	46.1	0

Regarding risk factors, the results showed that in the Sele river, the risk factor of PCBs for the sampling site were elevated at the mouth and at 500 m south, although in other sites, the risk value ranged from appreciable to low. Consequentially, based on the data obtained, the risk in the sediments of the Sele river was medium. Concerning the OCPs, in all analyzed samples, the ratio indicated a RQ < 1 for most of the pesticides. These data show that negative effects on the aquatic organism would rarely be observed [28,42,62,63].

4. Conclusions

This study analyzed the pollution characteristics, spatiotemporal variation, source identification and potential ecological risk of PCBs and OCPs in the Sele river; the input was also calculated of this watercourse into the Tyrrhenian Sea (Central Mediterranean Sea).

A higher amount of this contaminant was built in sediment samples than in their correspondent water bodies, DP and SPM, which suggests that suspension processes and sedimentation are principally in the Sele river. The data obtained showed that industrial procedure was reputed to be the principal source of PCBs; regarding the risk assessment, the risk factors of PCBs in sediment samples were elevated at the Sele river mouth and at 500 mt south, while in other sites, they are low. OCPs ratio, instead, was lower and showed an RQ <1 for most analytes. Thus, the pollution situation in the Sele river and its estuary should be monitored regularly to assess the ecological risk in time. These data improve our knowledge on the Sele river water quality and they would inform such things as environmental monitoring, sediment quality guidelines application and ecological risk assessments. Our expectation is that the important and significative activity of establishing a rich database for different pollution factors can be developed, and more emerging contaminants should be included in ecological risk assessments of river ecosystems. Furthermore, this study's results will help prevent future environmental water system contamination of the Sele river from PCBs and OCPs and strengthen prevention and pollution control measures against future risks. It would further help policymakers identify high-risk pollutants areas, improve environmental protection regulatory policy and sensitize the public to its importance. This study presents a novel result on the current status of water and sediment PCBs and OCPs levels in the area surrounding the Sele river. Therefore, the PCBs and OCPs levels in water and sediment from the Sele river should be further analyzed to ensure the contaminant levels reported in these areas are not being underestimated due to the continued increase in environmental activities.

Supplementary Materials: The following supporting information can be downloaded at: https://www.mdpi.com/article/10.3390/toxics10110662/s1, Table S1: Percentage of PCBs in the different seasons in the water samples (DP) from Sele River (Southern Italy); Table S2: Percentage of PCBs in the different seasons in the suspended particulate matter (SPM) from Sele River (Southern Italy), Table S3: Percentage of PCBs in the April month in the sediment samples (SED) from Sele River (Southern Italy); Table S4: Environmental data from sampling sites: Table S5: Percentage of OCPs in the different seasons in the water samples (DP) from Sele River (Southern Italy); Table S6: Percentage of OCPs in the different seasons in the suspended particulate matter (SPM) from Sele River (Southern Italy); Table S7: Percentage of OCPs in the April month in the sediment samples (SED) from Sele River (Southern Italy); Table S8: Description of concentration of PCBs in water dissolved phase (DP) samples from Sele River, (Southern Italy); Table S9: Description of concentration of PCBs in Suspended Particulate Matter (SPM) samples from Sele River, (Southern Italy): Table S10: Description of concentration of PCBs in Sediment (SED) samples from Sele River, (Southern Italy); Table S11: Description of concentration of OCPs in water dissolved phase (DP) samples from Sele River, (Southern Italy); Table S12: Description of concentration of OCPs in Suspended Particulate Matter (SPM) samples from Sele River, (Southern Italy); Table S13: Description of concentration of OCPs in Sediment (SED) samples from Sele River, (Southern Italy).

Author Contributions: Conceptualization, M.T. and P.M.; Data curation, P.M., A.M., E.D.R. and A.N.; Formal analysis, E.D.R.; Methodology, P.M. and E.D.R.; Project administration, P.M.; Resources, P.M., E.D.R. and A.N.; Software, E.D.R.; Supervision, P.M. and M.T.; Writing—original draft, P.M., E.D.R.; Writing—review & editing, P.M., A.M. and M.T. All authors have read and agreed to the published version of the manuscript.

Funding: This research received no external funding.

Institutional Review Board Statement: Not applicable.

Informed Consent Statement: Not applicable.

Data Availability Statement: The datasets obtained and analyzed in the current study are available from the corresponding author on reasonable request.

Conflicts of Interest: The authors declare that they have no competing interest.

References

1. Islam, M.S.; Ahmed, M.K.; Rakanuzzaman, M.; Habibullah-Al-Mamun, M.; Islam, M.K. Heavy metal pollutionin surface water and sediment: A preliminary assessment of an urbanriver in a developing country. *Ecol. Indic.* **2015**, *48*, 282–291. [CrossRef]
2. Xu, X.; Cao, Z.; Zhang, Z.; Li, R.; Hu, B. Spatial distribution and pollution assessment of heavy metals in the surface sediments of the Bohai and Yellow Seas. *Mar. Pollut. Bull.* **2016**, *110*, 596–602. [CrossRef] [PubMed]
3. Xu, X.; Guo, C.S.; Luo, Y.; Lv, J.P.; Zhang, Y.; Lin, H.X.; Wang, I.; Xu, J. Occurrence and distribution of antibiotics, antibiotic resistence genes in the urban rivers in Beijing, China. *Environ. Pollut.* **2016**, *213*, 833–840. [CrossRef]
4. Barhoumi, B.; Beldean-Galea, M.S.; Al-Rawabdeh, A.M.; Roba, C.; Martonos, I.M.; Balc, R.; Kahlaoui, M.; Touil, S.; Tedetti, M.; Driss, M.R.; et al. Occurrence, distribution and ecological risk of trace metals and organic pollutants in surface sediments from a Southeastern European river (Someşu Mic River, Romania). *Sci. Total Environ.* **2019**, *660*, 660–676. [CrossRef]
5. Minh, N.H.; Minh, T.B.; Kajiwara, N.; Kunisue, T.; Iwata, H.; Viet, P.H.; Cam Tu, N.P.; Tuyen, B.C.; Tanabe, S. Pollution sources and occurrences of selected persistent organic pollutants (POPs) in sediments of the Mekong River delta, South Vietnam. *Chemosphere* **2007**, *67*, 1794–1801. [CrossRef] [PubMed]
6. Montuori, P.; Triassi, M. Polycyclic aromatic hydrocarbons loads into the Mediterranean Sea: Estimate of Sarno River inputs. *Mar. Pollut. Bull.* **2012**, *64*, 512–520. [CrossRef]
7. Loqanathan, B.G.; Kannan, K. Global organochlorine contamination trends: An overview. *Ambio* **1994**, *23*, 187–191.
8. Muir, D.; Sverko, E. Analytical methods for PCBs and organochlorine pesticides in environmental monitoring and surveillance: A critical appraisal. *Anal. Bioanal. Chem.* **2006**, *386*, 769–789. [CrossRef]
9. Zhao, L.; Hou, H.; Zhou, Y.; Xue, N.; Li, H.; Li, L. Distribution and ecological risk of polychlorinated biphenyls and organochlorine pesticides in surficial sediments from Haihe River and Haihe Estuary Area, China. *Chemosphere* **2010**, *78*, 1285–1293. [CrossRef]
10. Wang, Q.; Shi, Y.; Hu, J.; Yao, Z.; Fang, X.; Dong, Y. Determination of dioxin-like polychlorinated biphenyls in soil and moss from Fildes Peninsula, Antarctica. *Chin. Sci. Bull.* **2012**, *57*, 992–996. [CrossRef]
11. Dimou, K.N.; Su, T.L.; Hires, R.I.; Miskewitz, R. Distribution of polychlorinated biphenyls in the Newark Bay Estuary. *J. Hazard. Mater.* **2006**, *136*, 103–110. [CrossRef] [PubMed]
12. Yu, Y.; Li, Y.; Shen, Z.; Yang, Z.; Mo, L.; Kong, Y.; Lou, I. Occurrence and possible sources of organochlorine pesticides (OCPs) and polychlorinated biphenyls (PCBs) along the Chao River, China. *Chemosphere* **2014**, *114*, 136–143. [CrossRef] [PubMed]
13. Montuori, P.; Aurino, S.; Garzonio, F.; Triassi, M. Polychlorinated biphenyls and organochlorine pesticides in Tiber River and Estuary: Occurrence, distribution and ecological risk. *Sci. Total Environ.* **2016**, *571*, 1001–1016. [CrossRef] [PubMed]
14. Montory, M.; Ferrer, J.; Rivera, D.; Villouta, M.V.; Grimalt, J.O. First report on organochlorine pesticides in water in a highly productive agro-industrial basin of the Central Valley, Chile. *Chemosphere* **2017**, *174*, 148–156. [CrossRef]
15. Kafilzadeh, F. Distribution and sources of polycyclic aromatic hydrocarbons in water and sediments of the Soltan Abad River, Iran. *Egypt. J. Aquat. Res.* **2015**, *41*, 227–231. [CrossRef]
16. Zheng, B.; Wang, L.; Lei, K.; Nan, B. Distribution and ecological risk assessment of polycyclic aromatic hydrocarbons in water, suspended particulate matter and sediment from Daliao River estuary and the adjacent area, China. *Chemosphere* **2016**, *149*, 91–100. [CrossRef]
17. Eremina, N.; Paschke, A.; Mazlova, E.A.; Schüürmann, G. Distribution of polychlorinated biphenyls, phthalic acid esters, polycyclic aromatic hydrocarbons and organochlorine substances in the Moscow River, Russia. *Environ. Pollut.* **2016**, *210*, 409–418. [CrossRef]
18. Chen, M.Y.; Yu, M.; Luo, X.J.; Chen, S.J.; Mai, B.X. The factors controlling the partitioning of polybrominated diphenyl ethers and polychlorinated biphenyls in the water-column of the Pearl River Estuary in South China. *Mar. Pollut. Bull.* **2010**, *62*, 29–35. [CrossRef]
19. Gómez-Gutiérrez, A.I.; Jover, E.; Bodineau, L.; Albaigés, J.; Bayona, J.M. Organic contaminant loads into the Western Mediterranean Sea: Estimate of Ebro river inputs. *Chemosphere* **2006**, *65*, 224–236. [CrossRef]
20. Guan, Y.F.; Wang, J.Z.; Ni, H.G.; Zeng, E.Y. Organochlorine pesticides and polychlorinated biphenyls in riverine runoff of the Pearl River Delta, China: Assessment of mass loading, input source and environmental fate. *Environ. Pollut.* **2009**, *157*, 618–624. [CrossRef]
21. Di Paola, G.; Alberico, I.; Aucelli, P.P.C.; Matano, F.; Rizzo, A.; Vilardo, G. Coastal subsidence detected by Synthetic Aperture Radar interferometry and its effects coupled with future sea-level rise: The case of the Sele Plain (Southern Italy). *J. Flood Risk Manag.* **2018**, *11*, 191–206. [CrossRef]
22. Arienzo, M.; Bolinesi, F.; Aiello, G.; Barra, D.; Donadio, C.; Stanislao, C.; Trifuoggi, M. The environmental assessment of an estuarine transitional environment, southern Italy. *J. Mar. Sci. Eng.* **2020**, *8*, 628. [CrossRef]

23. Diodato, N.; Fagnano, M.; Alberico, I. Geospatial and visual modeling for exploring sediment source areas across the Sele river landscape, Italy. *Ital. J. Agron.* **2011**, *6*, e14. [CrossRef]
24. Albanese, S.; De Vivo, B.; Lima, A.; Cicchella, D. Geochemical background and baseline values of toxic elements in stream sediments of Campania region (Italy). *J. Geochem. Explor.* **2007**, *93*, 21–34. [CrossRef]
25. Menghan, W.; Stefano, A.; Annamaria, L.; Claudia, C.; Antonio, C.; Wanjun, L.; Angela, D. Compositional analysis and pollution impact assessment: A case study in the Gulfs of Naples and Salerno. *Estuar. Coast. Shelf Sci.* **2015**, *160*, 22–32. [CrossRef]
26. Albanese, S.; De Vivo, B.; Lima, A.; Cicchella, D.; Civitillo, D.; Cosenza, A. Geochemical baselines and risk assessment of the Bagnoli brownfield site coastal sea sediments (Naples, Italy). *J. Geochem. Explor.* **2010**, *105*, 19–33. [CrossRef]
27. Montuori, P.; De Rosa, E.; Di Duca, F.; De Simone, B.; Scippa, S.; Russo, I.; Triassi, M. Polycyclic Aromatic Hydrocarbons (PAHs) in the Dissolved Phase, Particulate Matter, and Sediment of the Sele River, Southern Italy: A Focus on Distribution, Risk Assessment, and Sources. *Toxics* **2022**, *10*, 401. [CrossRef]
28. Montuori, P.; De Rosa, E.; Sarnacchiaro, P.; Di Duca, F.; Provvisiero, D.P.; Nardone, A.; Triassi, M. Polychlorinated biphenyls and organochlorine pesticides in water and sediment from Volturno River, Southern Italy: Occurrence, distribution and risk assessment. *Environ. Sci. Eur.* **2020**, *32*, 123. [CrossRef]
29. Montuori, P.; Cirillo, T.; Fasano, E.; Nardone, A.; Esposito, F.; Triassi, M. Spatial distribution and partitioning of polychlorinated biphenyl and organochlorine pesticide in water and sediment from Sarno River and Estuary, southern Italy. *Environ. Sci. Pollut. Res.* **2014**, *21*, 5023–5035. [CrossRef]
30. UNEP/MAP. *Guidelines for River (Including Estuaries) Pollution Monitoring Programme for the Mediterranean Region*; MAP Technical Reports Series No. 151; UNEP/MAP: Athens, Greece, 2004.
31. Jolliffe, I.T. *Principal Component Analysis*, 2nd ed.; Springer: Berlin/Heidelberg, Germany, 2002.
32. Van den Berg, M.; Birnbaum, L.S.; Denison, M.; De Vito, M.; Farland, W.; Feeley, M.; Fiedler, H.; Hakansson, H.; Hanberg, A.; Haws, L.; et al. The 2005 World Health Organization reevaluation of human and Mammalian toxic equivalency factors for dioxins and dioxin-like compounds. *Toxicol. Sci.* **2006**, *93*, 223–241. [CrossRef]
33. Birch, G.F. A review of chemical-based sediment quality assessment methodologies for the marine environment. *Mar. Pollut. Bull.* **2018**, *133*, 218–232. [CrossRef] [PubMed]
34. Hakanson, L. An ecological risk index for aquatic pollution control A sedimentological approach. *Water Res.* **1980**, *14*, 9751001. [CrossRef]
35. WHO. *Water Quality: Guidelines, Standards and Health*; IWA: London, UK, 2001.
36. USEPA (US Environmental Protection Agency). Regional Screening Levels for Chemical Contaminants at Superfund Sites. Regional Screening Table. User's Guide. 2012. Available online: https://www.epa.gov/risk/regional-screening-levels-rslsgeneric-tables (accessed on 2 August 2022).
37. Chen, C.; Zou, W.; Chen, S.; Zhang, K.; Ma, L. Ecological and health risk assessment of organochlorine pesticides in an urbanized river network of Shanghai, China. *Environ. Sci. Eur.* **2020**, *32*, 42. [CrossRef]
38. Xu, H.; Yang, H.; Ge, Q.; Jiang, Z.; Wu, Y.; Yu, Y.; Han, D.; Cheng, J. Long-term study of heavy metal pollution in the northern Hangzhou Bay of China: Temporal and spatial distribution, contamination evaluation, and potential ecological risk. *Environ. Sci. Pollut. Res. Int.* **2020**, *28*, 10718–10733. [CrossRef] [PubMed]
39. Srédlovà, K.; Cajthaml, T. Recent advances in PCB removal from historically contaminate environmental matrices. *Chemosphere* **2022**, *287*, 132096. [CrossRef] [PubMed]
40. Lin, T.; Nizzetto, L.; Guo, Z.; Li, Y.; Li, J.; Zhang, G. DDTs and HCHs in sediment cores from the coastal East China Sea. *Sci. Total Environ.* **2016**, *539*, 388–394. [CrossRef]
41. Čonka, K.; Chovancová, J.; Stachová Sejáková, Z.; Dömötörová, M.; Fabišiková, A.; Drobná, B.; Kočan, A. PCDDs, PCDFs, PCBs and OCPs in sediments from selected areas in the Slovak Republic. *Chemosphere* **2014**, *98*, 37–43. [CrossRef]
42. Syed, J.H.; Malik, R.N.; Li, J.; Chaemfa, C.; Zhang, G.; Jones, K.C. Status, distribution and ecological risk of organochlorines (OCs) in the surface sediments from the Ravi River, Pakistan. *Sci. Total Environ.* **2014**, *472*, 204–211. [CrossRef]
43. Bhattacharya, B.; Sarkar, S.K.; Mukherjee, N. Organochlorine pesticide residues in sediments of a tropical mangrove estuary, India: Implications for monitoring. *Environ. Int.* **2003**, *29*, 587–592. [CrossRef]
44. Barakat, A.O.; Khairy, M.; Aukaily, I. Persistent organochlorine pesticide and PCB residues in surface sediments of Lake Qarun, a protected area of Egypt. *Chemosphere* **2013**, *90*, 2467–2476. [CrossRef]
45. Ren, N.Q.; Que, M.X.; Li, Y.F.; Liu, L.Y.; Wang, X.N.; Xu, D.D.; Sverko, E.D.; Ma, J.M. Polychlorinated biphenyls in Chinese surface soils. *Environ. Sci. Technol.* **2013**, *41*, 3871–3876. [CrossRef] [PubMed]
46. Tian, Y.; Li, W.; Shi, G.; Feng, Y.; Wang, Y. Relationships between PAHs and PCBs, and quantitative source apportionment of PAHs toxicity in sediments from Fenhe reservoir and watershed. *J. Hazard. Mater.* **2013**, *248–249*, 89–96. [CrossRef]
47. Guo, W.; He, M.; Yang, Z.; Lin, C.; Quan, X.; Wang, H. Distribution of polycyclic aromatic hydrocarbons in water, suspended particulate matter and sediment from Daliao River watershed, China. *Chemosphere* **2007**, *68*, 93–104. [CrossRef]
48. Yang, Z.; Feng, J.; Niu, J.; Shen, Z. Release of polycyclic aromatic hydrocarbons from Yangtze River sediment cores during periods of simulated resuspension. *Environ. Pollut.* **2008**, *155*, 366–374. [CrossRef] [PubMed]
49. Wang, L.; Jia, H.; Liu, X.; Sun, Y.; Yang, M.; Hong, W.; Li, Y.F. Historical contamination and ecological risk of organochlorine pesticides in sediment core in northeastern Chinese river. *Ecotoxicol. Environ. Saf.* **2013**, *93*, 112–120. [CrossRef] [PubMed]

50. Doong, R.; Lee, S.; Lee, C.; Sun, Y.; Wu, S. Characterization and composition of heavy metals and persistent organic pollutants in water and estuarine sediments from Gao-ping River, Taiwan. *Mar. Pollut. Bull.* **2008**, *57*, 846–857. [CrossRef]
51. Salem, M.S.; Khaled, A.; Nemr, A.E. Assessment of pesticides and polychlorinated biphenyls in sediments of the Egyptian Mediterranean coast. *Egypt. J. Aquat. Res.* **2013**, *39*, 141–152. [CrossRef]
52. Peng, S.; Kong, D.; Li, L.; Zou, C.; Chen, F.; Li, M.; Cao, T.; Yu, C.; Song, J.; Jia, W.; et al. Distribution and sources of DDT and its metabolites in porewater and sediment from a typical tropical bay in the South China Sea. *Environ. Pollut.* **2020**, *267*, 115492. [CrossRef]
53. Alonso-Hernandez, C.M.; Mesa-Albernas, M.; Tolosa, I. Organochlorine pesticides (OCPs) and polychlorinated biphenyls (PCBs) in sediments from the Gulf of Batabano, Cuba. *Chemosphere* **2013**, *94*, 36–41. [CrossRef]
54. Kuranchie-Mensah, H.; Atiemo, S.M.; Palm, L.M.; Blankson-Arthur, S.; Tutu, A.O.; Fosu, P. Determination of organochlorine pesticide residue in sediment and water from the Densu river basin, Ghana. *Chemosphere* **2011**, *86*, 286–292. [CrossRef]
55. Aries, E.; Anderson, D.R.; Fisher, R. PCDD/F and Dioxin-like PCB emissions from iron ore sintering plants in the UK. *Chemosphere* **2006**, *65*, 1470–1480. [CrossRef] [PubMed]
56. Shibamoto, T.; Yasuhara, A.; Katami, T. Dioxin formation from waste incineration. In *Reviews of Environmental Contamination and Toxicology*; Springer: New York, NY, USA, 2007. [CrossRef]
57. Gioia, R.; Nizzetto, L.; Lohmann, R. Polychlorinated biphenyls (PCBs) in air and seawater of the Atlantic Ocean: Sources, trends and processes. *Environ. Sci. Technol.* **2008**, *42*, 1416–1422. [CrossRef] [PubMed]
58. Totten, L.A.; Gigliotti, C.L.; VanRy, D.A. Atmospheric concentrations and deposition of polychorinated biphenyls to the Hudson River Estuary. *Environ. Sci. Technol.* **2004**, *38*, 2568–2573. [CrossRef] [PubMed]
59. Gao, S.; Chen, J.; Shen, Z.; Liu, H.; Che, Y. Seasonal and spatial distributions and possible sources of polychlorinated biphenyls in surface sediments of Yangtze Estuary, China. *Chemosphere* **2013**, *91*, 809–816. [CrossRef]
60. Wang, M.; Wang, C.; Hu, X.; Zhang, H.; He, S.; Lv, S. Distributions and sources of petroleum, aliphatic hydrocarbons and polycyclic aromatic hydrocarbons (PAHs) in surface sediments from Bohai Bay and its adjacent river, China. *Mar. Pollut. Bull.* **2015**, *90*, 88–94. [CrossRef]
61. Quiroz, R.; Grimalt, J.O.; Fernández, P. Toxicity assessment of polycyclic aromatic hydrocarbons in sediments from European high mountain lakes. *Ecotoxicol. Environ. Saf.* **2010**, *73*, 559–564. [CrossRef]
62. Chen, Y.P.; Zhao, Y.; Zhao, M.M.; Wu, J.H.; Wang, K.B. Potential health risk assessment of HFRs, PCBs, and OCPs in the Yellow River basin. *Environ. Pollut.* **2021**, *275*, 116648. [CrossRef]
63. Dinc, B.; Çelebi, A.; Avaz, G.; Canlı, O.; Güzel, B.; Eren, B.; Yetis, U. Spatial distribution and source identification of persistent organic pollutants in the sediments of the Yeşilırmak River and coastal area in the Black Sea. *Mar. Pollut. Bull.* **2021**, *172*, 112884. [CrossRef]

toxics

Article

Polycyclic Aromatic Hydrocarbons (PAHs) in the Dissolved Phase, Particulate Matter, and Sediment of the Sele River, Southern Italy: A Focus on Distribution, Risk Assessment, and Sources

Paolo Montuori [1,*], Elvira De Rosa [1], Fabiana Di Duca [1], Bruna De Simone [1], Stefano Scippa [1], Immacolata Russo [1], Pasquale Sarnacchiaro [2] and Maria Triassi [1]

[1] Department of Public Health, "Federico II" University, Via Sergio Pansini no 5, 80131 Naples, Italy; elvira_derosa@libero.it (E.D.R.); fabianadiduca91@gmail.com (F.D.D.); desimonebruna7@gmail.com (B.D.S.); stefanoscippa923@gmail.com (S.S.); imrusso@unina.it (I.R.); triassi@unina.it (M.T.)

[2] Department of Law and Economics, "Federico II" University, Via Cinthia 26, 80126 Naples, Italy; sarnacch@unina.it

* Correspondence: pmontuor@unina.it

Abstract: The Sele River, located in the Campania Region (southern Italy), is one of the most important rivers and the second in the region by average water volume, behind the Volturno River. To understand the distribution and sources of polycyclic aromatic hydrocarbons (PAHs) in the Sele River, water sediment samples were collected from areas around the Sele plain at 10 sites in four seasons. In addition, the ecosystem health risk and the seasonal and spatial distribution of PAHs in samples of water and sediment were assessed. Contaminant discharges of PAHs into the sea were calculated at about 1807.9 kg/year. The concentration ranges of 16 PAHs in surface water (DP), suspended particulate matter (SPM), and sediment were 10.1–567.23 ng/L, 121.23–654.36 ng/L, and 331.75–871.96 ng/g, respectively. Isomeric ratio and principal component analyses indicated that the PAH concentrations in the water and sediment near the Sele River were influenced by industrial wastewater and vehicle emissions. The fugacity fraction approach was applied to determine the trends for the water-sediment exchange of 16 priority PAHs; the results indicated that fluxes, for the most part, were from the water into the sediment. The toxic equivalent concentration (TEQ) of carcinogenic PAHs ranged from 137.3 to 292.6 ngTEQ g^{-1}, suggesting that the Sele River basin presents a definite carcinogenic risk.

Keywords: polycyclic aromatic hydrocarbons; Sele River; fugacity; source; TEQ

check for updates

Citation: Montuori, P.; De Rosa, E.; Di Duca, F.; De Simone, B.; Scippa, S.; Russo, I.; Sarnacchiaro, P.; Triassi, M. Polycyclic Aromatic Hydrocarbons (PAHs) in the Dissolved Phase, Particulate Matter, and Sediment of the Sele River, Southern Italy: A Focus on Distribution, Risk Assessment, and Sources. *Toxics* **2022**, *10*, 401. https://doi.org/10.3390/toxics10070401

Academic Editor: Pierre-Marie Badot

Received: 13 June 2022
Accepted: 17 July 2022
Published: 19 July 2022

1. Introduction

Polycyclic aromatic hydrocarbons (PAHs) are a class of ubiquitous and persistent pollutants that are highly dangerous for humans, as they are carcinogenic and mutagenic. The extent of the potential risk and the distribution of PAHs in the environment is a public health issue [1,2]. With population development and economic growth, the input of PAHs intensified considerably in the 20th century; therefore, 16 PAHs have been identified as priority contaminants by the U.S. Environmental Protection Agency [3]. Seven of them, namely benz[a]anthracene, chrysene, benzo[b]fluoranthene, benzo[k]fluoranthene, benzo[a]pyrene, indo[1,2,3-cd]pyrene, and dibenzo[a,h]anthracene, are potentially carcinogenic to humans according to the International Agency for Research on Cancer [4,5]. In addition, four PAHs (benzo[a]pyrene, benz[a]anthracene, benzo[b]fluoranthene and chrysene) were recently defined as the main indicators of the presence of genotoxic and mutagenic PAHs in the environment and, in particular, in food [6]. The majority of the PAH load in the environment is from the combustion of organic matter (pyrolytic origin), which is usually released from human activities, such as coal combustion, petrol and diesel oil combustion, industrial processes, and home heating. Other types of non-anthropogenic sources, such as petrogenic and diagenetic origins, are relatively less abundant [7,8]. PAHs introduced into the aquatic environment move from the water into the sediment due to their chemical and physical properties; in particular, the high-molecular-weight PAHs, consisting of several aromatic rings, have a greater tendency to bind to the sediment. In contrast, low-molecular-weight (LMW) PAHs, consisting of few aromatic rings, degrade faster, and their concentrations in surface water and sediments are relatively low [9]. The partitioning of PAHs in water and sediment is one of the major processes controlling the toxicity of PAHs in aquatic environments [10]. Over the years, numerous studies have been conducted on important rivers in central and southern Italy to evaluate and estimate the PAH levels in the water, suspended particulate matter, and sediment; toxicity was also assessed to verify the harmful effects on the environment and the possible biological risks for living organisms in the watercourses.

Beginning with central Italy, the Tiber River has concentration ranges of 10.3 to 951.6 ng/L (DP + SPM) and 36.2 to 545.6 ng/g for the sediment, with a relatively low toxicity [11]. In southern Italy, we find the Volturno River, with concentration ranges from 256.0 to 1686.3 ng/L (DP + SPM) and 434.8 to 872.1 ng/g for the sediment, with a toxicity value that highlights an area possibly at risk [12]. In the Sarno River, on the other hand, ranges of 23.1 to 2670.4 ng/L (DP + SPM) and 5.3 to 678.6 ng/g for sediment are found, with toxicity values that do not indicate an area experiencing immediate biological effects [13]. Qu et al. [14] studied the Gulfs of Salerno and Naples, reporting concentrations for the sediment from 9.58 to 15.81 μg/kg for the Bagnoli area, 317 μg/kg for the Salerno area, and 768.0 μg/kg for the Gulf of Naples area, with significant toxicity and biological risk values. Campania is one of the most populated regions of Italy, with over half of its population concentrated in metropolitan areas such as Naples and Salerno. Currently, industrial activity, agricultural practices, and illegal waste disposal represent difficult problems in the effort to mitigate the high levels of contamination in the Campania plain [14,15]. The plain is dominated by the presence of numerous industrial activities, including dairies, canning, and chemical industries. In addition, there are many contaminated sites, both landfills and illegal disposal areas. There are also well-developed agricultural activities in the region, such as livestock farming (buffalo farms); the large-scale production of vegetables and fruits feeds the local food industry. In areas where mainly agricultural and livestock products are processed, the emission of waste with high amounts of organic and inorganic substances can impact ecological and environmental integrity [13,16]. The Gulf of Salerno is one of the main environments in which pollutants accumulate from the Campania Plain. The Sele River is an important river in the Campania region; it has a length of 64 km and is the second in the region and the south of Italy by average water volume, behind the Volturno River.

The current paper reports the concentrations of PAHs in the water and sediment of the Sele River in the Gulf of Salerno (central Mediterranean Sea), southern Italy. The specific objectives of the present study are to: (I) investigate the contamination levels and spatial distribution of PAHs, (II) identify their potential sources, and (III) estimate the environmental risk in this area.

2. Materials and Methods

2.1. Study Area

The Sele River basin (3236 km^2) is located on the western (i.e., Tyrrhenian) side of southern Italy and includes a large alluvial plain. The plain has a triangular surface area of about 400 km^2. It is delimited offshore by a narrow sandy coastal strip between the towns of Salerno (NW) and Agropoli (SE); landward, it is delimited to the north and northwest by the Lattari and Picentini Mountains and to the southeast by the Alburni Mountains and the Cilento Promontory (Figure 1) [17]. The climate in the Sele basin is of Mediterranean type, with important spatial variations in both erosive rainfall and temperature according to the elevation and the distance from the coast. The Mediterranean climate is characterized by mild temperatures. It is a particularly dry climate in summer and mild in winter. Rainfall is concentrated from autumn to spring, and in the driest month of the year is less than 30 mm, which is about a third of the wettest month. The lack of rainfall in the summer, with at least two consecutive months of drought, is a peculiarity of the Mediterranean climate. In other climate classifications, precipitation is concentrated in the hot season. In the Mediterranean climate, the sea contributes to determining the climate, which is warm temperate, with modest daily and annual temperature ranges (less than 21 °C); in fact, the sea retains the summer heat, accumulating and then releasing it during the winter. The combination of dry summers and rainy winters is a typical characteristic of the Mediterranean climate.

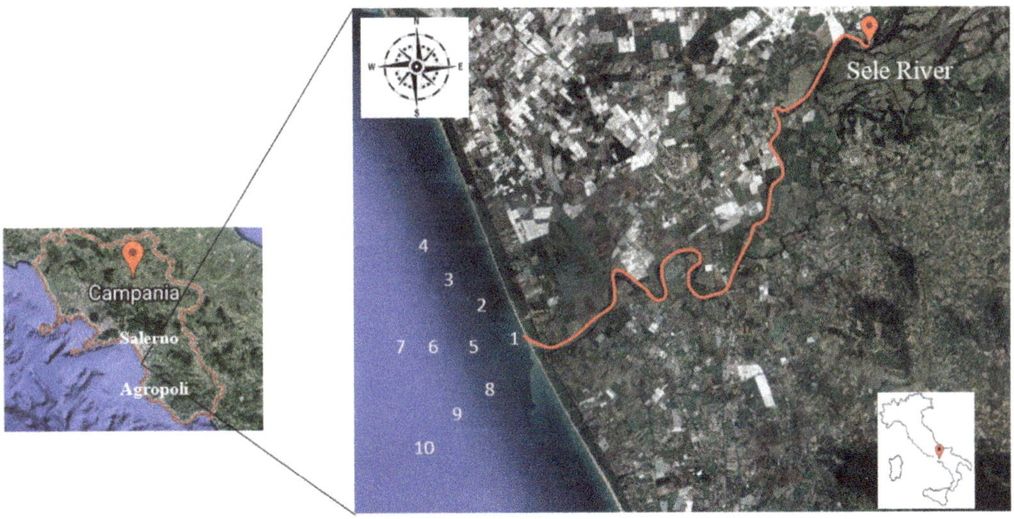

Figure 1. Map of the study area and sampling sites along the Sele River and estuary, southern Italy.

An increase in population density, high industrial pollution, the presence of road and railway networks, and an increasing influx of tourists to the city of Salerno have caused an increase in environmental pollution in this area [18].

2.2. Sampling

A total of 40 surface water samples and 10 sediment samples were collected in the summer, autumn, winter, and spring of 2020–2021 from 10 sampling locations along the Sele River (Table 1). For each season and at each sampling point, three sample aliquots were taken. This process was repeated in duplicate. The aliquots were transported to the laboratory and analyzed in triplicate to calculate the standard deviation and evaluate the repeatability of the method.

The first sampling point was the river mouth, with the purpose of assessing downstream pollution; in addition, nine other points were sampled at 500, 1000, and 1500 mt away from the river mouth to evaluate the impact of Sele River pollution on the Mediterranean Sea environment (Figure 1). The samples were collected in 2.5 L amber bottles, using 6 M of hydrochloric acid, from the surface layer at a depth of 0–50 cm from the sampling locations, while sediment samples were collected at 0–5 cm with a Van Veen Grab sampler and preserved in aluminum boxes. All samples were temporarily stored in refrigerated containers containing crushed ice until they were transported back to the laboratory and preserved at −20 °C until analysis.

Table 1. Description of the sampling sites and concentrations of PAHs in the water-dissolved phase (DP), suspended particulate matter (SPM), and sediment of the Sele River, southern Italy. (The values in brackets represent the values of PAH concentrations in SPM expressed in ng g^{-1} dry wt, after drying the filters in an air-heated oven and weighing them).

Sampling Location			**ΣPAHs**								
Site Number Identification	**Site Characteristics**	**Site Location**	**Dissolved Phase (ng L^{-1})**				**Particulate Phase (ng L^{-1}) (ng g^{-1} Dry wt)**				**Sediment (ng g^{-1} Dry wt)**
			Apr	**Jul**	**Nov**	**Feb**	**Apr**	**Jul**	**Nov**	**Feb**	**Apr**
1 (river water)	Sele river source	40°28′55″ N 14°56′33″ E	419.3	567.2	487.3	309.9	520.1 (41,364.1)	276.1 (28,122.3)	234.8 (19,865.6)	654.3 (23,487.2)	871.1
2 (sea water)	River mouth at 500 mt north	40°29′04″ N 14°56′14″ E	204.2	387.3	471.2	200.0	332.3 (30,542.6)	144.8 (18,657.1)	138.3 (6068.5)	381.0 (26,589.1)	712.4
3 (sea water)	River mouth at 500 mt central	40°29′12″ N 14°55′56″ E	226.5	552.3	408.2	331.8	233.9 (74,510.7)	182.3 (26,789.8)	128.3 (58,745.8)	277.7 (16,895.9)	724.3
4 (sea water)	River mouth at 500 mt south	40°29′20″ N 14°55′38″ E	487.3	560.2	509.1	334.1	504.2 (41,263.6)	261.2 (48,756.3)	181.2 (5986.8)	507.2 (47,596.2)	852.2
5 (sea water)	River mouth at 1000 mt north	40°28′55″ N 14°56′12″ E	309.5	497.3	424.4	121.9	370.3 (29,865.2)	125.1 (11,587.3)	204.5 (13,501.6)	589.9 (2843.2)	649.5
6 (sea water)	River mouth at 1000 mt central	40°28′55″ N 14°55′50″ E	227.3	498.3	529.3	249.7	328.7 (10,859.8)	214.7 (65,741.0)	190.2 (18,459.2)	461.1 (14,896.2)	708.1
7 (sea water)	River mouth at 1000 mt south	40°28′55″ N 14°55′28″ E	302.1	499.2	502.6	262.3	467.6 (36,587.2)	294.9 (24,189.2)	188.2 (10,453.2)	369.1 (4875.2)	744.3
8 (sea water)	River mouth at 1500 mt north	40°28′47″ N 14°56′16″ E	300.2	112.3	289.7	10.1	367.9 (19,845.5)	121.9 (10,354.3)	219.0 (16,181.1)	192.3 (5489.5)	331.7
9 (sea water)	River mouth at 1500 mt central	40°28′39″ N 14°55′56″ E	361.7	331.2	424.8	175.9	482.1 (86,412.3)	240.2 (66,587.4)	185.7 (58,476.5)	277.9 (13,489.2)	602.1
10 (sea water)	River mouth at 1500 mt south	40°28′30″ N 14°55′38″ E	471.0	489.3	509.1	207.1	545.8 (85,647.1)	387.3 (29,875.1)	173.2 (39,485.2)	451.7 (8746.2)	683.2

2.3. Extraction and Analysis

The samples collected were transported to the laboratory within 24 h, and they were filtered through 47 mm × 0.7 μm glass fiber filters (Whatman, Maidstone, UK) that had been heated at 400 °C overnight to separate the water from the suspended particulate matter (SPM). The dissolved phase PAHs were extracted from water samples using a solid-phase extraction (SPE) cartridge by Oasis HLB (6 mL, 500 mg; Waters, Milford, MA, USA), according to the method proposed by Liu et al. [19]. C18 SPE cartridges (to elute and concentrate) were pre-washed with dichloromethane (DCM) before conditioning with methanol and ultrapure water; then, 10 μL of surrogate standard (benzo[a]pyrene-d_{12} and indeno[1,2,3-cd]pyrene-d_{12}) was added to 1 L of the water sample before mixing. Following that, the water sample was passed through a column for concentration at a flow rate of 3 mL/min. Next, a vacuum pump was used to dry the column. The eluate was concentrated to 0.5 mL using a nitrogen flow before 10 μL of internal standard (Chrysene-d_{12}) was added, followed by GC/MS analysis.

SPM content was determined by gravimetry. First, the filter was dried in an air-heated oven (55 °C until constant weight) and equilibrated at room temperature in a desiccator. Filters were then spiked with three surrogate standards (10 ng of chrysene-d12, benzo[a]pyrene-d12, and indeno[1,2,3-cd]pyrene-d12) and extracted three times by sonication with 10 mL of dichloromethane-methanol (1:1) for 15 min. After extraction, the extracts were concentrated using a rotary evaporator. The volume of the extracts was adjusted to 0.5 mL and solvent-exchanged into hexane. Cleanup and fraction procedures were performed with open column chromatography (3 g of neutral alumina deactivated with 3% (*w*/*w*) Milli-Q water). Three fractions were collected: fraction I with 5.5 mL of hexane, fraction II with 6 mL of hexane:ethylacetate (9:1), and fraction III with 12 mL of ethylacetate. PAHs were eluted in fraction II, while fractions I and III contained other organic pollutants that were also detected in the samples. The sediment samples were air-dried in the dark for 5 days, crushed, sieved (250 μm particles were used as the sample), and divided into 5 g portions. The PAH concentrations in the sediment samples were calculated according to dry weight (ng/g dw) [20,21]. PAHs were extracted from the filters and sediment samples using a Soxhlet extractor (Table S1). As in [22], the samples were draped onto a filter paper, placed into the cellulose extraction thimble, and covered with cotton wool. The thimble was located inside the main Soxhlet chamber and fitted to a 250 mL round-bottomed flask containing methylene chloride (150 mL). A condenser was then attached. The samples were extracted for 24 h under reflux. The extracts were purified through a column composed of 1 g of sodium sulfate and 2.5 g (10% deactivated) of silica gel and eluted with 70 mL of a hexane:methylene chloride (7:3) solution. The extracts were evaporated to dryness, reduced to a final volume (500 μL) using flushing nitrogen gas, and chrysene-d_{12} was added as an internal standard. To evaluate the organic carbon normalized partition coefficients (Koc′), which estimate PAH attraction to sediment and define the sediment-water partitioning level, the total organic carbon (TOC) content of the sediments was analyzed using a TOC analyzer (TOC-VCPH, Shimadzu Corp., Kyoto, Japan).

2.4. *Instrumental Analysis*

All samples were analyzed on a gas chromatograph with a mass spectrometer detector (TRACE^{TM} 1310 Gas Chromatograph coupled to an ISQ^{TM} 7000 Single Quadrupole Mass Spectrometer, Thermo Scientific, Waltham, MA, USA) to determine PAHs with selected ion monitoring (SIM) (Table S2). A TG-5MS capillary column with 30 mm length × 0.25 mm inner diameter × 0.25 μm film thickness was used. The column temperature was programmed to rise from 60 °C to 200 °C for 2 min at 25 °C min^{-1}, then to 270 °C at 10 °C min^{-1} (maintained for 6 min), and finally, to 310 °C at 25 °C min^{-1} (maintained for 10 min). The mass spectrometer was operated in the electron ionization (EI) mode set at 70 eV, and the injector and detector temperatures were 280 °C and 300 °C, respectively (Table S3). Acquisition was carried out in the single ion monitoring mode (SIM) using two characteristic ions for each target analyte. Target analytes were identified and verified by comparing the retention times of the samples with standards and using the characteristic ions and their ratios for each target analyte. Furthermore, for the more highly concentrated samples, the identification of target analytes was confirmed in full-scan mode (m/z range from 60 to 350), and the analytes were quantified using the characteristic ions and their ratios for each target analyte. The concentrations of 16 PAHs were determined: naphthalene (Nap), acenaphthene (Ace), acenaphthylene (Acy), fluorine (Flu), phenanthrene (Phe), anthracene (Ant), fluoranthene (Fla), pyrene (Pyr), benz[a]anthracene (BaA), chrysene (Chr), benzo[b]fluoranthene (BbF), benzo[k]fluoranthene (BkF), benzo[a]pyrene (BaP), dibenz[a,h]anthracene (DahA), indeno[1,2,3-cd]pyrene (IcdP), and benzo[g,h,i]perylene (BghiP). PAH quantification was performed using a five-point calibration curve (5–25–100–500–1000 ng/L) for the 16 PAHs (Dr. Ehrenstorfer GmbH, Augsburg, Germany) (r^2 > 0.97), and chrysene-d_{12} was used as an internal standard. The quantification of individual compounds was determined by the comparison of peak areas with those of the recovery standards. The samples were analyzed in triplicate. For the water-dissolved phase samples (final concentration in water of 10 ng L^{-1}), after passage through a column, the eluate was concentrated to 0.5 mL using a nitrogen flow before 10 μL of internal standard (chrysene$_{d12}$) was added.

The PAH concentrations in the sediment samples were calculated according to dry weight (ng/g dw). PAHs were extracted from the filters and sediment samples using a Soxhlet extractor. The samples were extracted for 24 h under reflux. The extracts were evaporated to dryness, reduced to a final volume (500 μL) using flushing nitrogen gas, and chrysene$_{d12}$ was added as an internal standard.

The detection limit (LOD) was calculated as three times the noise in a blank sample chromatogram. In the water and SPM, LODs ranged from 1.3 to 1.6 ng L^{-1}; in sediment samples, they ranged from 1.5 to 1.9 ng g^{-1}. The quantification limits (LOQ) were in the range of 4.8–5.4 ng L^{-1} in the water and SPM samples and 5.1–6.3 ng g^{-1} in the sediment samples (Tables S4 and S5). A total of ten blanks were analyzed in the same manner as the samples; the PAHs in the blanks showed a concentration below the LOD. The recovery of PAHs in the standard checks and samples was between 70% and 130%, which met quality control requirements. For the effective and reproducible detection and quantification of low concentrations of PAHs in water, several parameters were determined, such as linear range (5–25–100–500–1000 ng/L), precision, limit of detection, and limit of quantification. The precision of the method was determined through repeatability studies and was expressed as relative standard deviation (RSD). The average of the results was used to estimate the precision of the method. The RSD was determined by analyzing one sample on the same day, with the same instrument, and by the same analyst under identical conditions.

2.5. Water-Sediment Partitioning

Water-sediment partitioning is an important environmental process that can be used to evaluate the equilibrium partition behavior of PAHs in aquatic environments [10,19]. The K_{OW} (octanol-water partition coefficient) is the coefficient expressing the lipophilicity or carbon affinity of a chemical, and it is related to the distribution coefficient so as to describe the fate of environmental pollutants such as PAHs [10]. Organic carbon normalized partition coefficients (Koc) estimate PAH attraction to sediment and define the sediment-water partitioning level [23–25]. In order to assess the behavior of PAHs in the Sele River area, in situ organic carbon coefficients (K_{oc}') were calculated by Equation (1) [26,27]:

$$K_{oc}' = C_S/(C_{aq} \times f_{oc}) \tag{1}$$

where C_s and C_{aq} are the PAH concentrations in the solid and liquid phases, respectively, and f_{oc} is the percentage of organic carbon in the sediment.

The difference between log K_{oc}' and the corresponding log K_{oc} indicates the equilibrium state of PAHs in an aquatic system [28].

If the average log K_{oc}' is lower than the corresponding K_{oc} and K_{ow}, PAHs are more absorbed into the sediment phase than exchanged into the water phase [27].

The movement of a chemical from one area to another is monitored by fugacity. For this reason, the exchange processes of PAHs between water and sediment were estimated by the fugacity fraction [2,29]. In Equation (2), ff is defined:

$$ff = K_{oc}'/(K_{oc}' + K_{oc}) \tag{2}$$

A value of $ff < 0.3$ indicates that PAHs are adsorbed into the sediment from water and that sediments act as a sink for PAHs. Values in the range $0.3 < ff < 0.7$ describe sediment-water equilibrium, and when $ff > 0.7$, a flux from sediment to water is predicted, and sediments act as a secondary emission source of PAHs [19,28].

2.6. Risk Assessment and Determination of Toxicity

2.6.1. Biological Adverse Effects

In sediment, PAHs can be very dangerous to life in the aquatic ecosystem and a source of pollutants that accumulate in the food chain [29].

In this study, the sediment quality guidelines (SQGs) were used to estimate the potentially toxic effects of contaminants in the sediment samples on animals and marine organisms [30].

The SQGs estimate the toxicity that these contaminants cause to the aquatic environment based on the following ranges: effects range low (ERL)/effects range median (ERM) and threshold effects level (TEL)/probable effects level (PEL) [31,32].

ERL and TEL classifications correspond to chemical amounts below which the probability of toxicity and other effects are low. In contrast, the ERM and PEL classifications represent a mid-range above which negative effects are likely to occur. ERL-ERM and TEL-PEL classifications represent a possible effects range within which adverse effects sometimes occur [13,33].

2.6.2. Toxicity Determination

Marine sediments are considered a contaminant pool of PAHs, and the potential toxicity of PAHs, in particular carcinogenic PAHs (C-PAHs), in the aquatic environment may threaten human health [34]. This study evaluated the potential impact of C-PAHs based on BaP toxic equivalency factors (TEFs). The toxic equivalent quantity (TEQ) of ΣPAHs was determined through the following equation:

$$TEQ_{PAHs} = \sum_i TEF_i \times C_{PAHi} \quad (3)$$

where TEF_i (toxic equivalency factor) is the toxic factor of each carcinogenic PAH relative to BaP and C_{PAHs} and represents the concentration of an individual carcinogenic PAH.

The TEF_s values determined by the U.S. EPA [22] for each carcinogenic PAH are as follows: 0.1 for BaA, 0.001 for Chr, 0.1 for BbF, 0.01 for BkF, 1 for BaP, 0.1 for IcdP, and 1 for DahA [35].

2.7. Identifying the Source of PAHs

PAHs mainly derive from industrial processes and incomplete combustion by various industrial activities, such as waste incineration, iron and aluminum production, cement manufacturing, dye manufacturing, and asphalt industries, as well as from vehicle emissions and other anthropogenic activities [36,37].

Identifying the possible sources of PAH pollution is an important objective for the institutions seeking to collect information on how to control the pollution caused by these pollutants.

The sources of PAHs, whether from fuel combustion (pyrolytic) or from crude oil (petrogenic) contamination, may be determined by the ratios of specific PAH compounds based on peculiarities in PAH composition and distribution as a function of the emission source. The diagnostic ratios of selected PAHs were utilized to distinguish PAHs from pyrogenic and petrogenic sources. For example, HMW/LMW PAHs, Flu/(Flu + Pyr), IcdP/(IcdP + BghiP), BaA/(BaA + Chr), Ant/(Ant + Phe), and BbF/BkF were applied for PAH source identification [38,39].

LMW contaminants are more common in samples containing petrogenic PAHs, and HMW contaminants are common in samples containing pyrogenic PAHs; this is because most of the HMW molecules are formed at higher temperatures [40,41].

In this study, the principal component analysis (PCA) technique was used to quantitatively explore PAH origins. PCA was used as a multivariate analytical tool to reduce a set of original variables (measured PAH content in the sediment samples) and to extract a small number of latent factors (principal components, PCs) for analyzing relationships among the observed variables. As a result of an effective ordination process, the first PC accounts for the greatest proportion of the original variance, while the second and subsequent PCs progressively explain smaller amounts of data variation [42,43].

3. Results and Discussion

3.1. PAH Distribution in Water, SPM, and Sediment

Analysis of samples collected from the Sele River showed the presence of various PAHs in surface water, SPM, and sediment; the mean concentrations of ΣPAHs were 10.1–567.2 ng/L, 121.9–654.3 ng/L, and 331.7–871.1 ng/g, respectively (Table 1). The high anthropogenic pressure of the city of Salerno is evident in the presence of large food facilities and a vast industrial zone; in particular, the environment surrounding the Sele River is characterized by industrial districts, urban areas, intensive cultivations, and agricultural crops [44,45].

The concentrations of total PAHs in the water-dissolved phase (DP) detected at 10 locations along the Sele River and its estuary ranged from 3.4 to 98.5 ng L^{-1} for two-ring PAHs (Nap), from 22.7 to 164.2 ng L^{-1} for three-ring PAHs (Acy, Ace, Flu, Phe, and Ant), from 1.2 to 24.2 ng L^{-1} for four-ring PAHs (Fla, Pyr, BaA, and Chr), from 9.4 to 37.2 ng L^{-1} for five-ring PAHs (BbF, BkF, BaP, and DahA), and from 17.6 to 44.5 ng L^{-1} for six-ring PAHs (BghiP and IcdP) (Table S6).The compositional pattern of PAHs in the dissolved phase indicates that two- and three-ring PAHs were abundant at all sampling sites, representing, on average, over 60% of all PAHs. The predominance of low-molecular-weight

PAHs (two-three-ring) in the water may be explained by their high water solubility and relatively high vapor pressures [46,47] (Figure S1).

The PAHs detected in SPM ranged from 3.2 to 62.3 ng L^{-1} for two-ring PAHs (Nap), from 21.2 to 58.3 ng L^{-1} for three-ring PAHs (Acy, Ace, Flu, Phe, and Ant), from 38.7 to 190.9 ng L^{-1} for four-ring PAHs (Fla, Pyr, BaA, and Chr), from 26.5 to 105.0 ng L^{-1} for five-ring PAHs (BbF, BkF, BaP, and DahA), and from 18.7 to 68.2 ng L^{-1} for six-ring PAHs (BghiP and IcdP) (Table S7). The compositional profiles of PAHs in SPM show that four-, five-, and six-ring PAHs were abundant at most sampling sites, accounting for 67% of ΣPAHs in SPM. Therefore, the higher PAH concentrations found in SPM may derive from PAH particles suspended in the air because the Sele River drainage basin passes through large agricultural areas and large industrial areas in southern Italy; these areas contain agri-food industries, chemical plants, and manufacturing industries. The emission of atmospheric particles from intensive agricultural activities and factories also causes serious air pollution, and the particulate-associated PAHs may be transported and deposited into the river [48,49] (Figure S1).

In sediment samples, the results ranged from 2.2 to 35.1 ng g^{-1} for two-ring PAHs (Nap), from 30.2 to 137.2 ng g^{-1} for three-ring PAHs (Acy, Ace, Flu, Phe, and Ant), from 59.2 to 241.2 ng g^{-1} for four-ring PAHs (Fla, Pyr, BaA, and Chr), from 191.3 to 490.3 ng g^{-1} for five-ring PAHs (BbF, BkF, BaP, and DahA), and from 11.2 to 109.4 ng g^{-1} for six-ring PAHs (BghiP and IcdP) (Table S8). In terms of individual PAHs in sediment, the composition characteristics were different from those in SPM and water (Figure S1). In sediment samples, the results showed the prevalence of four- and five-ring PAHs at most sites sampled, accounting for 36% and 42% of ΣPAHs in sediments, respectively. Liu et al. [50] reported a similar distribution of PAHs between water and sediment, confirming that HMW PAHs were mainly absorbed by sediment. This difference in distribution between water and sediment may be due not only to water solubility but also to bacterial degradation. In fact, the water solubility of PAHs probably decreases as the number of attached benzene rings increases, suggesting that HMW PAHs are less easily mobilized from solid substrates and dissolved into aquatic media than LMW PAHs and, as a result, they are less receptive to biodegradation. Instead, LMW PAHs have high solubility in water and greater benthic recycling, and were, therefore, more concentrated in the dissolved phase [51–53].

Such differences in pollutant composition between individual PAHs may be caused by different input methods and characteristics of PAHs. Firstly, river water receives direct PAH inputs from various sources, including wastewater discharge, runoff, atmospheric fallout, and so on. Secondly, low-molecular-mass PAHs gradually decrease as a result of degradation and adsorption, and only those PAHs that have relatively high molecular mass and are more resistant to degradation can resist such pressures to reach the sediment bed. Thirdly, water conditions, which change with the seasons, change the state of the water column process by mechanisms including dissolution, adsorption, desorption, degradation, and deposition [2,54,55].

Spatial distribution data showed that concentrations reached their peak values at sites near the river mouth, while the levels of PAHs at other sites decreased from location one (river mouth) to four (1500 mt). In the Tyrrhenian Sea, PAH concentrations ranged in general from high values near the river outflow to low values in offshore areas (Figure 2). At 500 mt of river outflow, the PAH concentrations were close to those at the Sele mouth (Figure 2). The concentrations at the sampling sites then decreased at 1000mt from the river outflow and more still at 1500 mt. From the Sele mouth, the PAH load moved into the Tyrrhenian Sea southward (Figure 2). As can be seen from the results obtained, the trend in the concentrations indicated a decrease from the mouth towards 1500 mt at sea. This may depend both on the flow of the river, which varies according to the season, and on the diluting effect of the sea. The seasonal variation in PAH concentrations depends on the hydrological conditions, which may cause dilution ratio variations [56,57]. Therefore, a high river flow rate resulted in a higher dilution ratio during the wet season floods and caused a decrease in the PAH concentrations in both the Sele River and its estuary. In the area of the sea where the sampling sites are located, the flow direction of the seawater moves south. The marine flow influences the concentrations and the distribution of the PAHs, which also change according to the seasons. This is influenced by the currents and characteristic winds of the Mediterranean Sea [58]. When the flow of the sea changes, several factors change that can contribute to altering the concentrations of the studied compounds: temperature, salinity, and often also "color" (more or less cloudy) [59]. These factors lead to changes in the density of the water. The occurrence of flow events implies a high presence of suspended solid matter of terrestrial origin, as well the resuspension of sediments caused by turbulence and the transport of the associated PAHs downstream. In contrast, when low-flow conditions predominate and previous flood events are long past, the settling of suspended matter and the associated storage of PAH particles in the sediment are favored [60]. The PAH sources present in the study area that may contain seawater are represented by the various industries present in the area and agricultural activities as well. The results showed that the PAH concentrations in DP decreased from July to February, in parallel with the increase in rainfall, which could cause dilution ratio variations. Therefore, the decrease in PAH concentrations moving from the Sele River mouth to the Mediterranean Sea was also affected by the high flow in the rainfall season, which resulted in an even higher dilution ratio. The lowest concentrations in SPM were recorded in the dry season (July) due to the decrease in flow and the greater stagnation of SPM, which led PAHs with a greater polarity to shift from SPM to DP (Figure 3).

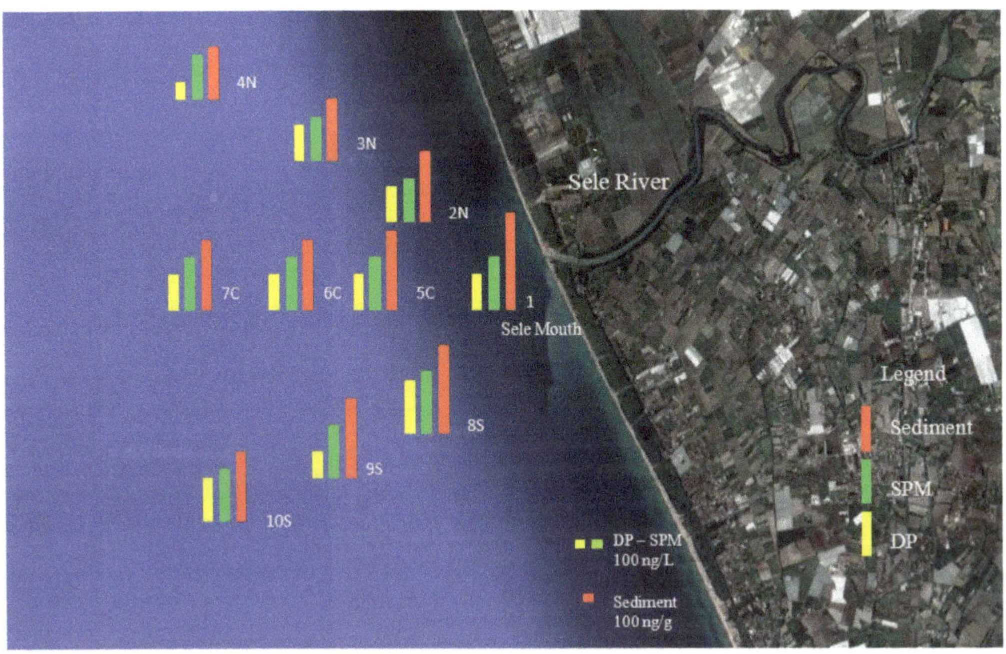

Figure 2. Spatial and temporal distributions of PAHs in the water-dissolved phase (DP, ng L^{-1}), suspended particulate matter (SPM, ng L^{-1}), and sediment (ng g^{-1} dry wt) of the Sele River and estuary, southern Italy.

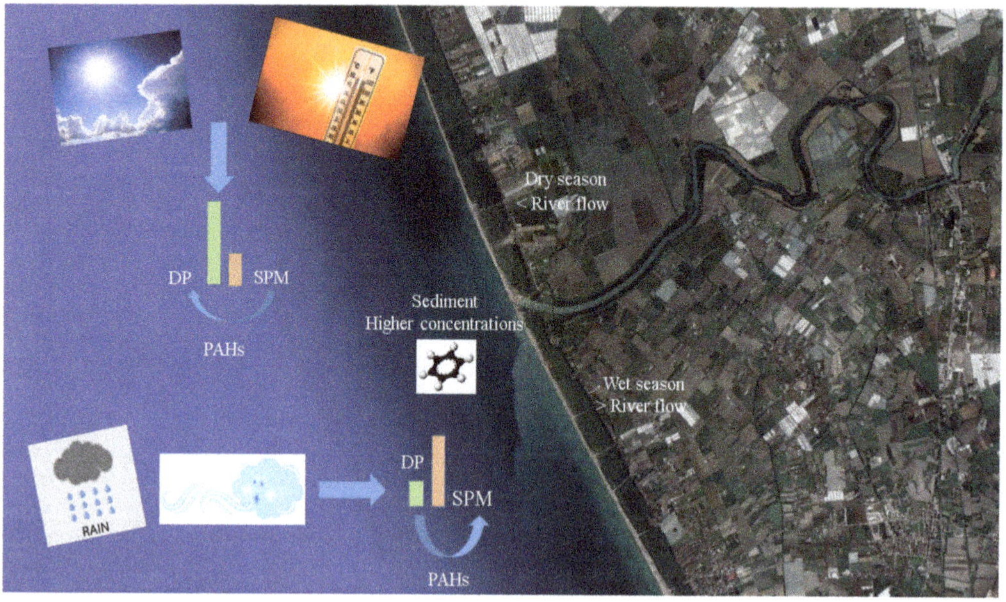

Figure 3. PAH distribution in water, suspended particulate matter, and sediment.

Based on these results, it can be concluded that the loads and migrations of PAHs between different phases at each sampling site of the Sele River were related to variations in the flow during rainy and dry seasons. Therefore, a high concentration of PAHs in sediments indicated that the contamination of PAHs in the Sele River and its estuary might be caused by the historical input of PAHs. The total load of PAHs that flowed into the Tyrrhenian Sea was evaluated to estimate the input of PAHs drained from rainwater outflow, tributary inflow, wastewater treatment plants, industrial effluent discharge, agricultural runoff, atmospheric deposition, and dredged material disposal. The total PAH loads contributed to the Tyrrhenian Sea from the Sele River were calculated considering the concentration values of the individual PAHs at the river mouth in the four months of sampling. The mean of the total concentrations was then multiplied by the annual average flow rate (m^3/year) of the Sele River. The load was calculated as about 1807.9 kg/year.

3.2. PAH Fugacity in the Aquatic System

Since water and sediment in aquatic ecosystems are subject to dynamic equilibration, it would be useful to identify the transport processes and fate of PAHs so that an estimation of the distribution between water and sediment could yield useful information. The values obtained in this study for the sediment-water equilibrium partitioning coefficient (log K_{oc}), in situ sediment-water distribution coefficient (log K_{oc}'), and fugacity fraction (ff) of PAHs at the 10 sampling sites are shown in Table 2.

Table 2. Comparison of log K_{oc} and log K'_{oc} for polycyclic aromatic hydrocarbons (PAHs) at the water-sediment interface and the fugacity fraction (*ff*) in the study area.

PAHs	log K_{oc} [a]	log K'_{oc} (Mean)	*ff*
Nap	3.11	3.25	0.05
Any	3.51	3.78	0.10
Ace	3.43	4.15	0.06
Flu	3.70	3.58	0.04

Table 2. *Cont.*

PAHs	log K_{OC} [a]	log K'_{OC} (Mean)	*ff*
Phe	3.87	4.22	0.06
Ant	3.40	4.00	0.06
Fla	3.70	4.79	0.09
Pyr	4.66	3.88	0.08
BaA	5.30	4.29	0.12
Chr	5.43	4.05	0.18
Bbf	5.36	1.21	0.27
Bkf	5.57	1.18	0.23
BaP	5.61	2.22	0.12
IcdP	6.64	0.41	0.28
DahA	6.22	2.10	0.10
Bghip	6.90	0.83	0.05

[a] Guo et al. [61].

The mean values of log K'_{oc} ranged from 0.41 to 5.99, but the average log K'_{oc} values for PAH compounds, except two-three-ring PAHs, were lower than their corresponding log K_{oc}. This indicates that these compounds were saturated in the water-dissolved phase, and their net flux was from the water into sediment. Overall, the difference between in situ log K'_{oc} and the corresponding log K_{oc} indicated non-steady-state conditions for the PAHs in the water-sediment system, but the differences between the log K'_{oc} and log K_{oc} values of HMW PAHs were relatively large, suggesting that the non-steady-state increased for HMW PAHs. In fact, the difference between log K_{oc}' and the corresponding log K_{oc} indicates the equilibrium state of PAHs in the aquatic system [28]. If the average log K_{oc}' is lower than the corresponding Koc and Kow, PAHs are more absorbed into the sediment phase than exchanged into the water phase [27]. Therefore, LMW PAHs were usually dominant in water, and they tend to be released from SPM to water; HMW PAHs were prevalent in sediment, and they tend to be adsorbed onto SPM from water [28,62].

The fugacity fraction ff was used to evaluate the equilibrium status of the organic pollutants and to better understand the interactions between the phases [27,63]. In the sediment-water of the Sele River, the ff values of the 16 PAHs were 0.04–0.28, i.e., lower than 0.3, causing a net flux of these PAHs from the water into the sediment.

3.3. Risk Assessment of PAHs

Several evaluation tools, such as sediment quality guidelines (SQGs) and the toxic equivalent quotient (TEQ), are frequently used for preliminary analysis and evaluation of the ecological risk faced by aquatic environments. These methods can rapidly and effectively evaluate the potential risk level to aquatic organisms induced by contaminant concentrations in the environmental medium.

In the Sele River, the obtained data showed concentrations of PAHs lower than the PEL and ERM values; however, for TEL and ERL values, not all concentrations were lower (Table 3). Moreover, the seasonal differences also influenced the risk assessment, which was higher for the compounds analyzed (DP and SPM) in July and lower in February, in relation to the concentrations found.

For individual compounds, TEL values were higher for Acy, Ace, and DahA for all samples; for Nap and Flu in 20% of samples; and for Bap in 70% of samples, indicating that adverse effects may occasionally exist. However, the mean concentrations of the detected PAHs were lower than their respective PEL values.

The amounts of individual PAHs did not exceed their respective ERM values, but the ERL values were exceeded for Ace in 50% of samples, for Flu in 20% of samples, and for DahA in all samples. The data obtained confirmed the presence of PAHs at some sites, showing that the environmental integrity of Sele River was at risk.

Table 3. A comparison of the TEL, PEL, ERL, and ERM guideline values (μg Kg^{-1}) for polycyclic aromatic hydrocarbons and the data found for the Sele River, southern Italy.

	PAHs																
	Nap	**Acy**	**Ace**	**Flu**	**Phe**	**Ant**	**Fla**	**Pyr**	**BaA**	**Chr**	**BbF**	**BkF**	**BaP**	**DahA**	**BghiP**	**IcdP**	**∑PAHs**
TEL [a]	34.6	5.87	6.71	21.2	86.7	46.9	113	153	74.8	108	-	-	88.8	6.22	-	-	1684
Percentage of samples over the TEL	20	100	100	40	0	0	0	0	0	0			70	100			0
PEL [a]	391	128	88.9	144	544	245	1494	1398	693	846	-	-	763	135	-	-	16770
Percentage of samples over the PEL	0	0	0	0	0	0	0	0	0	0			0	10			0
ERL [b]	160	44	16	19	240	85	600	665	261	384	-	-	430	63.4	-	-	4022
Percentage of samples over the ERL	0	0	50	20	0	0	0	0	0	0			0	100			0
ERM [b]	2100	640	500	540	1500	1100	5100	2600	1600	2800	-	-	1600	260	-	-	44792
Percentage of samples over the ERM	0	0	0	0	0	0	0	0	0	0			0	0			0

[a] Long et al. [64]. [b] MacDonald et al. [65].

In this study, the total TEQ_{PAHs} ranged from 137.3 to 292.6 ngTEQ/g; the highest values were measured at the river mouth and site eight, while all other sampling sites presented TEQ_{PAHs} values under the safe level.

Qu et al. [14] evaluated the PAH levels in the sediment of the Gulfs of Naples and Salerno, reporting TEQ_{PAHs} values ranging from 0.07 to 1425 ngTEQ/g; Arienzo et al. [16] studied the PAH levels in the sediment of the Gulf of Pozzuoli, with values between 1580 and 501.70 ngTEQ/g.

3.4. *Source Identification by PAH Diagnostic Ratios*

Data from this study highlighted a prevailing pattern of pyrolytic inputs of PAHs in the Sele River and its estuary. In effect, the results demonstrated that the Ant/(Ant + Phe) ratio was >0.1 in DP, SPM, and sediment (means of 0.40, 0.41, and 0.43, respectively), which assigned the origin of the PAHs to pyrogenic sources. Moreover, Flu/(Flu + Pyr) ratios allowed us to differentiate petroleum origins from combustion processes and make the distinction between such sources [66,67]. For Flu/(Flu + Pyr), low ratios (<0.40) indicate petroleum, intermediate ratios (0.40–0.50) indicate liquid fossil fuel combustion, and ratios >0.50 are characteristic of grass, wood, and coal combustion. In the Sele River and its estuary, a ratio value of Flu/(Flu + Pyr) > 0.5 was found in the dissolved phase, particulate matter, and sediment, indicating that combustion was the main source of pollution there (Figure 4a).

Ratio values of BaA/(BaA + Chr) > 0.35 and InP/(InP + BghiP) > 0.35 were found in the dissolved phase, particulate matter, and sediment, indicating a mixed source of petroleum and combustion (Figure 4b). The ratio results from the samples indicated that they were mainly contaminated by combustion. In general, atmospheric particles emitted from factories may be transported and deposited into the river. Moreover, industrial wastewater and vehicle emissions also suggest a pyrolytic origin for PAH pollution in the area. Among the pollutants evaluated in this study, Per was probably the most important diagenetic PAH found; therefore, the high concentration of this compound compared to the others could indicate a natural origin [55,68–70].

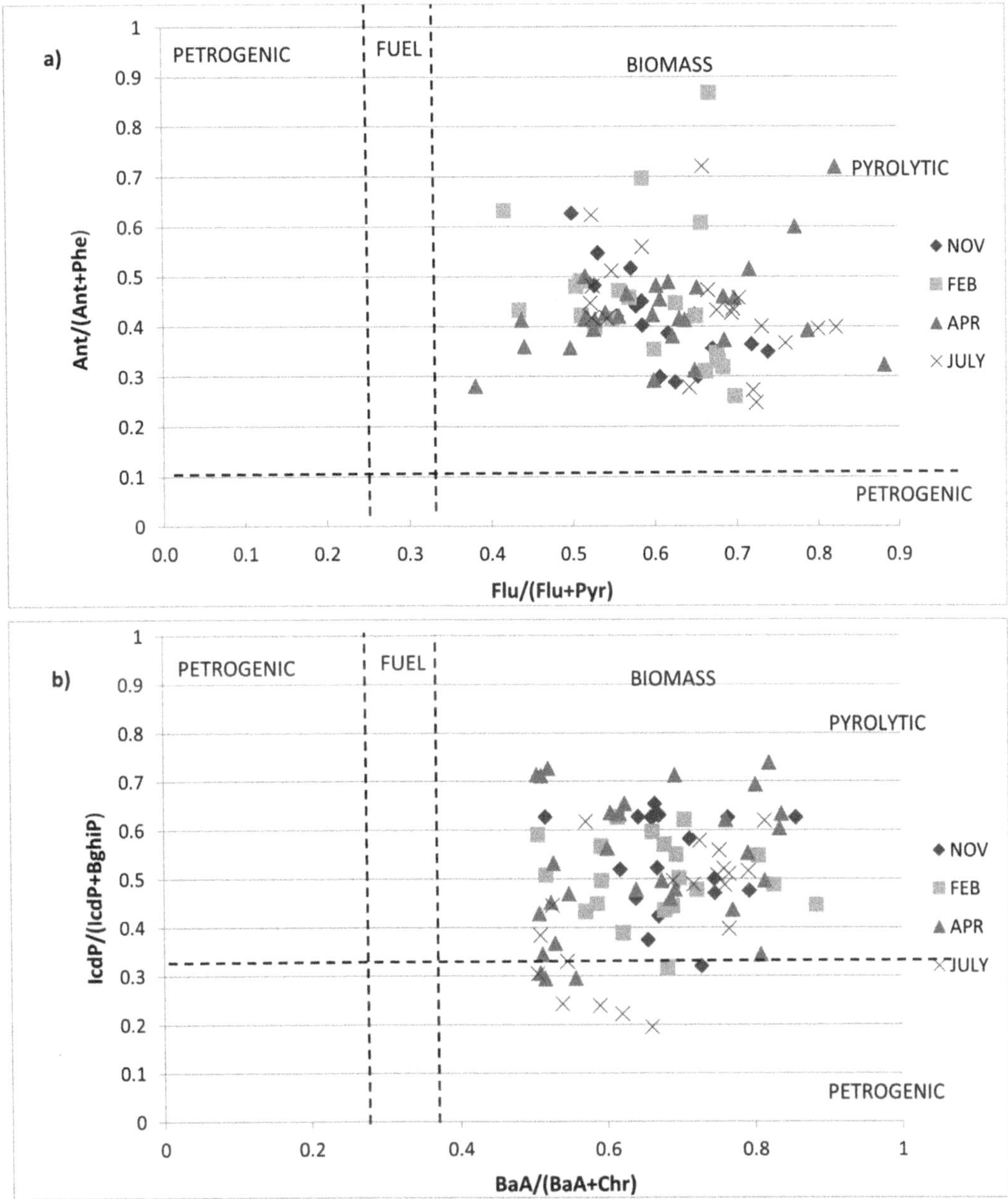

Figure 4. Cross plots of the values of (**a**) Flu/(Flu + Pyr) versus Ant/(Ant + Phe) and (**b**) BaA/(BaA + Chr) versus IcdP/(IcdP + BghiP) for all sample data from the Sele River and its estuary.

In fact, it has been indicated that amounts of Per above 10% of the total penta-aromatic isomers suggest a probable diagenetic input, whereas those samples in which Per accounts for less than 10% suggest a probable pyrolytic origin of the compound. In this study, the amount of Per detected in all sediment samples was very low (range 1.97–9.72 ng g^{-1}) and contributed less than 2% to the penta-aromatic isomers, indicating a pyrolytic origin of these pollutants. Differences in PAH spatial distributions in different periods are expected to be due to different sources of PAH inputs, water conditions, and the characteristics of individual PAHs. In the dry period, the river is stagnant, which weakens the transport of the pollutants from upstream to downstream, and the higher values at some sites may be the result of some highly local inputs. Special PAH ratios such as BaA/(BaA + Chr) and IcdP/(IcdP + BhiP) indicated that, in July, in dry season weather conditions, the PAHs found in the Sele River were primarily from petrogenic sources, while under wet weather season conditions, they were from pyrolytic sources.

PCA was used to quantitatively assess PAH origins, and molecular ratios between isomers were used: PAHs were represented by three PC factors (PC1, PC2, and PC3), and the extracted eigenvectors showed 47% for PC1 (Figure 5). This factor was mostly loaded by the four-ring PAHs Pyr, Flu, Chr, and BaA and the six-ring PAHs IcdP and BghiP. PAHs such as Pyr and Chr are indicators for coal burning, while Flu may indicate combustion. Similar behavior was observed for the ratio IcdP/(IcdP + BghiP, which indicates a mixed source of petroleum and combustion [71]. PC2 (21%) represented HMWs belonging to the five-ring PAHs BbF and BkF. High-molecular-weight PAHs such as these indicate pyrolysis and incomplete biomass combustion [72]. PC3, in contrast, contributed only 13% of the total load and represented the three-ring PAHs Phe and Ant (Figure 5). Therefore, the LMW/HMW ratio was low (<1 for each site), which is an indication of a pyrolytic origin of PAHs at these sites [73]. The PCA and diagnostic ratios indicate that the origins of contamination by PAHs in the Sele River were due to pyrolytic sources and combustion sources, such as gasoline burning and fuel and coal burning.

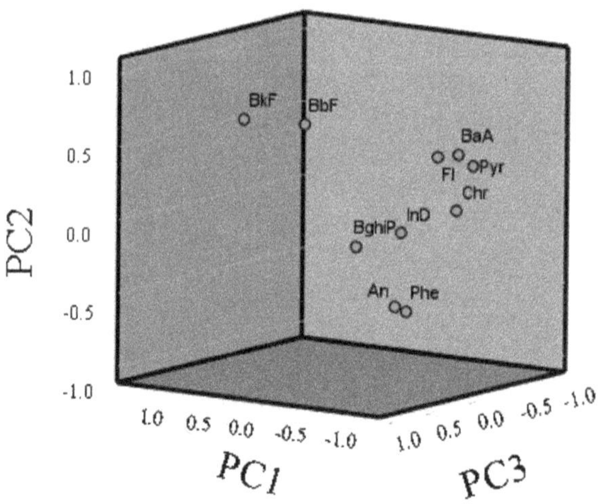

Figure 5. Principal component analysis (PCA) of PAH composition in samples from the Sele River estuary, southern Italy.

4. Conclusions

This paper offers important data on PAH concentrations and composition in the Sele River where it empties into the Tyrrhenian Sea (Central Mediterranean Sea), southern Italy, and presents the first comprehensive study of PAHs in water, SPM, and sediment in that area. The levels of LMW PAHs were particularly high in water samples, while the levels of HMW PAHs were predominant in sediment samples. A determination of the diagnostic ratio of PAHs revealed that the main PAH sources were pyrolytic and suggested that the majority of this pollution derived from vehicle traffic and combustion processes. The exchange of PAHs between water and sediment occurs in the direction of adsorption into the sediment from water. Regarding the risk assessment, the concentrations of many single PAHs at a number of sites were above ERL and/or TEL (and below ERM and/or PEL), which would on occasion yield negative environmental consequences. However, the toxic equivalent concentration (TEQ) of carcinogenic PAHs suggests that the Sele River basin presents a definite carcinogenic risk. Thus, the waters of the Sele River should be continuously monitored, as PAHs could lead to negative consequences for its aquatic ecosystems and organisms.

Supplementary Materials: The following are available online at https://www.mdpi.com/article/10.3390/toxics10070401/s1, Table S1: Characteristic Ions of the analyzed PAHs; Table S2: Individual PAHs recovery values; Table S3: Description of concentration of PAHs in sediment samples of the Sele River, southern Italy; Table S4: Description of concentration of PAHs in water dissolved phase (DP) samples of the Sele River, southern Italy; Table S5: Description of concentration of PAHs in suspended particulate matter (SPM) samples of the Sele River, southern Italy; Table S6: Parameter of GC-MS system; Table S7: Validation parameter values of PAHs in water samples and SPM samples; Table S8: Validation parameter values of PAHs in Sediment samples; Figure S1: Chromatograms obtained in different phase of Sele River. (a) PAHs chromatogram identified in a water sample (Dissolved phase) of Sele River. (b) PAHs chromatogram identified in a Suspended particulate matter (SPM) sample of Sele River. (c) PAHs chromatogram identified in a water sample (Dissolved phase) of Sele River. (d) PAHs chromatogram identified in a Suspended particulate matter (SPM) sample of Sele River. (e) PAHs chromatogram identified in Sediment sample of Sele River. (f) PAHs chromatogram identified in Sediment sample of Sele River.

Author Contributions: Conceptualization, M.T.; data curation, P.M., E.D.R., F.D.D., I.R. and P.S.; formal analysis, E.D.R., F.D.D., B.D.S., S.S. and I.R.; methodology, E.D.R., F.D.D., B.D.S. and S.S.; project administration, P.M.; resources, E.D.R., F.D.D., B.D.S., S.S. and I.R.; software, P.S.; supervision, P.M. and M.T.; writing—original draft, E.D.R. and F.D.D.; writing—review and editing, P.M. and M.T. All authors have read and agreed to the published version of the manuscript.

Funding: This research received no external funding.

Institutional Review Board Statement: Not applicable.

Informed Consent Statement: Not applicable.

Data Availability Statement: The datasets obtained and analyzed in the current study are available from the corresponding author on reasonable request.

Conflicts of Interest: The authors declare that they have no competing interests.

References

1. Zhang, J.; Liu, G.; Wang, R.; Huang, H. Polycyclic aromatic hydrocarbons in the water-SPM-sediment system from the middle reaches of Huai River, China: Distribution, partitioning, origin tracing and ecological risk assessment. *Environ. Pollut.* **2017**, *230*, 61–71. [CrossRef] [PubMed]
2. Chen, C.-F.; Ju, Y.-R.; Su, Y.-C.; Lim, Y.C.; Kao, C.-M.; Chen, C.-W.; Dong, C.-D. Distribution, sources, and behavior of PAHs in estuarine water systems exemplified by Salt River, Taiwan. *Mar. Pollut. Bull.* **2020**, *154*, 111029. [CrossRef] [PubMed]
3. USEPA (US Environmental Protection Agency). Regional Screening Levels for Chemical Contaminants at Superfund Sites. Regional Screening Table. User's Guide. 2012. Available online: https://www.epa.gov/risk/regional-screening-levels-rsls-generic-tables (accessed on 4 May 2022).
4. An, N.; Liu, S.; Yin, Y.; Cheng, F.; Dong, S.; Wu, X. Spatial distribution and sources of polycyclic aromatic hydrocarbons (PAHs) in the reservoir sediments after impoundment of Manwan Dam in the middle of Lancang River, China. *Ecotoxicology* **2016**, *25*, 1072–1081. [CrossRef] [PubMed]
5. Li, Q.; Wu, J.; Zhao, Z. Spatial and temporal distribution of Polycyclic Aromatic Hydrocarbons (PAHs) in sediments from Poyang Lake, China. *PLoS ONE* **2018**, *13*, e0205484. [CrossRef] [PubMed]
6. COMMISSION REGULATION (EU) No 835/2011 of 19 August 2011 amending Regulation (EC) No 1881/2006 as Regards Maximum Levels for Polycyclic Aromatic Hydrocarbons in Foodstuffs. Available online: https://eur-lex.europa.eu/LexUriServ/LexUriServ.do?uri=OJ:L:2011:215:0004:0008:En:PDF (accessed on 18 May 2022).
7. Liu, Y.; Zarfl, C.; Basu, N.B.; Cirpka, O.A. Turnover and legacy of sediment-associated PAH in a baseflow-dominated river. *Sci. Total Environ.* **2019**, *671*, 754–764. [CrossRef]
8. Zanardi-Lamardo, E.; Mitra, S.; Vieira-Campos, A.A.; Cabral, C.B.; Yogui, G.; Sarkar, S.K.; Biswas, J.K.; Godhantaraman, N. Distribution and sources of organic contaminants in surface sediments of Hooghly river estuary and Sundarban mangrove, eastern coast of India. *Mar. Pollut. Bull.* **2019**, *146*, 39–49. [CrossRef]

9. Castro-Jiménez, J.; Berrojalbiz, N.; Wollgast, J.; Dachs, J. Polycyclic aromatic hydrocarbons (PAHs) in the Mediterranean Sea: Atmospheric occurrence, deposition and decoupling with settling fluxes in the water column. *Environ. Pollut.* **2012**, *166*, 40–47. [CrossRef]
10. Sun, C.; Zhang, J.; Ma, Q.; Chen, Y.; Ju, H. Polycyclic aromatic hydrocarbons (PAHs) in water and sediment from a river basin: Sediment-water partitioning, source identification and environmental health risk assessment. *Environ. Geochem. Health* **2016**, *39*, 63–74. [CrossRef]
11. Montuori, P.; Aurino, S.; Garzonio, F.; Sarnacchiaro, P.; Nardone, A.; Triassi, M. Distribution, sources and ecological risk assessment of polycyclic aromatic hydrocarbons in water and sediments from Tiber River and estuary, Italy. *Sci. Total Environ.* **2016**, *566–567*, 1254–1267. [CrossRef]
12. Montuori, P.; De Rosa, E.; Di Duca, F.; Provvisiero, D.P.; Sarnacchiaro, P.; Nardone, A.; Triassi, M. Estima-tion of Polycyclic Aromatic Hydrocarbons Pollution in Mediterranean Sea from Volturno River, Southern It-aly: Distribution, Risk Assessment and Loads. *Int. J. Environ. Res. Public Health* **2021**, *18*, 1383. [CrossRef]
13. Montuori, P.; Triassi, M. Polycyclic aromatic hydrocarbons loads into the Mediterranean Sea: Estimate of Sarno River inputs. *Mar. Pollut. Bull.* **2012**, *64*, 512–520. [CrossRef] [PubMed]
14. Qu, C.; Li, J.; Albanese, S.; Lima, A.; Wang, M.; Sacchi, M.; Molisso, F.; De Vivo, B. Polycyclic aromatic hydrocarbons in the sediments of the Gulfs of Naples and Salerno, Southern Italy: Status, sources and ecological risk. *Ecotoxicol. Environ. Saf.* **2018**, *161*, 156–163. [CrossRef] [PubMed]
15. Qu, C.; Albanese, S.; Lima, A.; Li, J.; Doherty, A.L.; Qi, S.; De Vivo, B. Residues of hexachlorobenzene and chlorinated cyclodiene pesticides in the soils of the Campanian Plain, southern Italy. *Environ. Pollut.* **2017**, *231*, 1497–1506. [CrossRef] [PubMed]
16. Arienzo, M.; Albanese, S.; Lima, A.; Cannatelli, C.; Aliberti, F.; Cicotti, F.; Qi, S.; De Vivo, B. Assessment of the concentrations of polycyclic aromatic hydrocarbons and organochlorine pesticides in soils from the Sarno River basin, Italy, and ecotoxicological survey by Daphnia magna. *Environ. Monit. Assess.* **2015**, *187*, 52. [CrossRef] [PubMed]
17. Diodato, N.; Fagnano, M.; Alberico, I. Geospatial and visual modeling for exploring sediment source areas across the Sele river landscape, Italy. *Ital. J. Agron.* **2011**, *6*, e14. [CrossRef]
18. Albanese, S.; De Vivo, B.; Lima, A.; Cicchella, D.; Civitillo, D.; Cosenza, A. Geochemical baselines and risk assessment of the Bagnoli brownfield site coastal sea sediments (Naples, Italy). *J. Geochem. Explor.* **2010**, *105*, 19–33. [CrossRef]
19. Liu, Q.; Xu, X.; Wang, L.; Lin, L.; Wang, D. Simultaneous determination of forty-two parent and halogenated polycyclic aromatic hydrocarbons using solid-phase extraction combined with gas chromatography-mass spectrometry in drinking water. *Ecotoxicol. Environ. Saf.* **2019**, *181*, 241–247. [CrossRef]
20. Kafilzadeh, F.; Shiva, A.H.; Malekpour, R. Determination of polycyclic aromatic hydrocarbons (PAHs) in water and sediments of the Kor River, Iran. *Middle-East J. Sci. Res.* **2011**, *10*, 1–7.
21. Lin, L.; Dong, L.; Meng, X.; Li, Q.; Huang, Z.; Li, C.; Li, R.; Yang, W.; Crittenden, J. Distribution and sources of polycyclic aromatic hydrocarbons and phthalic acid esters in water and surface sediment from the Three Gorges Reservoir. *J. Environ. Sci.* **2018**, *69*, 271–280. [CrossRef]
22. USA-Environmental Protection Agency (US EPA). *Method 3540C: Soxhlet Extraction*; USA Environmental Protection Agency: Washington, DC, USA, 1996.
23. Ashayeri, N.Y.; Keshavarzi, B. Geochemical characteristics, partitioning, quantitative source apportionment, and ecological and health risk of heavy metals in sediments and water: A case study in Shadegan Wetland, Iran. *Mar. Pollut. Bull.* **2019**, *149*, 110495. [CrossRef]
24. Gou, Y.; Zhao, Q.; Yang, S.; Wang, H.; Qiao, P.; Song, Y.; Cheng, Y.; Li, P. Removal of polycyclic aromatic hydrocarbons (PAHs) and the response of indigenous bacteria in highly contaminated aged soil after persulfate oxidation. *Ecotoxicol. Environ. Saf.* **2019**, *190*, 110092. [CrossRef] [PubMed]
25. He, Y.; Yang, C.; He, W.; Xu, F. Nationwide health risk assessment of juvenile exposure to polycyclic aromatic hydrocarbons (PAHs) in the water body of Chinese lakes. *Sci. Total Environ.* **2020**, *723*, 138099. [CrossRef] [PubMed]
26. Fakhradini, S.S.; Moore, F.; Keshavarzi, B.; Lahijanzadeh, A. Polycyclic aromatic hydrocarbons (PAHs) in water and sediment of Hoor Al-Azim wetland, Iran: A focus on source apportionment, environmental risk assessment, and sediment-water partitioning. *Environ. Monit. Assess.* **2019**, *191*, 233. [CrossRef]
27. Zhao, Z.; Gong, X.; Zhang, L.; Jin, M.; Cai, Y.; Wang, X. Riverine transport and water-sediment exchange of polycyclic aromatic hydrocarbons (PAHs) along the middle-lower Yangtze River, China. *J. Hazard. Mater.* **2020**, *403*, 123973. [CrossRef] [PubMed]
28. Ashayeri, N.Y.; Keshavarzi, B.; Moore, F.; Kersten, M.; Yazdi, M.; Lahijanzadeh, A.R. Presence of polycyclic aromatic hydrocarbons in sediments and surface water from Shadegan wetland-Iran: A focus on source apportionment, human and ecological risk assessment and Sediment-Water Exchange. *Ecotoxicol. Environ. Saf.* **2018**, *148*, 1054–1066. [CrossRef]
29. Wenning, R.J.; Ingersoll, C.G. *Summary of the SETAC Pellston Workshop on Use of Sediment Quality Guide-Lines and Related Tools for the Assessment of Contaminated Sediments*; Society of Environmental Toxicology and Chemistry (SETAC): Pensacola, FL, USA, 2002. Available online: http://www.setac.org/files/SQGSummary.pdf (accessed on 18 May 2022).
30. Mogashane, T.M.; Mujuru, M.; McCrindle, R.I.; Ambushe, A.A. Quantification, source apportionment and risk assessment of polycyclic aromatic hydrocarbons in sediments from Mokolo and Blood Rivers in Limpopo Province, South Africa. *J. Environ. Sci. Health Part A* **2019**, *55*, 71–81. [CrossRef]

31. Zaghden, H.; Tedetti, M.; Sayadi, S.; Serbaji, M.M.; Elleuch, B.; Saliot, A. Origin and distribution of hydrocarbons and organic matter in the surficial sediments of the Sfax-Kerkennah channel (Tunisia, Southern Mediterranean Sea). *Mar. Pollut. Bull.* **2017**, *117*, 414–428. [CrossRef]
32. Yuan, H.M.; Li, T.G.; Ding, X.G.; Zhao, G.M.; Ye, S.Y. Distribution, Sources Analysis and Eco-Toxicological Risk Assessment of Polycyclic Aromatic Hydrocarbons (PAHs) in Surface Soils in the Northern Yellow River Delta, China. *Adv. Mater. Res.* **2013**, *726–731*, 750–756. [CrossRef]
33. Liu, X.; Chen, Z.; Xia, C.; Wu, J.; Ding, Y. Characteristics, distribution, source and ecological risk of polycyclic aromatic hydrocarbons (PAHs) in sediments along the Yangtze River Estuary Deepwater Channel. *Mar. Pollut. Bull.* **2019**, *150*, 110765. [CrossRef]
34. Pozo, K.; Perra, G.; Menchi, V.; Urrutia, R.; Parra, O.; Rudolph, A.; Focardi, S. Levels and spatial distribution of polycyclic aromatic hydrocarbons (PAHs) in sediments from Lenga Estuary, central Chile. *Mar. Pollut. Bull.* **2011**, *62*, 1572–1576. [CrossRef]
35. Křůmal, K.; Mikuška, P. Mass concentrations and lung cancer risk assessment of PAHs bound to PM1 aerosol in six industrial, urban and rural areas in the Czech Republic, Central Europe. *Atmos. Pollut. Res.* **2019**, *11*, 401–408. [CrossRef]
36. Gupte, A.; Tripathi, A.; Patel, H.; Rudakiya, D.; Gupte, S. Bioremediation of Polycyclic Aromatic Hydrocarbon (PAHs): A Perspective. *Open Biotechnol. J.* **2016**, *10*, 363–378. [CrossRef]
37. Mojiri, A.; Zhou, J.L.; Ohashi, A.; Ozaki, N.; Kindaichi, T. Comprehensive review of polycyclic aromatic hydrocarbons in water sources, their effects and treatments. *Sci. Total Environ.* **2019**, *696*, 133971. [CrossRef] [PubMed]
38. Tongo, I.; Ezemonye, L.; Akpeh, K. Levels, distribution and characterization of Polycyclic Aromatic Hydrocarbons (PAHs) in Ovia river, Southern Nigeria. *J. Environ. Chem. Eng.* **2017**, *5*, 504–512. [CrossRef]
39. Gdara, I.; Zrafi, I.; Balducci, C.; Cecinato, A.; Ghrabi, A. Seasonal occurrence, source evaluation and ecological risk assessment of polycyclic aromatic hydrocarbons in industrial and agricultural effluents discharged in Wadi El Bey (Tunisia). *Environ. Geochem. Health* **2018**, *40*, 1609–1627. [CrossRef]
40. Haiba, N.S.A. Polycyclic Aromatic Hydrocarbons (PAHs) in the River Nile, Egypt: Occurrence and Distribution. *Polycycl. Aromat. Compd.* **2017**, *39*, 425–433. [CrossRef]
41. Lima, E.A.R.; Neves, P.A.; Patchineelam, S.R.; da Silva, J.F.B.R.; Takiyama, L.R.; Martins, C.C.; Lourenço, R.A.; Taniguchi, S.; Elias, V.O.; Bícego, M.C. Anthropogenic and natural inputs of polycyclic aromatic hydrocarbons in the sediment of three coastal systems of the Brazilian Amazon. *Environ. Sci. Pollut. Res.* **2021**, *28*, 19485–19496. [CrossRef]
42. Triassi, M.; Nardone, A.; Giovinetti, M.C.; De Rosa, E.; Canzanella, S.; Sarnacchiaro, P.; Montuori, P. Ecological risk and estimates of organophosphate pesticides loads into the Central Mediterranean Sea from Volturno River, the river of the "Land of Fires" area, southern Italy. *Sci. Total Environ.* **2019**, *678*, 741–754. [CrossRef]
43. Pearson, K. On lines and planes of closest fit to systems of points in space. *Lond. Edinb. Dublin Philos. Mag. J. Sci.* **1901**, *2*, 559–572. [CrossRef]
44. Alberico, I.; Amato, V.; Aucelli, P.P.C.; Di Paola, G.; Pappone, G.; Rosskopf, C.M. Historical and recent changes of the Sele River coastal plain (Southern Italy): Natural variations and human pressures. *Rend. Lince* **2011**, *23*, 3–12. [CrossRef]
45. Giordano, L.; Alberico, I.; Ferraro, L.; Marsella, E.; Lirer, F.; Di Fiore, V. A new tool to promote sustainability of coastal zones. The case of Sele plain, southern Italy. *Rend. Lince* **2013**, *24*, 113–126. [CrossRef]
46. Yu, H.; Liu, Y.; Han, C.; Fang, H.; Weng, J.; Shu, X.; Pan, Y.; Ma, L. Polycyclic aromatic hydrocarbons in surface waters from the seven main river basins of China: Spatial distribution, source apportionment, and potential risk assessment. *Sci. Total Environ.* **2020**, *752*, 141764. [CrossRef] [PubMed]
47. Kim, L.; Jeon, H.-J.; Kim, Y.-C.; Yang, S.-H.; Choi, H.; Kim, T.-O.; Lee, S.-E. Monitoring polycyclic aromatic hydrocarbon concentrations and distributions in rice paddy soils from Gyeonggi-do, Ulsan, and Pohang. *Appl. Biol. Chem.* **2019**, *62*, 18. [CrossRef]
48. Yang, H.-H.; Lai, S.-O.; Hsieh, L.-T.; Hsueh, H.-J.; Chi, T.-W. Profiles of PAH emission from steel and iron industries. *Chemosphere* **2002**, *48*, 1061–1074. [CrossRef]
49. Guo, W.; He, M.; Yang, Z.; Lin, C.; Quan, X.; Wang, H. Distribution of polycyclic aromatic hydrocarbons in water, suspended particulate matter and sediment from Daliao River watershed, China. *Chemosphere* **2007**, *68*, 93–104. [CrossRef]
50. Liu, X.; Wang, H.; Wei, L.; Liu, J.; Reitz, R.D.; Yao, M. Development of a reduced toluene reference fuel (TRF)-2,5-dimethylfuran-polycyclic aromatic hydrocarbon (PAH) mechanism for engine applications. *Combust. Flame* **2016**, *165*, 453–465. [CrossRef]
51. Gong, G.; Zhao, X.; Wu, S. Effect of natural antioxidants on inhibition of parent and oxygenated polycyclic aromatic hydrocarbons in Chinese fried bread youtiao. *Food Control* **2018**, *87*, 117–125. [CrossRef]
52. Kumar, A.; Schimmelmann, A.; Sauer, P.E.; Brassell, S.C. Distribution and sources of polycyclic aromatic hydrocarbons (PAHs) in laminated Santa Barbara Basin sediments. *Org. Geochem.* **2017**, *113*, 303–314. [CrossRef]
53. Ferraro, A.; Massini, G.; Miritana, V.M.; Panico, A.; Pontoni, L.; Race, M.; Rosa, S.; Signorini, A.; Fabbricino, M.; Pirozzi, F. Bioaugmentation strategy to enhance polycyclic aromatic hydrocarbons anaerobic biodegradation in contaminated soils. *Chemosphere* **2021**, *275*, 130091. [CrossRef]
54. Liu, Z.; He, L.; Lu, Y.; Su, J.; Song, H.; Zeng, X.; Yu, Z. Distribution, source, and ecological risk assessment of polycyclic aromatic hydrocarbons (PAHs) in surface sediments from the Hun River, northeast China. *Environ. Monit. Assess.* **2015**, *187*, 290. [CrossRef]
55. Abdel-Shafy, H.I.; Mansour, M.S.M. A review on polycyclic aromatic hydrocarbons: Source, environmental impact, effect on human health and remediation. *Egypt. J. Pet.* **2016**, *25*, 107–123. [CrossRef]

56. Tzoraki, O.; Karaouzas, I.; Patrolecco, L.; Skoulikidis, N.; Nikolaidis, N.P. Polycyclic Aromatic Hydrocarbons (PAHs) and Heavy Metal Occurrence in Bed Sediments of a Temporary River. *Water Air Soil Pollut.* **2015**, *226*, 421. [CrossRef]
57. Chizhova, T.; Koudryashova, Y.; Prokuda, N.; Tishchenko, P.; Hayakawa, K. Polycyclic Aromatic Hydro-carbons in the Estuaries of Two Rivers of the Sea of Japan. *Int. J. Environ. Res. Public Health* **2020**, *17*, 6019. [CrossRef] [PubMed]
58. Sicre, M.A.; Fernandes, M.B.; Pont, D. Poly-aromatic hydrocarbon (PAH) inputs from the Rhône River to the Mediterranean Sea in relation with the hydrological cycle: Impact of floods. *Mar. Pollut. Bull.* **2008**, *56*, 1935–1942. [CrossRef]
59. Cao, Y.; Xin, M.; Wang, B.; Lin, C.; Liu, X.; He, M.; Lu, S. Spatiotemporal distribution, source, and ecological risk of polycyclic aromatic hydrocarbons (PAHs) in the urbanized semi-enclosed Jiaozhou Bay, China. *Sci. Total Environ.* **2020**, *717*, 137224. [CrossRef]
60. Zhang, H.; Sun, L.; Sun, T.; Li, H.; Luo, Q. Spatial distribution and seasonal variation of polycyclic aromatic hydrocarbons (PAHs) contaminations in surface water from the Hun River, Northeast China. *Environ. Monit. Assess.* **2013**, *185*, 1451–1462. [CrossRef] [PubMed]
61. Guo, W.; He, M.; Yang, Z.; Lin, C.; Quan, X.; Men, B. Distribution, partitioning and sources of polycyclic aromatic hydrocarbons in Daliao River water system in dry season, China. *J. Hazard. Mater.* **2009**, *164*, 1379–1385. [CrossRef]
62. Chen, Y.; Sun, C.; Zhang, J.; Zhang, F. Assessing 16 Polycyclic Aromatic Hydrocarbons (PAHs) in River Basin Water and Sediment Regarding Spatial-Temporal Distribution, Partitioning, and Ecological Risks. *Pol. J. Environ. Stud.* **2018**, *27*, 579–589. [CrossRef]
63. Akhbarizadeh, R.; Moore, F.; Keshavarzi, B.; Moeinpour, A. Aliphatic and polycyclic aromatic hydrocarbons risk assessment in coastal water and sediments of Khark Island, SW Iran. *Mar. Pollut. Bull.* **2016**, *108*, 33–45. [CrossRef]
64. Long, E.R.; Macdonald, D.D.; Smith, S.L.; Calder, F.D. Incidence of adverse biological effects within ranges of chemical concentrations in marine and estuarine sediments. *Environ. Manag.* **1995**, *19*, 81–97. [CrossRef]
65. Macdonald, D.D.; Carr, R.S.; Calder, F.D.; Long, E.R.; Ingersoll, C.G. Development and evaluation of sedi-ment quality guidelines for Florida coastal waters. *Ecotoxicology* **1996**, *5*, 253–278. [CrossRef] [PubMed]
66. Yunker, M.B.; Macdonald, R.W.; Vingarzan, R.; Mitchell, R.H.; Goyette, D.; Sylvestre, S. PAHs in the Fraser River basin: A critical appraisal of PAH ratios as indicators of PAH source and composition. *Org. Geochem.* **2002**, *33*, 489–515. [CrossRef]
67. Ekpo, B.O.; Oyo-Ita, O.E.; Oros, D.R.; Simoneit, B.R.T. Distributions and sources of polycyclic aromatic hydrocarbons in surface sediments from the Cross River estuary, S.E. Niger Delta, Nigeria. *Environ. Monit. Assess.* **2011**, *184*, 1037–1047. [CrossRef]
68. Tobiszewski, M.; Namieśnik, J. PAH diagnostic ratios for the identification of pollution emission sources. *Environ. Pollut.* **2012**, *162*, 110–119. [CrossRef]
69. Lin, B.-S.; Brimblecombe, P.; Lee, C.-L.; Liu, J.T. Tracing typhoon effects on particulate transport in a submarine canyon using polycyclic aromatic hydrocarbons. *Mar. Chem.* **2013**, *157*, 1–11. [CrossRef]
70. Pérez-Fernández, B.; Viñas, L.; Franco, M.; Bargiela, J. PAHs in the Ría de Arousa (NW Spain): A consideration of PAHs sources and abundance. *Mar. Pollut. Bull.* **2015**, *95*, 155–165. [CrossRef] [PubMed]
71. Tavakoly Sany, S.; Hashim, R.; Salleh, A.; Rezayi, M.; Mehdinia, A.; Safari, O. Polycyclic aromatic hydrocar-bons in coastal sediment of Klang Strait, Malaysia: Distribution pattern, risk assessment and sources. *PLoS ONE* **2014**, *9*, e9490. [CrossRef]
72. Jiang, Y.F.; Wang, X.T.; Wang, F.; Jia, Y.; Wu, M.H.; Sheng, G.Y.; Fu, J.M. Levels, composition profiles and sources of polycyclic aromatic hydrocarbons in urban soil of Shanghai, China. *Chemosphere* **2009**, *75*, 1112–1118. [CrossRef]
73. Khabouchi, I.; Khadhar, S.; Driouich Chaouachi, R.; Chekirbene, A.; Asia, L.; Doumenq, P. Study of organic pollution in superficial sediments of Meliane river catchment area: Aliphatic and polycyclic aromatic hydro-carbons. *Environ. Monit. Assess.* **2020**, *192*, 283. [CrossRef]

MDPI
St. Alban-Anlage 66
4052 Basel
Switzerland
www.mdpi.com

Toxics Editorial Office
E-mail: toxics@mdpi.com
www.mdpi.com/journal/toxics

www.ingramcontent.com/pod-product-compliance
Lightning Source LLC
Chambersburg PA
CBHW040748250926
51850CB00007BB/1442
* 9 7 8 3 7 2 5 8 0 9 0 6 6 *